Benchmark Papers in Ecology

Series Editor: Frank B. Golley
University of Georgia

Volume

1. CYCLES OF ESSENTIAL ELEMENTS / *Lawrence R. Pomeroy*
2. BEHAVIOR AS AN ECOLOGICAL FACTOR / *David E. Davis*
3. NICHE: Theory and Application / *Robert H. Whittaker and Simon A. Levin*
4. ECOLOGICAL ENERGETICS / *Richard G. Wiegert*
5. ECOLOGICAL SUCCESSION / *Frank B. Golley*
6. PHYTOSOCIOLOGY / *Robert P. McIntosh*
7. POPULATION REGULATION / *Robert H. Tamarin*
8. PATTERNS OF PRIMARY PRODUCTION IN THE BIOSPHERE / *Helmut F. H. Lieth*
9. SYSTEMS ECOLOGY / *H. H. Shugart and R. V. O'Neill*
10. TROPICAL ECOLOGY / *Carl F. Jordan*
11. DISPERSAL AND MIGRATION / *William Z. Lidicker, Jr. and Roy L. Caldwell*

Related Titles in BENCHMARK PAPERS IN BEHAVIOR Series

Volume

1. HORMONES AND SEXUAL BEHAVIOR / *Carol Sue Carter*
2. TERRITORY / *Allen W. Stokes*
3. SOCIAL HIERARCHY AND DOMINANCE / *Martin W. Schein*
4. EXTERNAL CONSTRUCTION BY ANIMALS / *Nicholas E. Collias and Elsie C. Collias*
5. IMPRINTING / *Eckhard H. Hess and Slobodan B. Petrovich*
6. PSYCHOPHYSIOLOGY / *Stephen W. Porges and Michael G. H. Coles*
7. SOUND RECEPTION IN FISHES / *William N. Tavolga*
8. VERTEBRATE SOCIAL ORGANIZATION / *Edwin M. Banks*
9. SOUND PRODUCTION IN FISHES / *William N. Tavolga*
10. EVOLUTION OF PLAY BEHAVIOR / *Dietland Müller-Schwarze*
11. PARENTAL BEHAVIOR IN BIRDS / *Rae Silver*
12. CRITICAL PERIODS / *John Paul Scott*
13. THERMOREGULATION / *Evelyn Satinoff*
14. LEARNING IN ANIMALS / *Robert W. Hendersen*
15. MAMMALIAN SEXUAL BEHAVIOR / *Donald A. Dewsbury*
16. BEHAVIOR-GENETIC ANALYSIS / *Jerry Hirsch and Terry R. McGuire*

Benchmark Papers in Ecology / 11

A BENCHMARK® Books Series

DISPERSAL AND MIGRATION

Edited by
WILLIAM Z. LIDICKER, JR.
and
ROY L. CALDWELL
University of California, Berkeley

Hutchinson Ross Publishing Company

Stroudsburg, Pennsylvania

Copyright ©1982 by **Hutchinson Ross Publishing Company**
Benchmark Papers in Ecology, Volume 11
Library of Congress Catalog Card Number: 82-9326
ISBN: 0-87933-435-5

All rights reserved. No part of this book covered by the copyrights hereon may be reproduced or transmitted in any form or by any means—graphic, electronic, or mechanical, including photocopying, recording, taping, or information storage and retrieval systems—without written permission of the publisher.

84 83 82   1 2 3 4 5
Manufactured in the United States of America.

LIBRARY OF CONGRESS CATALOGING IN PUBLICATION DATA
Main entry under title:
Dispersal and migration.
   (Benchmark papers in ecology ; 11)
   Includes indexes.
   1. Animal migration—Addresses, essays, lectures.
2. Animal populations—Addresses, essays, lectures.
3. Zoogeography—Addresses, essays, lectures.
I. Lidicker, William Zander, 1932-   . II. Caldwell, Roy L. III. Series.
QL754.D53   1982   591.52'5   82-9326
ISBN 0-87933-435-5   AACR2

Distributed worldwide by Van Nostrand Reinhold Company Inc., 135 W. 50th Street, New York, NY 10020.

CONTENTS

Series Editor's Foreword	ix
Preface	xi
Contents by Author	xiii
Introduction	1

PART I: DOES DISPERSAL HAVE A GENETIC BASIS?

Editors' Comments on Papers 1 Through 10 — 8

1. SAKAI, K., T. NARISE, and S. IYAMA: Migration Studies in Several Wild Strains of *Drosophila melanogaster*
   Nat. Inst. Genet. (Japan) Ann. Rep., No. 7, pp. 73-75 (1957) — 17

2. NARISE, T.: Studies on Competition in Plants and Animals, X. Genetic Variability of Migratory Activity in Natural Populations of *Drosophila melanogaster*
   Jpn. J. Genet. **37**:451-461 (1962) — 19

3. BERTHOLD, P., and U. QUERNER: Genetic Basis of Migratory Behavior in European Warblers
   Science **212**:77-79 (1981) — 30

4. HOWARD, W. E.: Innate and Environmental Dispersal of Individual Vertebrates
   Am. Midl. Nat. **63**:152-161 (1960) — 33

5. JOHNSTON, R. F.: Population Movements of Birds
   Condor **63**:386-389 (1961) — 43

6. SOUTHWOOD, T. R. E.: Migration of Terrestrial Arthropods in Relation to Habitat
   Biol. Rev. **37**:171-175, 190-194, 198-211 (1962) — 47

7. OGDEN, J. C.: Artificial Selection for Dispersal in Flour Beetles (Tenebrionidae: *Tribolium*)
   Ecology **51**:130-133 (1970) — 71

8. CALDWELL, R. L., and J. P. HEGMANN: Heritability of Flight Duration in the Milkweed Bug *Lygaeus kalmii*
   Nature **223**:91-92 (1969) — 75

9. RITTE, U., and B. LAVIE: The Genetic Basis of Dispersal Behavior in the Flour Beetle *Tribolium castaneum*
   Can. J. Genet. Cytol. **19**:717-722 (1977) — 77

10. HILBORN, R.: Similarities in Dispersal Tendency Among Siblings in Four Species of Voles *(Microtus)*
    Ecology **56**:1221-1225 (1975) — 83

Contents

PART II: WHAT CONDITIONS MOTIVATE ORGANISMS TO MOVE?

Editors' Comments on Papers 11 Through 16   90

11   LIDICKER, W. Z., JR.: The Role of Dispersal in the Demography of Small Mammals   102
Small Mammals: Their Productivity and Population Dynamics, F. B. Golley, K. Petrusewicz, and L. Ryszkowski, eds., Cambridge University Press, London, 1975, pp. 103–128

12   GRANT, P. R.: Dispersal in Relation to Carrying Capacity   134
Nat. Acad. Sci. (USA) Proc. **75:**2854-2858 (1978)

13   JOHNSON, C. G.: Physiological Factors in Insect Migration by Flight   139
Nature **198:**423-427 (1963)

14   EVANS, F. C.: Studies of a Small Mammal Population in Bagley Wood, Berkshire   144
J. Anim. Ecol. **11:**194-197 (1942)

15   NARISE, T.: The Mode of Migration of *Drosophila ananassae* Under Competitive Conditions   148
Stud. Genet., (Univ. Texas) No. 4, pp. 121-131 (1966)

16   HERRNKIND, W.: Queuing Behavior of Spiny Lobsters   159
Science **164:**1425-1427 (1969)

PART III: WILL MOVEMENTS BE SUCCESSFUL?

Editors' Comments on Papers 17 Through 20   164

17   JÄRVINEN, O., and K. VEPSÄLÄINEN: Wing Dimorphism as an Adaptive Strategy in Water-Striders *(Gerris)*   174
Hereditas **84:**61-68 (1976)

18   KENNEDY, J. S.: A Turning Point in the Study of Insect Migration   182
Nature **189:**785-791 (1961)

19   RAINEY, R. C.: Weather and the Movements of Locust Swarms: A New Hypothesis   189
Nature **168:**1057-1060 (1951)

20   METZGAR, L. H.: An Experimental Comparison of Screech Owl Predation on Resident and Transient White-footed Mice *(Peromyscus leucopus)*   194
J. Mammal. **48:**387-391 (1967)

PART IV: WHAT ARE THE CONSEQUENCES OF DISPERSAL AND MIGRATION?

Editors' Comments on Papers 21 Through 26   200

21   GILLESPIE, J. H.: The Role of Migration in the Genetic Structure of Populations in Temporally and Spatially Varying Environments I. Conditions for Polymorphism   207
Am. Nat. **109:**127-135 (1975)

22   DINGLE, H.: Migration Strategies of Insects   216
Science **175:**1327-1335 (1972)

23	LOMNICKI, A.: Individual Differences between Animals and the Natural Regulation of their Numbers *J. Anim. Ecol.* **47**:461-475 (1978)	225
24	TAYLOR, L. R., and R. A. J. TAYLOR: Aggregation, Migration and Population Mechanics *Nature* **265**:415-421 (1977)	240
25	DETHIER, V. G., and R. H. MacARTHUR: A Field's Capacity to Support a Butterfly Population *Nature* **201**:728-729 (1964)	246
26	BAILEY, E. D.: Immigration and Emigration as Contributory Regulators of Populations through Social Disruption *Can. J. Zool.* **47**:1213-1215 (1969)	248

PART V: THE EVOLUTION OF DISPERSAL AND MIGRATION

Editors' Comments on Papers 27 Through 32 — 252

27	DINGLE, H.: Ecology and Evolution of Migration *Animal Migration, Orientation, and Navigation,* S. A. Gauthreaux, Jr. ed., Academic, New York, 1980, pp. 1-2 and 78-83.	259
28	GILL, D. E.: Effective Population Size and Interdemic Migration Rates in a Metapopulation of the Red-spotted Newt, *Notophthalmus viridescens* (Rafinesque) *Evolution* **32**:839-849 (1978)	267
29	HAMILTON, W. D., and R. M. MAY: Dispersal in Stable Habitats *Nature* **269**:578-581 (1977)	278
30	LIDICKER, W. Z., JR.: Emigration as a Possible Mechanism Permitting the Regulation of Population Density below Carrying Capacity *Am. Nat.* **96**:29-33 (1962)	282
31	VAN VALEN, L.: Group Selection and the Evolution of Dispersal *Evolution* **25**:591-598 (1971)	287
32	HANSEN, T. A.: Larval Dispersal and Species Longevity in Lower Tertiary Gastropods *Science* **199**:885-887 (1978)	295

Author Citation Index — 299
Subject Index — 305
About the Editors — 311

SERIES EDITOR'S FOREWORD

Ecology—the study of interactions and relationships between living systems and environment—is an extremely active and dynamic field of science. The great variety of possible interactions in even the most simple ecological system makes the study of ecology compelling but difficult to discuss in simple terms. Further, living systems include individual organisms, populations, communities, and ultimately the entire biosphere; there are thus numerous subspecialties in ecology. Some ecologists are interested in wildlife and natural history, others are intrigued by the complexity and apparently intractable problems of large ecological systems, and still others apply ecological principles to the problems of man and the environment. This means that a Benchmark Series in Ecology could be subdivided into innumerable subvolumes that represented these diverse interests. However, rather than take this approach, I have tried to focus on general patterns or concepts that are applicable to two particularly important levels of ecological understanding: the population and the community. I have taken the dichotomy between these two as my major organizing concept in this series.

In a field that is rapidly changing and evolving, it is often difficult to chart the transition of single ideas into cohesive theories and principles. In addition, it is not easy to make judgments as to the benchmarks of the subject when the theoretical features of a field are relatively young. These twin problems—the relationship between interweaving ideas and the elucidation of theory, and the youth of the subject itself—make development of a Benchmark series in the field of ecology difficult. Each of the volume editors has recognized this inherent problem, and each has acted to solve it in his or her unique way. Their collective efforts will, we anticipate, provide a survey of the most important concepts in the field.

The Benchmark series is especially designed for libraries of colleges, universities, and research organizations that cannot purchase the older literature of ecology because of costs, lack of staff to select from the hundreds of thousands of journals and volumes, or from the unavailability of the reference materials. For example, in developing countries where a science library must be developed *de novo,* I have seen where the Benchmark series can provide the only background literature available to the students and staff. Thus, the intent of the series is to provide an authoritative selection of literature, which can be read in the original form, but that is cast

Series Editor's Foreword

in a matrix of thought provided by the editor. The volumes are designed to explore the historical development of a concept in ecology and point the way toward new developments, without being a historical study. We hope that even though the Benchmark Series in Ecology is a library oriented series and bears an appropriate cost, it will also be of sufficient utility that many professionals will place it in their personal library. In a few cases the volumes have even been used as textbooks for advanced courses. Thus, we expect that the Benchmark Series in Ecology will be useful not only to the student who seeks an authoritative selection of original literature, but also to the professional who wants to quickly and efficiently expand his or her background in an area of ecology outside his special competence. The present volume should be particularly appealing to the professional ecologist and advanced student because the authors have provided short reviews of various aspects of the field, including extensive bibliographies.

Dispersal and migration is a topic that was of major importance in the past, forming a key element in natural history studies of animals. Then the topic went into a period of decline while ecologists focused on other issues. Recently, this situation has changed again because dispersal is a fundamental element in genetic exchange among members of populations and therefore, in evolutionary ecology. It is also increasingly recognized as a critical variable in population demography. This Benchmark book on dispersal has been developed by two ecologists who have personally contributed theoretically and practically to this revival of interest in dispersal. William Z. Lidicker, Jr., is professor of zoology, curator of mammals, and associate director of the Museum of Vertebrate Zoology, University of California, Berkeley. Lidicker is a mammalogist with extensive experience in all phases of mammalian ecology and life history. Roy L. Caldwell is an associate professor in zoology at the University of California, Berkeley. Caldwell's research has been mainly directed toward invertebrates, including insects, and he has extensive tropical and marine experience as well. Thus, the editors bring to this topic a very broad as well as deep experience with dispersal and migration in the context of animal population dynamics, behavior, and evolution.

FRANK B. GOLLEY

PREFACE

In this Benchmark volume, we attempt to provide a useful collection of papers and commentary that will give some coherence to an important, but currently diffuse and semantically confusing, field of investigation. The subject of animal movements is currently being pursued vigorously on many fronts, and is fundamentally important to future advances in ecological, ethological, and evolutionary theory, as well as to applied disciplines such as natural resource management, pest control, and public health. However, much of the research in this area, both past and present, predictably remains unburdened by a synthetic overview of the field. We hope this volume will be a significant contribution toward such a synthesis.

Although our objectives are integrative and synthetic, practical considerations have dictated that some important topics related to movements be omitted. Perhaps most obtrusive of these neglected subjects is that of plant dispersal. Other topics from which we have necessarily steered away include navigation, orientation, physiological substrates of movement, and those spatial and temporal patterns of movements that are confined to home ranges. What is left is comprehensive enough! Moreover, other Benchmark volumes treat related topics and can serve as useful adjuncts to this one. In the ecology series, there is *Behavior as an Ecological Factor* (David, 1974), and *Population Regulation* (Tamarin, 1978). Among those in the behavior series, we call attention to *Territory* by Stokes (1974) and *Vertebrate Social Organization* by Banks (1977). The genetics series offers *Genetics and Social Structure* (Ballonoff, 1974), and *Demographic Genetics* (Weiss and Ballonoff, 1976).

Papers have been selected for this volume to illustrate our views on the conceptual history of the field as well as on current directions that research is taking. We feel that a historical perspective is essential to any synthetic effort of this sort. Moreover, we wish to emphasize, particularly to students, the long developmental time-frame characterizing many of our concepts. Ideas have a way of recurring periodically, each time being clothed in a patina of new jargon that tends to obfuscate underlying intellectual continuity. We have also given special consideration to valuable papers published in obscure places and/or often overlooked by researchers in the field. We bring to this effort complementary backgrounds in vertebrates/ecology/evolution

Preface

and invertebrates/behavior/physiology, and we hope a truly synergistic interaction has occurred.

WILLIAM Z. LIDICKER, JR.
ROY L. CALDWELL

REFERENCES

Ballonoff, P. A., 1974, *Genetics and Social Structure: Mathematical Structuralism in Population Genetics and Social Theory,* Dowden, Hutchinson & Ross, Stroudsburg, Pa., 504p.

Banks, E. M., 1977, *Vertebrate Social Organization,* Dowden, Hutchinson & Ross, Stroudsburg, Pa., 432p.

Davis, D. E., 1974, *Behavior As an Ecological Factor,* Dowden, Hutchinson & Ross, Stroudsburg, Pa., 408p.

Stokes, A. W., 1974, *Territory,* Dowden, Hutchinson & Ross, Stroudsburg, Pa., 416p.

Tamarin, R. H., 1978, *Population Regulation,* Dowden, Hutchinson & Ross, Stroudsburg, Pa., 416p.

Weiss, K. M., and P. A. Ballonoff, 1976, *Demographic Genetics,* Dowden, Hutchinson & Ross, Stroudsburg, Pa., 414p.

CONTENTS BY AUTHOR

Bailey, E. D., 248
Berthold, P., 30
Caldwell, R. L., 75
Dethier, V. G., 246
Dingle, H., 216, 259
Evans, F. C., 144
Gill, D. E., 267
Gillespie, J. H., 207
Grant, P. R., 134
Hamilton, W. D., 278
Hansen, T. A., 295
Hegmann, J. P., 75
Herrnkind, W., 159
Hilborn, R., 83
Howard, W. E., 33
Iyama, S., 17
Järvinen, O., 174
Johnson, C. G., 139
Johnston, R. F., 43

Kennedy, J. S., 182
Lavie, B., 77
Lidicker, W. Z., Jr., 102, 282
Łomnicki, A., 225
MacArthur, R. H., 246
May, R. M., 278
Metzgar, L. H., 194
Narise, T., 17, 19, 148
Ogden, J. C., 71
Querner, U., 30
Rainey, R. C., 189
Ritte, U., 77
Sakai, K., 17
Southwood, T. R. E., 47
Taylor, L. R., 240
Taylor, R. A. J., 240
Van Valen, L., 287
Vepsäläinen, K., 174

DISPERSAL AND MIGRATION

INTRODUCTION

Although there may be "no place like home," most animals leave their places of birth sometime during their lifetimes. They may venture out into the unknown at any age; as a larva, as a just weaned or hatched juvenile, or as an adult, or at any stage of life. Moreover, biologists are gradually realizing that animal movements are just as important a component of life histories as more conventional parameters, such as age at first reproduction, fecundity, or mortality schedules. Until recently the relevant dynamics of populations were assumed to revolve around birth and death rates. Movements by individuals into and out of populations were considered rare—perhaps even pathological—or at best were inconsequential attempts to escape conditions that were no longer supportive. There were exceptions to this generalization, but these consisted mainly of the spectacular seasonal migrations exhibited by some species of vertebrates, especially birds, and a few insects (Williams, 1958; Orr, 1970). This low-key, but widespread view of animal movements was succintly summarized by Charles Elton in 1927 (p. 148): "The idea with which we have to start is. . . that animal dispersal is on the whole a rather quiet, humdrum process, and that it is taking place all the time as a result of the normal life of animals."

The situation has changed dramatically. There is a growing awareness among biologists that movements of animals away from their home areas can no longer be ignored. They may, in fact, be of very great importance and interest. Dispersal and migratory movements are increasingly appreciated as important components in population dynamics, social behavior, and genetic structure of populations. Moving individuals carry their genes with them, and if they are successful in reproducing in a new area, "gene flow" has been accomplished. Clearly, the interplay between demographic, social, and genetic processes, all tied together and modulated by movement patterns is at the heart of microevolutionary dynamics. We approach the development of this book in a spirit of enthusiasm for the integral

Introduction

importance of animal movements in understanding life history strategies, population dynamics, and evolutionary processes.

The roots of our knowledge of animal movements are deep and diversified. Investigations in this field have been pursued from the provincial perspectives of the physiologist, behaviorist, ecologist, wildlife manager, pest-control specialist, and evolutionist, to name only the most obvious. What we feel is now needed is a synthesis of this diverse array of information, techniques, and attitudes in an evolutionary context. We intend that this volume will be a contribution in this desired direction. It has been our plan to select papers of historical interest as well as those that represent important conceptual advances in the field and/or interesting points of view. Finally, we have selected a cross-section of recent papers that give the reader some idea of current activity in this rapidly advancing field.

We should point out what portions of this large field of endeavor we do not intend to cover. The omission of plant dispersal is probably our most embarrassing area of neglect. Although many aquatic plant species and even some terrestrial ones are mobile as adults or are dispersed by vegetative propagation, the typical plant is sessile. Dispersal is, nevertheless, a common and important process in plants, taking place through such disseminules as spores, pollen, and seeds. The reader is referred to *Principles of Dispersal in Higher Plants* by van der Pijl (1972) and general interest papers on plants by Allard, Baker, Ehrendorfer, Harper, Heiser, Mulligan, and Stebbins all of which appear in Baker and Stebbins (1965) for an introduction to the botanical side of this story. In defense of our omission, we can only offer the view that plant dispersal will ultimately be found to conform to the general principles developed in this volume. It has also been necessary to avoid the large and complex fields of navigation and orientation. Although these disciplines are vital aspects of any complete understanding of animal movement, they would massively over-extend what is already too large a chunk of intellectual activity for one review volume. Furthermore, four ambitious reviews have appeared recently that consider these areas in detail (Street, 1976; Baker, 1978; Schmidt-Koenig and Keeton, 1978; Able, 1980). Another subject not covered in any detail is the physiological and anatomical substrates of mobility in vagrants (see reviews by Rankin, 1978; Blem, 1980; Meier and Fivizzani, 1980). Moreover, much animal movement occurs within the confines of home ranges, having social, ecological, and evolutionary implications completely different from those we are concerned with here and hence will be neglected in our discussions. These somewhat arbitrary boundaries to our subject still leave us with extensive subject matter. Our focus will be on those movements of animals that take an individual away from its place of birth, current home, or feeding site.

As is often the case with good science, we must begin by tangling with definitions. Typically, in a rapidly growing intellectual field and especially in one that emerges from a variety of traditions in biology, the terminology is in a fluid state. This condition is advantageous in that too firm definitions too early tend to solidify concepts prematurely. However, intelligent communication requires that we organize our thoughts, and hence concepts, around a few key working definitions. While these definitions will inevitably promote disagreement, we hope also to promote improved communication, our most important consideration.

We view animal movements as not easily sorted into discrete categories. Rather they more realistically seem to fall on a continuous spectrum or axis, ranging from accidental or idiosyncratic dispersal on one extreme, through various kinds of adaptive dispersal, to round-trip types of migration indulged in by entire populations on the other extreme. The essence of *dispersal* is that an individual leaves its living place. There is thus at minimum a leaving component and a traveling component to dispersing. Generally, dispersers can be presumed to be seeking other suitable home sites so that there is an arriving component as well. Dispersal carries no implication of two-way movement by an individual or its immediate descendants. A net movement of zero is not a necessary feature of dispersal. Likewise, successful gene flow is not a necessary requisite of dispersal. Finally, dispersal should not be confused with *dispersion*, which refers to the spatial arrangement of individuals and is not a process but a property of populations.

Migration, on the other hand, is dispersal plus an implied returning to the original area (if not the same home site) by an individual or its immediate descendants. Such movements tend to be seasonally organized and may involve a large fraction of the individuals in a population. Breeding may or may not occur in the alternative home site. Except for mortality losses, a zero-net movement thus occurs with respect to any one home site. Recently, several authors, and particularly those dealing with insects, have begun to use the term "migration" to apply to any adaptive movement (see Papers 22 and 27; Baker, 1978). Most researchers working with vertebrates continue to restrict the use of the term "migration" to back and forth movements. We feel that because of the historical and continued lay use of the term in this way, it is best to continue this tradition and to confine migration as a special subset of dispersal. Note that we also are excluding the definition of "migration" as used by the paleontologist and biogeographer to describe large-scale extensions of range by species or faunal groups.

In general, as we move along our hypothetical dispersal-migration axis from accidental dispersal toward migration, we encounter behaviors that are increasingly complex and firmly integrated into an individual's genotype. However, particular cases may be difficult to classify and considerable variation may occur even within a particular species or population. A special case of dispersal, falling somewhere near the middle of the axis, is that of *nomadism*. Individuals exhibiting this type of behavior can be thought of as daily (or extremely frequent) dispersers. That is, they leave, move, and arrive on a continuing basis. Another behavior that is difficult to classify is that of the exploratory excursions often indulged in by many individuals in numerous species. Such behavior is similar to dispersal in that there are leaving and travel phases. Leaving is only temporary, however, as a new home site is not established. The behavior is also somewhat like migration in that there is a round-trip aspect. Motivationally, exploration is probably also closely related to dispersal and may actually precede the latter. Moreover, an exploratory excursion may turn into dispersal if an individual fails to return for any reason. These examples further emphasize the continuous nature of behaviors on the dispersal-migration axis.

We now come to a group of terms that we would like to restrict to the realm of population biology. The first of these is the word "population" itself. This term is used in a great variety of contexts, and we feel it should continue to be used in its most general sense, namely, to refer to a group (two or more) of conspecific individuals. The group boundary is determined arbitrarily by the investigator or observer, and no functional integration among the individuals is necessarily implied. Of course, an astute observer will choose boundaries that have particular significances for his or her purposes.

The term "deme" refers to a population having some form of functional integration. Such integration can be either genetic, demographic, social, or some combination of these. A genetic deme, for example, refers to a population whose members are both potentially interbreeding and separated by a partial or complete barrier from other such groups. In other words, there is a genetic discontinuity of some sort marking demic boundaries. Similarly, groups of individuals that share the same density regulating processes can be thought of as demographic units (or demes if one wants to be etymologically redundant) (see Lidicker, 1978, p. 134). Finally, population integrated on social bases and separated by social discontinuity from other groups are social units or demes. One of the virtually unexplored areas of population biology is the consideration of the extent to which these three sorts of demes either coincide or fail to do so.

Finally, we come to the terms "emigration" and "immigration." We define these as one-way movements out of or into populations respectively. Dispersal thus becomes emigration if an individual leaves the population under study and hence is lost to it. If a disperser subsequently enters another population, it becomes an immigrant (gained to that population). On the other hand, a disperser can leave home, travel, and establish a new home site all within a defined population, in which case no immigration or emigration has occurred. Moreover, emigration is not necessarily followed by immigration. It may instead be followed by death, nomadism, or *colonization,* which occurs when residency is established in unoccupied habitat. Lastly, it should be obvious from these definitions that what may be emigration or immigration to one investigator may only be dispersal to another. Because each investigator may define population boundaries differently, a dispersal event may be emigration to, say, a social behaviorist, but merely dispersal to a demographer. It is important that both a general term such as dispersal be available, as well as terms such as emigration and immigration with meanings specifically suitable for demographic analysis.

We have organized the papers in this volume into five sections reflecting what we believe to be logical development of the topic. First, we consider the evidence that dispersal and migratory movements can in fact have a genetic basis. This consideration is essential for establishing the claim that our hypothetical dispersal-migration axis has genetic determination of the behavior as a major variable. A genetic basis for these behaviors is also essential for any consideration of their evolution. Second, we consider circumstances leading to the leaving component of dispersal and migration. What conditions motivate organisms to move? In this context, we will also consider a balance sheet of advantages and disadvantages for individuals contemplating risks and benefits of dispersal or migration. Having launched the dispersers on their way, we are next interested in factors that influence the success of their travels and chances for re-establishment. Fourth, we consider the consequences of dispersal, both genetic and ecological, leading, in turn, into our final synthetic chapter on the evolution of dispersal and migratory behaviors.

REFERENCES

Able, K. P., 1980, Mechanisms of Orientation, Navigation, and Homing, in *Animal Migration, Orientation, and Navigation,* S. A. Gauthreaux, Jr., ed., Academic, New York, pp. 283-373.

Baker, H. G., and G. L. Stebbins, 1965, *The Genetics of Colonizing Species,* Academic, New York, 588p.

Baker, R. R., 1978, *The Evolutionary Ecology of Animal Migration,* Holmes and Meier, New York, 1024p.

Introduction

Blem, C. R., 1980, The Energetics of Migration, in *Animal Migration, Orientation, and Navigation,* S. A. Gauthreaux, Jr., ed., Academic, New York, pp. 175-224.

Elton, C., 1927, *Animal Ecology,* MacMillan, New York, 207p.

Lidicker, W. Z., Jr., 1978, Regulation of Numbers in Small Mammal Populations—Historical Reflections and a Synthesis, in *Populations of Small Mammals Under Natural Conditions,* D. P. Snyder, ed., Pymatuning Laboratory of Ecology, University of Pittsburgh, Pittsburgh, pp.122-141.

Meier, A. H., and A. J. Fivizzani, 1980, Physiology of Migration, in *Animal Migration, Orientation, and Navigation,* S. A. Gauthreaux, Jr., ed., Academic, New York, pp. 225-282.

Orr, R. T., 1970, *Animals in Migration,* MacMillan, New York, 303p.

Rankin, M. A., 1978, Hormonal Control of Insect Migratory Behavior, in *Evolution of Insect Migration and Diapause,* H. Dingle ed., Springer-Verlag, New York, pp. 5-32.

Schmidt-Koenig, K., and W. T. Keeton, eds., 1978, *Animal Migration, Navigation, and Homing,* Springer-Verlag, New York, 480p.

Street, P., 1976, *Animal Migration and Navigation,* Scribner, New York, 144p.

van der Pijl, L., 1972, *Principles of Dispersal in Higher Plants,* 2nd ed., Springer-Verlag, New York, 162p.

Williams, C. B., 1958, *Insect Migration,* Collins, London, 235p.

Part I

DOES DISPERSAL HAVE A GENETIC BASIS?

Editors' Comments on Papers 1 Through 10

1 **SAKAI, NARISE, and IYAMA**
  Migration Studies in Several Wild Strains of Drosophila melanogaster

2 **NARISE**
  Studies on Competition in Plants and Animals, X. Genetic Variability of Migratory Activity in Natural Populations of Drosophila melanogaster

3 **BERTHOLD and QUERNER**
  Genetic Basis of Migratory Behavior in European Warblers

4 **HOWARD**
  Innate and Environmental Dispersal of Individual Vertebrates

5 **JOHNSTON**
  Population Movements of Birds

6 **SOUTHWOOD**
  Excerpts from *Migration of Terrestrial Arthropods in Relation to Habitat*

7 **OGDEN**
  Artificial Selection for Dispersal in Flour Beetles (Tenebrionidae: Tribolium)

8 **CALDWELL and HEGMANN**
  Heritability of Flight Duration in the Milkweed Bug Lygaeus kalmii

9 **RITTE and LAVIE**
  The Genetic Basis of Dispersal Behavior in the Flour Beetle Tribolium castaneum

10 **HILBORN**
  Similarities in Dispersal Tendency Among Siblings in Four Species of Voles (Microtus)

In a most basic sense, anything an organism does is done in the context of its genetic constitution. An existing behavior may not be directed by an individual's genotype, but that behavior must at least be permitted by it. For example, an animal may behave so as to be caught and eaten by a predator. Such behavior is not expected to be directed by the prey's genotype, but certainly its genetics are compatible with such an event. Is being caught by a predator genetically based behavior? Although this may seem a trivial point, it illustrates the profound difficulties associated with questions concerning the genetic basis of behaviors, especially those that are complex. The simple fact that an organism possesses locomotory organs necessitates a genetic basis for all its active movements, including any that we choose to define as dispersal or migration. Similarly, organisms tend to be genetically programmed to avoid noxious stimuli and uninhabitable places. From these kinds of fundamental genetic involvement in movements, there exists the potential for a continuous spectrum of increasingly sophisticated genetic control of behaviors associated with dispersal and migration.

Furthermore, the control of dispersal and migration may be affected through genes governing a wide variety of behavioral, physiological, and anatomical factors. Some of these factors may have an obvious and direct bearing on movement. Others will not. For example, dispersal in some insects has been shown to be influenced by polymorphisms for wing length—an obvious and direct mechanism. On the other hand, control may be exerted much more subtly such as through variation in behavioral responses to stimuli eliciting takeoff; differences in the ability to lay down fuel reserves; and variation in behavioral, physiological, and in some cases even anatomical responses to social stimuli such as crowding or the presence of mates. Also, these genetically based differences affecting dispersal and migration are often only realized under very specific sets of environmental conditions, making their discovery and study most challenging.

The extent to which an organism's genotype is involved in determining dispersal and migratory behavior is of fundamental importance. If these kinds of behaviors are to evolve, a genetic basis for distinguishing among individuals must exist. Moreover, if our postulated dispersal-migration axis approximates reality, it depends on the corollary of increasing genetic control along its length. Finally, the discipline of behavioral genetics, of which we are concerned here only with a small subset, has its own intrinsic interest and merits.

In Part I we provide papers illustrating various levels of genetic involvement in dispersal and migration. Some adaptations characterize all members of a species or at least some large fraction thereof.

The existence of *disseminules,* or life-history stages specifically adapted for dispersal, illustrates this point. Plants provide the most numerous and diverse examples, but even among animals there are often fertilized eggs, larvae, or other ontogenetic stages particularly adapted for dispersal. In addition to age bias, dispersal can often be sex biased, and this form of sexism similarly illustrates species-specific dispersal adaptations that clearly have a genetic basis. Having firm genetic roots are seasonal migrations, sometimes involving thousands of miles and many navigational skills, indulged in by all members of a species. Although there may be a learned component in some cases, genetic determination clearly predominates (see reviews by Orr, 1970; Schmidt-Koenig, 1975; Emlen, 1975; Leggett, 1977; Gwinner, 1977). There is extensive literature documenting these species-specific adaptations for dispersal and migration. We conclude confidently that the question of a genetic basis for these kinds of species-specific adaptations must be answered in the affirmative. Additional evidence is provided by well-documented cases, especially among birds and insects that colonize oceanic islands and then proceed to lose their adaptations for flight. If selection can operate against dispersal, it can also operate to improve dispersal abilities and tendencies.

If age- and sex-biased dispersal are under genetic control, what factors are critical in the evolution of such patterns? These issues are of considerable potential interest but rarely have been addressed. Is it, for example, most important that dispersers have the highest possible survival value under environmental hardships, or is this irrelevant in view of a disperser's low probability of successfully reestablishing itself in a new location? On the other hand, should dispersers carry substantial potential for future reproduction (called "reproductive value" by Fisher, 1958) or should such valuable individuals be protected against the risks of dispersal (Paper 22)? Under what circumstances should one sex be vagile and the other not? Clearly, the answers to these questions will relate to the kind of dispersal involved and its motivating circumstances (see Part II).

A few examples will illustrate these points. Ghiselin (1974) discusses at length the adaptive significance of various dispersal life-history stages, particularly of invertebrates. Often in insects pre-reproductives predominate among dispersers (Johnson, 1969; Waloff and Bakker, 1963; Paper 22; Foster, 1978). On the other hand, the maximum dispersal rate in flour beetles, *Tribolium castaneum,* occurs when reproductive potential is greatest (Ritte and Agur, 1977), and the spectacular migrations of the Monarch butterfly are adult phenomena (Urquhart, 1960). While long-distance seasonal migrations are well-known among birds, it is less appreciated that juveniles of nonmigratory birds often exhibit a dispersal phase (Grinnell, 1922; Pinowski, 1965,1967; Weise and Meyer, 1979). Among mammals, juvenile males often represent the dispersal class (see review by Brown, 1966), and this situation is best illustrated by the Family Sciuridae (squirrels). A typical case is represented by the Richardson's Ground Squirrel, *Spermophilus richardsonii,* in which young males disperse while young females take over territories of closely related females (Yeaton, 1972). This pattern is no more than a generalization, however, as even in the squirrel family there is an exception in the Least Chipmunk, *Eutamias minimus,* where more females than males were observed to move long distances (Meredith, 1974). Moreover, young male predominance is often a plurality only, since all other age and sex groups may contribute significantly to dispersal. In fact pregnant females are not uncommonly found dispersing in some groups such as microtine rodents (see page 110 of Paper 11). Northcutt et al. (1974) even report a pregnant mink, *Mustela vison,* dispersing over saltwater. In birds, on the other hand, females are more often the highly dispersive sex (Greenwood, 1980), most known exceptions to this generalization being among the water birds.

Of greater interest than cases of species-specific dispersal adaptations is the question of whether conspecific individuals can vary in their genetic propensity to move. Of course intraspecific genetic variation is to be expected as an intermediate step necessarily preceding genetic fixation. Nevertheless, the potential existence of such variation poses interesting ecological and evolutionary questions. We will first examine interpopulation, subspecific, or strain differences in dispersal behavior, and then consider evidence for intrapopulation (polymorphic) variation in such traits.

Although little appreciated, there is in fact a considerable body of evidence for interpopulational variation in dispersal and migratory behavior, and the cases are rapidly accumulating. One of the foremost contributors to this field has been Takashi Narise, formerly with the National Institute of Genetics (Japan) and now at the Josai Dental University, Sakado. In a series of ingenious experiments he has demonstrated strain differences in dispersal movements in *Drosophila melanogaster* and *D. ananassae* (See Papers 1 and 2, as well as Sakai et al., 1958; Narise et al., 1960). Then he went on to show that the dispersal tendency of a particular strain depends also on which other strains are in competition with it (Paper 15). Moreover, the larger the difference in genetic background between the two strains in competition, the more intense is the stimulatory effect on dispersal (Narise, 1969). This latter result is particularly interesting in view of the opposite relation prevailing in interspecific competition where more similar species tend to compete more strongly.

Naylor (1961), Żyromska-Rudzka (1964), and Prus (1966) all report strain differences in dispersal in flour beetles (*Tribolium*). Another insect example, now classic, is the report by Brown (1951) that water boatmen (Corixidae) from temporary bodies of water are much more dispersible than are those from permanent water. More recently, Gilbert and Singer (1973) have discovered what appear to be genetically based and adaptively appropriate differences in dispersal rates between colonies of the butterfly, *Euphydryas editha*, living in different habitats.

Among vertebrates, there are too many examples of interpopulation variation in dispersal for a complete review here. In both birds and mammals, examples are known where some populations of a species are migratory and others are not. A terrestrial bird example is the Orange-crowned Warbler, *Vermivora celata*, which is migratory except for the subspecies from the California Channel Islands (Johnson, 1972). A particularly well-documented case in an aquatic bird is reported by Coulson and Brazendale (1968). They give evidence that different breeding colonies of the cormorant, *Phalacocorax carbo*, in the eastern North Atlantic have different dispersal potentials, different directions of dispersal, and different tendencies to cross large bodies of water and/or land. A good example from mammals is provided by the Free-tailed Bat, *Tadarida brasiliensis*, in which some subspecies are migratory and others are not (Cockrum, 1969). As far as strain differences in dispersal are concerned, Incerti and Pasquali (1967) have made an extensive survey of exploratory behavior in wild, semiwild, and domestic strains of house mice, *Mus musculus*. They found many significant differences attributable to genetic variation as well as to sex and reproductive condition. Winston and Lindzey (1964) report that two albino strains of domestic mice were the slowest of six strains tested in escaping from a water bath, and that this difference persisted even among hybrids between the two albino strains. Paper 3 documents the genetic basis for interpopulation variation in migratory behavior in the European warbler, *Sylvia atricapilla*. This paper is particularly significant in that it is one of the very few studies to confront directly the issue of genetic variability in the control of migration rather than simply implying such control from indirect evidence.

We now come to the most intriguing question of all, and that is whether or not dispersal polymorphisms exist within populations. This possibility raises some particularly important and difficult questions about the evolutionary origin and maintenance of such conditions, issues that will be addressed to some extent in Part V.

The notion of two types of dispersers coexisting in a given population is not a new one. As far back as 1949, the respected mammalogist, W. H. Burt, ventured the opinion that "an inherent desire to wander was differentially developed among individuals" (p.26). Andrewartha and Birch (1954) also refer to "an innate tendency for dispersal." In 1960, Howard published a paper on innate and environmental dispersal that is still widely cited (Paper 4). He suggested that many studies support the idea that populations consist of some individuals who are sedentary and some who go considerable distances. The latter were innate (genetically programmed) dispersers, and that as a result of this situation, dispersal distances were bimodal. In addition to "innate dispersal," Howard described "environmental dispersal," which was over a short distance and stimulated by poor conditions, social rejection, and so on. Finally, he prophetically outlined Krebs's modification of the Chitty hypothesis (Krebs, 1978) for explaining microtine cycles in claiming that "the higher the population density, the greater the number of innate dispersers" (p. 158). In 1961 Johnston reached similar conclusions for birds (Paper 5), and in 1962 Lidicker (Paper 30) provided some arguments for how natural selection could favor the evolution of genetic dispersers. In 1962 Southwood (Paper 6) published a particularly useful review of movements in arthropods and provided many examples of dispersal polymorphisms, a portion of which is reprinted here, including his detailed treatment of one representative group, the Coleoptera.

In the last several decades numerous investigators have searched for evidence of genetically based dispersal variability within populations. Some failed; for example, Blackwell and Ramsey (1972) failed to find any correlation between exploratory activity in the Old-field Mouse, *Peromyscus polionotus,* and several biochemical characters. Similar results were obtained by Riggs (1979) for *Microtus californicus*. Some early efforts in both field (Łomnicki, 1969) and laboratory (Ogden, Paper 7), while not conclusive, were suggestive of such genetic variation. Ogden's paper is a moderately successful attempt to select artificially for improved dispersal behavior in flour beetles (*Tribolium*). Over the past 15 years enough evidence has accumulated to permit the conclusion that genetic variation for dispersal tendencies, including sometimes true polymorphism, is widespread among animals. Two papers are reprinted here to illustrate these findings among insects. Caldwell and Hegmann (Paper 8) document substantial variation in flight duration in the milkweed bug, *Lygaeus kalmii,* and report heritabilities for this behavior of .20 and .41 for male and female parents respectively. Ritte and Lavie (Paper 9) report on remarkably successful experiments for selection of dispersal behavior in *Tribolium castaneum*. Both low and high dispersal lines were produced with the difference between them being attributed to the genotype at a single sex-linked locus.

Among mammals, the evidence is mixed. Two papers that failed to find correlations between dispersal behavior and selected biochemical traits have already been mentioned (Blackwell and Ramsey, 1972; Riggs, 1979). In contrast, Myers and Krebs (1971) describe genetic (biochemical), behavioral, and reproductive differences between dispersing and resident voles of the species *Microtus pennsylvanicus* and *M. ochrogaster*. In this study, dispersers were defined as individuals trapped on plots from which resident voles had been previously removed. Pickering et al. (1974) also report biochemical differences of a similar sort on a very small sample of *Microtus pennsylvanicus*. French et al. (1968) describe a bimodal distribution of dispersal distances in the pocket mouse, *Perognathus formosus*. Hilborn (Paper 10) used a unique approach to arrive at the fascinating conclusion that groups of siblings in four species of *Microtus* are more similar in their dispersal tendencies than are nonsiblings.

In our enthusiasm for supporting genetic involvement in dispersal and migratory movements, we should not forget that there are many cases of movements in which genetic factors operate only in the most general permissive sense. These are the movements toward the left end of our dispersal-migration axis. Some of these are individual affairs—accidental or idiosyncratic. Others, however, may involve organized movements of entire populations or even species. Much more will be said about these types of movements in Part II, but here we would like to mention as examples the mass movements (irruptions) that occur erratically in many kinds of animals. Svärdson (1957) provides a review of such movements in birds and mentions some mammalian examples. Holman and Hill (1961) describe a mass unidirectional movement of the water snake *Natrix sipedon*. Many other examples can be found in Orr (1970, pp. 4-13). No direct genetic control and no polymorphisms would seem to be involved in these cases. Not even all cases of dispersal polymorphism are entirely genetic, however, as illustrated by the migratory locust, *Locusta migratoria* (Kennedy, 1956, 1961) and tent caterpillars, *Malacosoma pluviale* (Wellington, 1965).

REFERENCES

Andrewartha, H. G., and L. C. Birch, 1954, *The Distribution and Abundance of Animals*, University of Chicago, Chicago, 782p.

Blackwell, T. L., and P. R. Ramsey, 1972, Exploratory Activity and Lack of Geneotypic Correlates in *Peromyscus polionotus*, *J. Mammal.* **53:** 401-403.

Brown, E. S., 1951, The Relation Between Migration-rate and Type of Habitat in Aquatic Insects, with Special Reference to Certain Species of Corixidae, *Zool. Soc. London Proc.* **121**:539-545.

Brown, L. E., 1966, Home Range and Movements of Small Mammals, *Zool. Soc. London, Symp. No. 18,* pp. 111-142.

Burt, W. H., 1949, Territoriality, *J. Mammal.* **30**:25-27.

Cockrum, E. L., 1969, Migration in the Guano Bat, *Tadarida brasiliensis,* in *Contributions in Mammalogy,* J. K. Jones, Jr., ed., Museum of Natural History, University of Kansas, Lawrence, pp. 303-336.

Coulson, J. C., and M. G. Brazendale, 1968, Movements of Cormorants Ringed in the British Isles and Evidence of Colony-specific Dispersal, *Br. Birds* **61**:1-21.

Emlen, S. T., 1975, Migration: Orientation and Navigation, in *Avian biology,* vol. 5, D. S. Farner and J. R. King, eds., Academic, New York, pp. 129-219.

Fisher, R. A., 1958, *The Genetical Theory of Natural Selection,* 2nd rev. ed., Dover, New York, 291p.

Foster, W. A., 1978, Dispersal Behaviour of an Intertidal Aphid, *J. Anim. Ecol.* **47**:653-659.

French, N. R., T. Y. Tagami, and P. Hayden, 1968, Dispersal in a Population of Desert Rodents, *J. Mammal.* **49**:272-280.

Ghiselin, M. T., 1974, *The Economy of Nature and the Evolution of Sex,* University of California, Berkeley, 346p.

Gilbert, L. E., and M. C. Singer, 1973, Dispersal and Gene Flow in a Butterfly Species, *Am. Nat.* **107**:58-72.

Greenwood, P. J., 1980, Mating Systems, Philopatry and Dispersal in Birds and Mammals, *Anim. Behav.* **28**:1140-1162.

Grinnell, J., 1922, The Role of the "Accidental," *Auk* **39**:373-380.

Gwinner, E., 1977, Circannual Rhythms in Bird Migration, *Annu. Rev. Ecol. Syst.* **8**:381-405.

Holman, J. A., and W. H. Hill, 1961, A Mass Unidirectional Movement of *Natrix sipedon pictiventris, Copeia* **1961**:498-499.

Incerti, G., and A. Pasquali, 1967, Esperimenti sulle attivita' esploratorie in topi domestici e selvatici *(Mus musculus), Ist. Lombardo Accad. Sci. Lett. Rendiconti,*ser. B, **101**:19-46.

Johnson, C. G., 1969, *Migration and Dispersal of Insects by Flight,* Methuen, London, 763p.

Johnson, N. K., 1972, Origin and Differentiation of the Avifauna of the Channel Islands, California, *Condor* **74**:295-315.

Kennedy, J. S., 1956, Phase Transformation in Locust Biology, *Biol. Rev.* **31**:349-370.

Kennedy, J. S., 1961, Continuous Polymorphism in Locusts, *R. Entomol. Soc. Symp.* **1**:80-90.

Krebs, C. J., 1978, A Review of the Chitty Hypothesis of Population Regulation, *Can. J. Zool.* **56**:2463-2480.

Leggett, W. C., 1977, The Ecology of Fish Migration, *Annu. Rev. Ecol. Syst.* **8**:285-308.

Łomnicki, A., 1969, Individual Differences among Adult Members of a Snail Population, *Nature* **223**:1073-1074.

Meredith, D. H., 1974, Long Distance Movements by Two Species of Chipmunks *(Eutamias)* in Southern Alberta, *J. Mammal.* **55**:466-469.

Myers, J. H., and C. J. Krebs, 1971, Genetic, Behavioral, and Reproductive Attributes of Dispersing Field Voles *Microtus pennsylvanicus* and *Microtus ochrogaster, Ecol. Monogr.* **41:**53-78.

Narise, T., 1969, Migration and Competition in *Drosophila*. II. Effect of Genetic Background on Migratory Behavior of *Drosophila melanogaster, Jpn. J. Genet.* **44:**297-302.

Narise, T., K. Sakai, and S. Iyama, 1960, Mass Migrating Activity in Inbred Lines Derived from Four Wild Populations of *Drosophila melanogaster, Nat. Inst. Genet. (Japan), Ann. Report No. 10,* pp. 28-29.

Naylor, A. F., 1961, Dispersal in the Red Flour Beetle *Tribolium castaneum* (Tenebrionidae), *Ecology* **42:**231-237.

Northcott, T. H., N. F. Payne, and E. Mercer, 1974, Dispersal of Mink in Insular Newfoundland, *J. Mammal.* **55:**243-248.

Orr, R. T., 1970, *Animals in Migration,* MacMillan, New York, 303p.

Pickering, J., L. L. Getz, and G. S. Whitt, 1974, An Esterase Phenotype Correlated with Dispersal in *Microtus, Ill. Acad. Sci. Trans.* **67:**471.

Pinowski, J., 1965, Dispersal of Young Tree Sparrows (*Passer m. montanus* L.), *Acad. Pol. Sci. Bull.,* cl. II, **13:**509-514.

Pinowski, J., 1967, Experimental Studies on the Dispersal of Young Tree Sparrows, *Ardea* **55:**241-248.

Prus, T., 1966, Emigrational Ability and Surface Numbers of Adult Beetles in 12 Strains of *Tribolium confusum* Duval and *T. castaneum* Herbst (Coleoptera, Tenebrionidae), *Ekol. Pol.,* ser. A, **14:**547-588.

Riggs, L. A., 1979, Experimental Studies of Dispersal in the California Vole, *Microtus californicus,* Ph.D. dissertation, University of California, Berkeley, 237p.

Ritte, U., and Z. Agur, 1977, Variability for Dispersal Behavior in a Wild Population of *Tribolium castaneum, Tribolium Inf. Bull.* **20:**122-131.

Sakai, K., T. Narise, Y. Hiraizumi, and S. Iyama, 1958, Studies on Competition in Plants and Animals. IX. Experimental Studies on Migration in *Drosophila melanogaster, Evolution* **12:**93-101.

Schmidt-Koenig, K., 1975, *Migration and Homing in Animals,* Springer-Verlag, New York, 99p.

Svärdson, G., 1957, The "Invasion" Type of Bird Migration, *Br. Birds* **50:**314-343.

Urquhart, F. A., 1960, *The Monarch Butterfly,* University Press, Toronto, 361p.

Waloff, N., and K. Bakker, 1963, The Flight Activity of Miridae (Heteroptera) Living on Broom, *Sarothamnus scoparius* (L.) Wimm., *J. Anim. Ecol.* **32:**461-480.

Weise, C. M., and J. R. Meyer, 1979, Juvenile Dispersal and Development of Site-fidelity in the Black-capped Chickadee, *Auk* **96:**40-55.

Wellington, W. G., 1965, Some Maternal Influences on Progeny Quality in the Western Tent Caterpillar Malacosoma pluviale (Dyar), *Can. Entomol.* **97:**1-14.

Winston, H. D., and G. Lindzey, 1964, Albinism and Water Escape Performance in the Mouse, *Science* **144:**189-191.

Yeaton, R. I., 1972, Social Behavior and Social Organization in Richardson's Ground Squirrel (*Spermophilus richardsonii*) in Saskatchewan, *J. Mammal.* **53:**139-147.

Żyromska-Rudzka, H., 1964, The Migration Tendency of Two Strains of *Tribolium castaneum* (Hbst)-Preliminary Observations, *Ekol. Pol.,* ser. A, **12:**379-388.

Migration Studies in Several Wild Strains of Drosophila melanogaster.

(By Kan-Ichi SAKAI, Takashi NARISE and Shinya IYAMA)

Six wild strains of *Drosophila melanogaster* were collected from different localities in some parts of Japan. They were propagated as populations for two generations in the laboratory and were tested for their migrating activity. In the original tube, 80, 120 and 160 flies were placed, and the number of migrated flies from the original tube was counted after 2 days. The experiment was replicated four times. Analysis of variance of data is presented in Table 1, which shows that local strains are different with regard to migrating activity.

Table 1. Analysis of variance of number of migrated flies for three densities of original population.

Source	d.f.	Mean square
Population density	2	308.98*
Local strains	5	489.31**
Density × strain	10	44.46
Replication	3	113.17
Error	51	67.48

*, ** Significant at the 5% and the 1% level, respectively.

The percentage of migrated flies in six wild strains is presented in Table 2.

Table 2. Percentage of migrated flies from the original tubes at three levels of population density.

Strains	Migrating activity (%)			
	80+)	120+)	160+)	Average
Tateba	39.4	40.9	42.8	41.03
Iguro	44.3	55.3	54.7	51.43
Kama	48.8	55.2	58.8	54.27
Te-sima	50.1	59.3	57.7	55.70
Isima	55.6	60.7	55.7	57.33
Katsunuma	55.0	55.8	64.8	58.53

+) The number of flies placed at the start of the experiment in the original tube: Population density.

The migration in these wild strains was incomparably more active than in the inbred laboratory strain, Samarkand, in which the migration hardly occurred unless the number of flies in the original tube exceeded 150.

The next question to be answered was: Is there a threshold in the size of population below which the burst of migration rarely occurs? A series of experiments each with 4 replications were again conducted making the size of population in the original tube at 10, 20, 40 and 60 flies. Analysis of variance of the data obtained in these experiments showed that the effect of population size as well as the effect of interaction between densities and strains were highly significant.

The migration percentage in relation to various sizes of population in the original tube is graphically illustrated in Fig. 1.

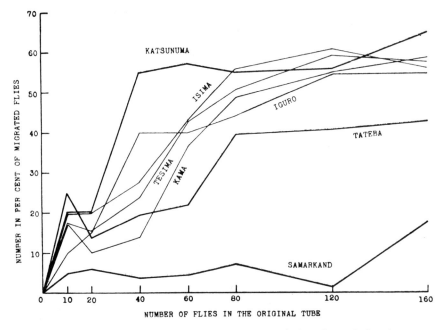

Fig. 1. Relation between migration percentage and size of population, in one laboratory stock, Samarkand, and six wild strains.

Fig. 1 shows that the thresholds in the size of population responsible for the burst of mass migration differ from strain to strain: the critical size of population was 60~80 in Tateba- and Kama-strain, 60 in Tesima- and Isima-strain, 40 in Iguro- and Katsunuma-strain,

2

Reprinted from *Jpn. J. Genet.* 37:451-461 (1962)

Studies on Competition in Plants and Animals, X. Genetic Variability of Migratory Activity in Natural Populations of *Drosophila melanogaster*[1]

Takashi NARISE

National Institute of Genetics, Misima

Introduction

It is generally accepted that the main factors responsible for organic evolution are mutation, selection, random genetic drift, competition and migration. Of those five, the first three have long been subjected to theoretical and experimental studies by many workers (Huxley 1948). Regarding competition, Sakai (1955) has carried out a series of experimental works with plants, and succeeded to demonstrate that competition in plants was controlled by genetic factors. Migration, however, has failed to capture evolutionists' attention which it deserves, except for a few theoretical studies conducted by Wright (1931) and Kimura (1953). It was 1958 that Sakai and his colleagues succeeded in opening a new gate for the problem with *D. melanogaster*. According to them, migratory behavior of the flies could be partitioned into two components: the mass-migration which is effected by the pressure of population density and the random-migration caused by random movement of individual flies. It was also found that wild populations of *D. melanogaster* involved variation with regard to activities for both kinds of migration. This paper will describe the result of some experiments conducted for the purpose of estimating genetic variability in four wild populations of *D. melanogaster* in succession of the experiments above described.

Material and Method

Strains of *D. melanogaster* used in this study include, beside one laboratory stock, Samarkand, four wild populations collected in Japan in 1956. They are named according to their native habitats as follows: Katsunuma (KN), Tateba-sima (TB), Iguro-sima (IG) and Te-sima (TS). In the inbreeding experiment, a certain number of inbred lines were grown by the continued full-sib matings for a number of generations. During the course of continuous inbreeding, migratory activity was examined in each generation. The apparatus, which we call population-tubes, for

---

[1] Contributions from the National Institute of Genetics, Japan, No. 442.

testing migratory activity are those described in the previous paper (Sakai *et al.* 1958). Measurement of migratory activity was conducted by counting flies migrated from the central tube to three connected tubes after definite time interval. Actually, the migratory activity was measured in three ways: the general, the mass- and the random-migratory activity. The general migratory activity involved the other two, and was measured by counting migrated flies in per cent of total number initially introduced into the central tube. The mass-migratory activity was measured by determining the lowest density of population which caused mass-migration of flies. In the present experiment, the tested sizes of original population were made at 10, 20, 40, 60, 80 and 120 for TB and KN line-groups, and 10, 20, 80 and 120 for TS and IG line-groups. Counting of migrated flies was conducted 6 hours after connection of fresh tubes. The random-migratory activity was measured 24, 30 and 48 hours after the connection of fresh tubes by counting number of migrated flies from the central tube in which 80 flies were initially introduced.

Materials for the selection experiment were the IG and TS strains which had been propagated without any artificial selection in our laboratory since their introduction. The experiment was conducted by selecting in each generation 20 to 30 with regard to high migratory activity from 120 flies in a combined set of 10 tubes. The selection was continued up to the 21st generation with intermittent tests for the effect of selection. The latter test consisted of measurement of the general migratory activity. At the final stages of selection, the selected lines were tested for their random- and mass-migratory activities.

It should be stated that all the experiments and tests above mentioned were conducted in the dark kept at 25°C. The fly population introduced in the central tube always consisted of the same number of males and females. Every test was replicated 10 to 30 times.

Experimental Results

1. The inbreeding experiment

The number of inbred lines obtained by the consecutive inbreeding through 20 generations was seven in KN and five in TB population. The general migratory activity of these 12 inbred lines is shown in Table 1 and in Fig. 1.

The graph in Fig. 1 is drawn taking the respective mean of each tested generation as zero. Deviation of various lines from the mean is expressed in terms of the corresponding standard deviation. As seen in Table 1, the general migratory activity of C-3 line of the TB population was increasing generation after generation, while A-1 line decreasing. Similar behavior was also found in lines derived from the KN population. These facts indicate that the general migratory activity is a genetic character, on which the long-continued inbreeding effected separation

Table 1. Migratory activity of 12 inbred lines of *D. melanogaster* derived from two wild populations by twenty consecutive full-sib matings

Inbred lines		Migratory activity measured for 20 inbred generations								
		1.2	5.6	9.10	11.12	13.14	15.16	17.18	19.20	Mean
KN	A-2	50.63	28.05	40.63	37.71	49.38	61.88	53.55	51.05	45.643
	B-1	53.13	60.21	48.34	63.96	64.58	46.58	53.01	56.46	58.344
	C-3	42.71	25.83	44.79	56.88	57.71	65.00	57.92	52.92	49.606
	D-2	44.79	55.04	30.84	36.25	27.09	31.46	26.25	37.92	41.132
	D-3	71.46	80.80	45.21	48.75	52.50	39.59	35.21	45.42	52.853
	F-2	27.92	37.09	35.21	27.71	23.34	39.38	23.34	48.13	37.650
	H-1	22.88	18.34	23.96	25.83	25.84	16.25	22.92	27.71	27.194
Mean		44.79	43.62	38.43	42.44	42.90	45.45	38.77	45.66	44.632
TB	A-1	50.42	28.34	25.00	22.09	18.54	25.83	24.17	22.29	26.002
	B-3	18.75	50.63	36.46	42.30	46.25	41.67	38.84	42.71	38.929
	C-2	24.59	49.58	43.54	56.46	41.88	51.04	53.13	46.67	47.981
	C-3	27.50	49.79	43.96	66.25	57.29	77.50	18.13	59.38	55.601
	F-1	27.83	37.09	21.88	18.34	14.59	22.29	24.17	35.21	25.827
Mean		29.82	43.09	34.17	41.09	35.71	43.67	39.69	41.25	38.868
Grand mean		38.55	43.40	36.65	41.88	39.96	44.71	39.22	43.82	42.320

of lines differing from each other with respect to the activity. The result of analysis of variance of data collected during the last ten generations is presented in Table 2.

Table 2. Analysis of variance of general migratory activity from 11th to 20th generation of inbreeding of twelve inbred lines derived from two wild populations

Source of variation	d. f.	m. s.
Generation	9	3583.9197**
Line	11	6281.7897**
Between populations	1	693.0178
Within population	10	6840.6669**
Generation × line	99	239.3942
Error	240	267.9908

** Significant at the 1% level.

It will have to be mentioned here that the general migratory activity of flies is not subject to inbreeding depression. No difference was also found between populations while intra-population variation was highly significant statistically. A

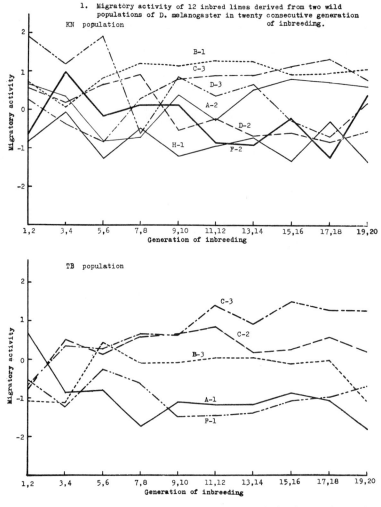

Fig. 1. Migratory activity of 12 inbred lines derived from two wild populations of *D. melanogaster* in twenty consecutive generations of inbreeding.

similar comparison among inbred lines was also made in two other populations, IG and TS. In this comparison, 16 IG-lines and 24 TS-lines were tested for their general migratory activity with five replications. The result of this test is presented in Table 3.

Table 3 again shows that segregation of inbred lines differing from each other with respect to migratory activity has taken place with no sign of inbreeding depression. All these data clearly indicate that the four natural populations of *D. melanogaster* were highly heterogeneous with regard to the migratory activity.

i) Mass-migratory activity of inbred lines: Observation on general migratory activity described above involved mass- and random-migratory activities together.

Table 3. Migratory activity of inbred lines derived from TS and IG populations in 30th generation of inbreeding

TS		IG	
Line No.	Migratory activity	Line No.	Migratory activity
2	35.50	1	34.25
6	35.25	2	34.75
8	29.50	4	31.25
9	41.35	6	14.25
11	61.35	8	40.25
12	37.00	14	49.25
13	62.50	25	41.50
14	46.35	28	50.20
18	17.50	30	56.25
20	43.25	32	65.25
22	49.50	35	54.25
23	12.50	40	16.00
28	63.50	41	41.25
30	38.50	48	45.35
31	55.25	49	55.35
32	53.75		
34	31.50		
36	55.50		
39	56.50		
40	61.75		
41	43.75		
42	44.75		
43	64.25		
44	60.50		
Mean	45.8771	Mean	41.4625
Original population	43.5000	Original population	36.5000

It was therefore necessary to reexamine those inbred lines for the two components of migratory activity separately. Two experiments were thus conducted, one with KN and TB populations, and the other with IG and TS populations. Analysis of variance of the data shows that variation among populations and also among inbred lines within same populations was statistically significant. Observations on the mass-migratory activity of various inbred lines derived from the four wild populations are presented in Table 4.

In Table 4, we find no evidence of inbreeding depression in the mass-migratory activity. Inter-line variability is very apparent and the correlation between the early and the late generations looked very high.

ii) Random-migratory activity of inbred lines: The same lines were again

Table 4. Mass-migratory activity of inbred strains in 30th and 50th inbred generations in TS and IG, and 20th and 40th in KN and TB populations

Population											
IG			TS			KN			TB		
Line	I-30*	I-50	Line	I-30	I-50	Line	I-20	I-40	Line	I-20	I-40
1	40		2	150		A-2	40		A-1	120	120
2	40	40	6	120		A-3	40		A-5	40	
4	40		8	120		A-4	40		B-1	80	
6	120	120	9	80		B-1	20	20	B-3	40	
8	60	60	11	80		C-2	40		B-4	60	60
14	60	50	12	80		C-3	60	50-60	C-2	40	
16	40		13	40	40	D-2	40	40	C-3	40	30
25	80		14	80		D-3	20	20	D-1	40	40
28	20	20	18	120	120	E-2	40		D-5	80	80
30	30		20	80		E-3	60	60	F-1	60	
32	30	30	22	60		F-2	40	40	F-4	60	50-60
35	40		23	120		G-3	60				
40	40		28	80		H-1	40				
41	40		30	120	120	H-2	40				
48	40		31	40	40						
49	80		32	40							
			34	40							
			36	40							
			39	80							
			40	40							
			41	60	60						
			42	80							
			43	40							
			44	80	80						
Mean	50.00			61.30			41.23			60.00	
Original population	40			60-80			40			80	

* I-30 means that inbreeding was continued 30 generations.

used for the test of random-migratory activity. The test was replicated ten times in IG and TS and five in KN and TB populations. In this test, counting of migrated flies was made three times from 24 hours after the commencement of the experiment to 48 hours. The statistical test of linearity for every inbred line showed that the increase of migrated flies was linearly proportional to the number of hours during which the migration was allowed to occur. Result of the statistical test of differences among regression coefficients of various lines is presented in Table 5.

Table 5 shows that the difference among regression coefficients of various inbred lines was highly significant. Since the regression coefficient is the direct

Table 5. Regression analysis of number of migrated flies on time in different lines derived from four wild populations of *D. melanogaster*

Source of variation	TS		IG		TB		KN	
	d.f.	m.s.	d.f.	m.s.	d.f.	m.s.	d.f.	m.s.
Differences among regressions	23	1221.6381**	15	717.1236**	10	1554.3364**	13	2680.2072**
Error	672	256.2822	448	225.4645	143	100.4880	118	168.1967

** Significant at the 1% level.

measure of the random-migratory activity, we could conclude that inbred lines were variable with regard to this activity. Random-migratory activities of different inbred lines are presented in Table 6.

Table 6. Random-migratory activity of inbred lines in 30th and 50th inbred generations in TS and IG, and 20th and 40th in KN and TB populations

					Population						
	IG			TS			KN			TB	
Line	I-30*	I-50	Line	I-30	I-50	Line	I-20	I-40	Line	I-20	I-40
1	0.6827		2	0.3926		A-2	0.1468	0.2067	A-1	0.4864	0.3830
2	0.8413		6	1.1321	1.2772	A-3	1.1759		A-5	1.2308	
4	1.0016	0.9872	8	0.5817		A-4	0.6372	0.6137	B-1	1.1074	
6	0.7179		9	0.9615		B-1	1.3333		B-3	0.7037	0.7039
8	0.4103	0.6538	11	0.3622		C-2	1.4183		B-4	1.0724	
14	0.5362		12	0.5000		C-3	1.1619		C-2	0.9279	
16	0.8686		13	-0.0208	0.1052	D-2	1.4243		C-3	0.9976	0.9616
25	1.0894	1.2266	14	0.8429	0.8538	D-3	1.2981		D-1	0.8273	0.7163
28	0.5498		18	0.2232		E-2	1.2720		D-5	0.9796	
30	0.3494	0.4932	20	0.8413		E-3	1.2244		F-1	1.2595	1.2628
32	0.4167	0.5083	22	0.5929		F-2	1.5401	1.4871	F-4	1.0897	
35	0.3157		23	0.3734		G-3	0.8353	0.8894			
40	0.7131		28	0.1442		H-1	1.5184	1.4089			
41	0.9439		30	0.4183		H-2	0.9199				
48	0.9695	0.9217	31	0.5112							
49	0.7243		32	0.6042							
			34	0.4183	0.6540						
			36	0.8814							
			39	0.0401							
			40	1.1410	1.1484						
			41	0.4455							
			42	0.4247							
			43	0.5497							
			44	0.4054	0.4971						
Mean	0.6962			0.5322			1.2305			0.9711	
Original population	0.7676			0.7243			0.7952			0.8604	

* I-30 means that inbreeding was continued 30 generations.

It is found in this table that inbreeding intensity has nothing to do with the random-migratory activity.

We should here remember that the Samarkand strain was very inactive with regard to either mass-migration or random-migration. On the basis of the inbreeding experiments described above, we may assume that the inactive migration of the Samarkand strain may not be due to the effect of inbreeding depression, but rather due to genetic segregation.

2. Selection experiment

So far, data have been presented to show that natural populations of Drosophila flies are genetically heterogeneous in respect of migratory activity. It may then be of interest to find to what extent the migratory activity could be an objective of selection. The selection experiment to be described here concerns with the selection of flies for high general migratory activity. A selection toward low migratory activity was not conducted because of a technical difficulty.

Table 7. Migratory activity of selected lines in different generations

Population	Generation	3	6	9	12	15	18	21
IG	Migratory activity	28.3750	29.6667	29.3750	35.4167	45.4583	34.2917	39.2083
	Standard deviation	8.1913	10.6721	6.0271	7.2467	6.0394	6.9145	5.7621
TS	Migratory activity	30.5205	35.7750	39.7833	38.4167	42.8750	36.2083	41.9583
	Standard deviation	5.6916	8.9926	13.2583	7.5354	5.3645	5.2346	6.6281

Table 8. The analysis of variance of effect of selection on migratory activity in TS and IG populations

Source of variation	d. f.	m. s.
Between populations	1	1420.8481
Between generations within population	12	912.8921**
Between replications within generation within population	406	133.8501

** Significant at the 1% level.

Result of the selection experiment and the analysis of variance of the data are presented in Tables 7 and 8.

As seen in Table 8, variation between generations is highly significant. These two tables in combination indicate that general migratory activity can be generally improved by selection, though the effect may vary among different populations.

A marked response to selection was detected between the 12th and the 15th generation, after which the response reached a plateau despite of further selection. A problem then arises whether those high general migratory activities attained by selection might be due to improvement in the mass-migratory activity or in the random one or both. In order to answer this question, both migratory activities were examined in the 21st generation of selection. The results of these experiments are summarized in Table 9.

Table 9. Mass and random-migratory activity of selected lines in the 21st generation of selection in IG and TS populations

Population		IG	TS
Lowest level of population density causing mass-migration	Selected line	30	30
	Original line	40	60-80
Intensity of random-migration per hour (%)	Selected line	1.0237	0.9795
	Original line	0.7676	0.7234

From Table 9, it is found that the selected lines are more active than their original strains in either mass or random-migratory activity. The analysis of variance of the data for mass-migratory activity is shown in Table 10.

Table 10. The analysis of variance of the mass-migratory activity in the 21st generation of selection

Source of variation	d. f.	m.s.
Between strains	1	970.0316
Between densities	3	3435.7688**
Strains × Densities	3	442.0658**
Error	112	86.2861

** Significant at the 1% level.

Table 10 shows that variation between fly densities and that due to interaction between strains and fly density are significant so far as mass-migratory activity is concerned. It means that strains differ among themselves with regard to fly density causing mass-migration. It is of interest to notice that the difference between

Table 11. Test of significance of difference between random-migratory activities of two selected line groups, IG and TS in the 21st generation of selection

Source of variation	d. f.	m. s.
Difference	1	260.6901
Error	104	417.9273

selected lines from two wild strains, which were initially apparently different, was not at all significant as seen in Table 11.

Thus, it can be shown that the two kinds of migratory activity are successfully improved by selection, providing an additional evidence toward the hypothesis of their genetically controlled nature.

Discussion and Conclusion

In the previous paper, Sakai *et al.* (1958) have pointed out that the migratory activity would be under genetic control. They found that wild strains were very active in migration in comparison with the laboratory strain, Samarkand. What could have brought about such a difference between wild and laboratory strains? Loss of hybrid vigor due to long continued inbreeding might have been responsible for the very low migratory activity of the Samarkand strain, or either a possible inconscious selection for it or random genetic drift might have been the cause. In the inbreeding experiment described in this paper, natural populations could be separated by the method of full-sib breeding into inbred lines which are different from each other with regard to migratory activity. Of interest in these experiments was to find no definite relation between inbreeding intensity and the migratory activity. In the selection experiment, migratory activity was proved to be subject of selection. Consequently, it seems reasonable to assume that the low migratory activity of the Samarkand strain would not be an effect of inbreeding, but it might be either an effect of artificial selection for low activity effected in the laboratory or an effect of random genetic drift.

It is now clear that the migratory activity is a character which is under genetic control. It is also concluded that natural populations involve various genotypes with respect to the character. It has been shown in the previous paper as well as in the present one that a general migratory activity involves two activities, one the mass and the other random, and they are independently inherited. The correlation coefficients between the mass- and the random-migratory activity measured among inbred lines ranged between $+0.2042$ and -0.5249 in four different wild strains.

Though we are not aware at present of what role could the migration of animals play in the maintenance of genetic variability in wild population, it may deserve more attention than it did by the geneticists interested in evolution of animal species.

Summary

1) A number of inbred lines derived from four wild strains of *D. melanogaster* were tested for their general, mass and random-migratory activities. It

was found that these wild strains were without exception quite heterogeneous in respect of the three kinds of migratory activities mentioned above.

2) The same wild strains were subjected to a selection experiment for higher general migratory activity. The selection proved to be effective and the selected lines were found to be very active in either mass- or random-migration.

3) The findings described above give an evidence that the migratory activities of Drosophila flies are genetic characters, a good deal of variation of which is concealed in wild populations.

Acknowledgement

The author is particularly indebted to Dr. Kan-Ichi Sakai for his helpful guidance and valuable suggestions during this experiment and a critical reading of the manuscript.

The author is also thankful to Dr. Shin-ya Iyama who was kind enough to conduct the inbreeding experiment on the author's behalf for so long duration as two years between 1957 and 1959 when latter was abroad on leave.

Literature Cited

Huxley, J. 1948 Evolution, The Modern Synthesis. George Allen and Unwin Ltd. London.
Kimura, M. 1953 "Stepping stone" model of population. Ann. Rep. Nat. Inst. Genet, Japan **3**: 62-63.
Sakai, K. I. 1955 Competition in plants and its relation to selection. Cold Spr. Harb. Symp. **20**: 137-157.
————, T. Narise, Y. Hiraizumi and S. Iyama 1958 Studies on competition in plants and animals, IX. Experimental studies on migration in *D. melanogaster*. Evolution **12**: 93-101.
Wright, S. 1931 Evolution in Mendelian population. Genetics **16**: 97-159.

GENETIC BASIS OF MIGRATORY BEHAVIOR IN EUROPEAN WARBLERS

Peter Berthold and Ulrich Querner

Vogelwarte Radolfzell, Max-Planck-Institut für Verhaltensphysiologie, D-7760 Radolfzell-Möggingen, West Germany

Abstract. *The seasonal course and magnitude of migratory restlessness recorded in four populations of the blackcap* Sylvia atricapilla *differ in a population-specific fashion that is related to the distance of travel. Experimentally produced hybrids of an exclusively migratory European population and a partially migratory African population showed intermediate migratory restlessness and an intermediate percentage of birds displaying restlessness compared to the two parental stocks. These results demonstrate the genetic basis of migratory behavior in this avian species and support the hypothesis that partial migration of populations is due to polymorphism.*

In Old World warblers of the genera *Phylloscopus* and *Sylvia* there are close relations between migratory restlessness (nocturnal hopping by captive birds at the time of migration) and several aspects of actual migration. In particular, the amount of migratory restlessness measured in the laboratory is closely correlated with the length of the migratory route. Thus it is possible that young, inexperienced birds migrate to their species- or population-specific winter quarters as a result of an endogenous program for migration that is controlled by or linked with an endogenous annual rhythm (*1*).

These observations can be explained by assuming that the patterns of migratory behavior are genetically fixed (*2*). Alternatively, environmental conditions (such as differences in photoperiod, temperature, or food supply) under which individuals grow up may be responsible for the differences in migratory performance (*3*). The genetic hypothesis would be supported if F_1 hybrids of birds from two populations that migrate over different distances showed intermediate patterns of migratory behavior (*4*).

The blackcap *Sylvia atricapilla*, which has a wide distribution from the Cape Verde Islands off the West African coast to approximately 65°N in Europe, is a suitable species for the necessary cross-breeding experiment. Blackcaps of different populations differ sharply in their migratory performance and in the percentage of individuals that display migratory restlessness when caged. Moreover, migratory activity can be easily measured as nocturnal activity in this nocturnally migrating species (*5*).

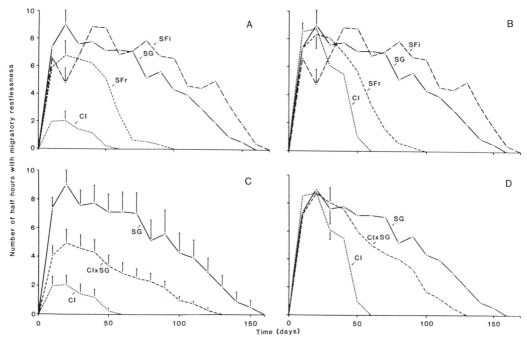

Fig. 1. (A and B) Time course of nocturnal migratory restlessness of groups of *Sylvia atricapilla* from three European and one African populations. Mean values for 10-day periods are given. Abbreviations for sources of the birds: *SFi*, southern Finland (60°N); *SG*, southern Germany (47°N); *SFr*, southern France (43°N); and *CI*, Canary Islands (28°N). (A) Data for all the birds in each group, including those birds that did not show migratory restlessness (N = 26, 25, 24, and 26 for *SFi*, *SG*, *SFr*, and *CI* birds, respectively). Significant differences between groups are as follows (Mann-Whitney U test): for *SFi* birds compared to *SFr* and *CI* birds, $P < .001$; for *SFr* birds compared to *SG* birds, $P < .01$; and for *SG* birds compared to *CI* birds, $P < .001$. (B) Data for the birds that showed restlessness (N = 26, 25, 20, and 6). All curves are normalized so that they start on the first night of restlessness for each bird. To maintain clarity, standard errors are given for representative examples. (C and D) Time course of migratory restlessness of hybrids ($CI \times SG$) compared with that of the *SG* and *CI* parental stocks. (C) Data for all the birds in each group, including those birds that did not show migratory restlessness (N = 25, 32, and 26 for *SG*, $CI \times SG$, and *CI* birds, respectively). Standard errors are given for all the mean values. (D) Data for the birds that showed migratory restlessness (N = 25, 18, and 6); standard errors are given for representative examples.

Twenty-four to 26 nestlings from each of three European populations (Finnish, German, and French) and one African population were hand-raised in the spring of 1976 and 1977 and their migratory restlessness was recorded in the following autumns (Fig. 1, A and B). The degree of migratory restlessness was greatest in the Finnish birds, with progressively less in the German, French, and African birds. One hundred percent of the Finnish and German birds, 80 percent of the French birds, and 23 percent of the African birds exhibited restlessness [$P < .001$ for the African birds compared to the Finnish and German birds (chi-square test)]. These results are in accordance with the distances of north-south travel in the free-living populations (6).

In 1978 and 1979 we used the African and German birds of the above experiment for a cross-breeding experiment. We successfully hand-raised 32 hybrids used for their parents. Twenty-eight of the 32 hybrids were derived by crossing migratory German and nonmigratory African birds; four were derived from a migratory German parent and a migratory African parent.

The hybrids displayed an amount of migratory restlessness intermediate between that of the parental populations of Africa ($P < .001$; Mann-Whitney U test) and Germany ($P < .01$) (Fig. 1, C and D). Also, the mean number of restless nights, the range of the amount of restlessness, and the coefficient of variation for restlessness were all intermediate in the hybrids (7).

There is consistent agreement between the proportion of migrants and nonmigrants in various blackcap populations and the proportion of birds in those populations which display migratory restlessness in captivity. Therefore and used the same method to record their migratory restlessness as had been the migratory or sedentary habits of different individuals constituting the partially migratory populations of blackcaps may be due to polymorphism (5). In our cross-breeding experiment all the German parental birds and 23 percent of the African ones displayed migratory restlessness. If polymorphism is the basis of the partially migratory behavior, the introduction of genes from the exclusively migratory German population into the partially migratory African population should increase the percentage of F_1 hybrids displaying restlessness. We observed restlessness in 56 percent of the hybrids. This value is intermediate between, and significantly different from, those obtained for the German parental population ($P < .001$, chi-square test) and the African parental population ($P < .025$). In addition, the expression of morphological features and juvenile development associated with migration was also intermediate in the hybrids (8).

References and Notes

1. E. Gwinner, *Annu. Rev. Ecol. Syst.* **8**, 381 (1977).
2. E. Mayr, *Animal Species and Evolution* [Belknap (Harvard Univ. Press), Cambridge, Mass., 1965].
3. T. I. Blyumental and V. R. Dolnik. *Tr. Vses. Soveshch.* (1965), p. 319; E. Gwinner, P. Berthold, H. Klein, *J. Ornithol.* **113**, 1 (1972).
4. K. Immelmann, *Einführung in die Verhaltensforschung* (Parey, Berlin, 1976), pp. 192-197; A. Jacquard, *The Genetic Structure of Populations* (Springer-Verlag, New York, 1974), pp. 86-98; C. Stern. *Principles of Human Genetics* (Freeman, San Francisco, 1973), pp. 443-467.
5. P. Berthold. *Experientia* **34**, 1451 (1978).
6. The birds were taken from their nests at 5 to 7 days of age, transferred to the Vogelwarte Radolfzell in southern Germany (47°46'N, 09°00'E), and hand-raised under simulated local light conditions standardized according to an arbitrary birth date of 10 May for all birds. Beginning at 50 days of age the birds were maintained under constant conditions [a cycle of 12.5 hours of light (400 lux) and 11.5 hours of darkness (0.01 lux); 20° ± 1.5°C] until the end of the season of migratory restlessness. Then they were transferred to natural southern German light conditions to allow them to achieve their normal spring breeding condition.
7. The German, F_1 hybrid, and African birds averaged 74, 53, and 35 nights of restlessness, respectively, and displayed restlessness during 0 to 2009, 0 to 991, and 0 to 534 total half-hours. Respective coefficients of variation: 145, 112, and 98 percent.
8. P. Berthold and U. Querner, in preparation.
9. We thank G. J. Kenagy for his especially extensive review of our report and the Finnish, French, Spanish, and German governments for permission to collect the birds.

4 December 1980

Innate and Environmental Dispersal of Individual Vertebrates

WALTER E. HOWARD
Field Station Administration, University of California, Davis

Little is known about the rates of dispersal of the young of wild populations of vertebrates, because it is difficult to recover marked individuals after they have dispersed to breed elsewhere. Nevertheless, growing evidence suggests that the observed dispersal patterns may be governed by the laws of heredity as well as being influenced by population pressure.

Dispersal of an individual vertebrate is the movement the animal makes from its point of origin to the place where it reproduces or would have reproduced if it had survived and found a mate. The significant role of an individual's dispersal is the greatest distance its genetic characteristics are transmitted, rather than the greatest distance the animal may have migrated or otherwise traveled away from the place where it was conceived, hatched or born. Most, but not all vertebrate animals make some sort of a dispersal movement at about the time they attain puberty, regardless of their actual age. Individuals that make innate dispersals, as defined by the hypothesis advanced herein, are predisposed at birth to disperse beyond the confines of their parental home range. They ignore available and suitable niches and voluntarily disperse into strange and sometimes even unfavorable habitats (Fig. 1). In contrast, environmental dispersal is defined as the movement an animal makes away from its birthplace in response to crowded conditions (mate selection, territoriality, lack of suitable homesites, or parental ejection). Environmental dispersers are assumed to have inherited a homing tendency. Environmental dispersal is a density-dependent factor, whereas innate dispersal is independent of density, but both are presumed to be inherited traits.

Even though the hypothesis that vertebrates are predisposed at birth to make either innate or environmental dispersal movements still needs to be substantiated, it seems worth while and desirable to present this concept at this time as a stimulus to further research on this neglected aspect of population ecology. In the literature there are many statements implying that vertebrates disperse because of population pressure factors, but there are relatively few that support the innate dispersal hypothesis.

Burt (1949) wrote, in regard to rodents and other mammals, "If . . . there is an inherent desire to wander, it most certainly is not developed to the same degree in all individuals of a species." Blair (1953) theorized that the dispersal of rodents may result from either population pressure or "an inherent tendency to disperse, stimulated by physiologic changes as the animal becomes sexually active." Dice and Howard (1951) provide some support for an innate and an environmental dispersal trait, for they found that dispersal distances of prairie deermice (*Peromyscus maniculatus*) are nonrandom — that apparently there is an innate stimulus which might motivate certain individuals but not others to leave the vicinity of their birthplaces. Johnston (1956) similarly found that salt-marsh song sparrows did not have a random dispersal pattern. As with the deermice, too many birds aggregated close by, too few moved intermediate distance (350 to 650 meters), and too many moved beyond 650 meters. He says that both his data and those of Mrs. Nice (Ohio song sparrow) have dispersal curves that differ from the expected significantly at the one percent level.

In an intensive study of valley quail by Howard and Emlen (1942), the birds were marked individually with colored bands, and practically every bird in three of six coveys studied was readily observed each day from an automobile. In this quail population there was an interchange of a few members between coveys prior to the spring nesting. At this time, additional, unmarked birds appeared from more distant coveys, and some resident birds disappeared. In every case when a strange bird attempted to merge with a new covey, it was repeatedly attacked for about a month, and ostracized by the members of the same sex in the covey. These ostracized birds had not been driven from their original coveys; they left voluntarily, and patiently awaited acceptance into a different covey. At that time of the year the birds were pairing,

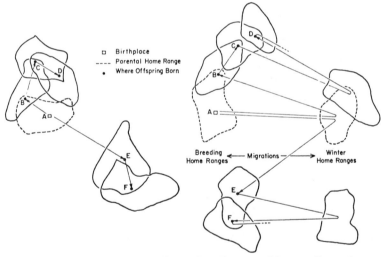

Fig. 1.—A schematic drawing illustrating distance of innate dispersal movements (shortest line between A-E) and of environmental dispersal changes of home ranges (the shortest distance between A-B, B-C, C-D, and E-F) for non-migratory (left) and migratory (right) animals.

although still in covey formation. Individuals of both sexes made dispersal movements, even though there were unpaired members of the opposite sex in the coveys they left. Perhaps "shuffling" in bobwhite quail is a similar phenomenon and is also composed of innate dispersers.

Von Haartman (1951) made a study in Finland of the tendency of pied flycatchers to return to their birthplaces. According to a review of von Haartman's article by Huntington (1951), about three times as many young females as young males did not return to their birthplaces. Many of these probably made innate dispersals. Of the birds that returned the second and following years after banding, the percentage was about the same for both sexes. Von Haartman considers the homing, or philopatric, quality to be hereditary rather than due to greater success in competition for territory, because combat is unusual in the females, and unsuccessful attempts to settle were much more common in males than in females. Furthermore, the dates of arrival and egg-laying of philopatric females were not significantly earlier than those of others. Apparently the nonphilopatric adult females lack a genetically controlled factor that makes their home-loving sisters try to settle in a familiar place. If that is so, then the nonphilopatric individuals may likewise possess a genetically controlled factor for innate dispersal. Of course, innate dispersal and philopatry may not be the hereditary consequence of each other, but von Haartman does consider the differences, *i.e.*, the variation in dispersal distance, to be inherited. He suggests that philopatry may be a recessive character, because six (3%) offspring of 196 banded philopatric females returned to the area, whereas only three (0.5%) offspring of 632 nonphilopatric females returned. This information is too meager to permit drawing any definite conclusions, but it provides some support for the hypothesis that philopatry, hence, also innate and environmental dispersal, are genetically determined.

There are numerous examples indicating that a certain percentage of both males and females (it will vary with species) is likely to make extensive dispersal movements. Only one case will be cited here. In Fitch's (1948) careful investigation of ground squirrels on an 80-acre plot, he found that 40.9 percent of young females survived to maturity, whereas only 31.1 percent of the males survived. He attributes this difference to an "extensive movement of young males off the area rather than actual mortality."

Introduced animals, like the English sparrow and starling in North America and rabbits in Australia and New Zealand, spread their range far more rapidly than population pressure factors could have demanded. It seems more likely that certain young individuals inherit an instinct to disperse considerable distances, and that is why the range of introduced species is often greatly extended by the time the earlier occupied habitats become crowded.

In 1905, three female and two male muskrats were released in two small natural ponds in Dobrisch, 40 km southwest of Prague.

Nine years later the resulting population in Bohemia was estimated at two million (Mohr, 1933). It takes very little calculation to realize that such a large population could not have resulted if young muskrats had not dispersed until population pressure in each favorable habitat caused them to disperse to the next available homesite. On the contrary, as the result of the presumed innate dispersal urge, there must have been a rapid spreading of individuals into new areas long before the populations they left behind had become very large. By the time the populations in the earlier occupied habitats approached overcrowding, which also probably resulted in increased mortality and decreased fecundity, the range of the species had probably already been extended many miles.

Beer and Meyer (1951) report on muskrats as follows: "The bulk of the reports on movements in the Madison area are received in March with a few in February and quite a few in April. This agrees with the period of increasing gonadotropic activity of the pituitary and the rapid development of the gonads. We believe that these early spring movements may be induced by an increased irritability of the animals or to an urge within the animal to move and that the latter is probably the closest to being correct, since the animals found in the early part of the spring movement have few, if any, wounds from fighting. The group that is found moving from April through July are usually severely wounded from fighting. These movements should be classified as force-out movements induced by population pressures."

Davenport (1915) stated that the nomadic impulse in man is an instinct that is sex-linked. Even though there seems to be little doubt that the dispersal trait is subject to the laws of heredity, additional evidence is necessary before one can be certain that the innate dispersal impulse is sex-linked, sex-modified, recessive, polygenic, or just how it operates.

At the San Joaquin Experimental Range, California, the author captured dispersing rodents in seven funnel traps located along drift fences of hardware cloth (patterned after traps used by Imler, 1945). With the nine species of rodents concerned, results indicate that the dispersal trait might be sex-linked, for about two males were captured for each female. Out of 766 trapped individuals that were carefully sexed, an average of 63 percent were males. It is interesting to note how close all these species independently approached this average. The percentage of male rodents captured between June, 1949, and July, 1952, is as follows (total number of individuals appears in parentheses): *Citellus beecheyi* 71% (7 individuals); *Thomomys bottae* 62% (158); *Perognathus inornatus* 68% (71); *Dipodomys heermanni* 61% (18); *Reithrodontomys megalotis* 63% (65); *Peromyscus maniculatus* 63% (51); *Peromyscus boylei* and *P. truei* 62% (26); and *Microtus californicus* 63% (370). If no locally established rodents or environmental dispersers had been captured, the figure might have been even closer to two males for each female. The traps were examined almost daily for four years and, at that time, it was obvious that nearly all the individuals captured were virgins in the act of dispersing.

In an attempt to measure the degree of dispersal among prairie deermice (*Peromyscus maniculatus*) living on 300 acres of grassland, the birthplace, distance of dispersal, and subsequent breeding site were learned for 155 (77 males and 78 females) of the young deermice (Howard, 1949). An arbitrary figure of 500 feet, which is beyond the boundary of most parental home ranges, is used to separate innate dispersal from environmental dispersal.

The over-all pattern of dispersal observed in the deermice seemed to indicate that: 1) the availability of surplus home sites (at least artificial ones) had little influence on degree of dispersal; 2) virgin mice made their extensive dispersal, if they ever made one, when they attained puberty, regardless of whether at 4.5 weeks or 25 weeks old (reflecting the season when born); 3) even though various proportions of the sexes were found in the same nest during the breeding season (up to two pairs were present together in the same nest boxes), there was no tendency for deermice to aggregate in the presumably more favorable habitats; 4) there was apparently very little trial and error searching for suitable niches, for dispersers were never found at intermediate points, suggesting that the dispersal urge for each individual was satisfied in a day or two; 5) individuals did not make more than one extensive dispersal, even if they later lost their mate, although there were minor shifts of nest sites, usually with each new

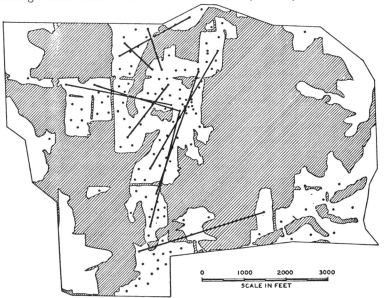

Fig. 2.—The lines indicate ten examples of extensive, innate dispersal movements made by prairie deermice on the Edwin S. George Reserve, Michigan. The grassland areas are unshaded and locations of nest boxes buried in field are marked by dots. (From Howard, 1949.)

litter; 6) when no suitable mates were present, some dispersers remained unmated for several months during the breeding season without moving, and adults did not move whenever an uneven sex ratio developed; and 7) there was no evidence suggesting that the parents forced the young to disperse, for quite frequently the mother would merely abandon the young and seek a new nest when a new litter was due.

With regard to the 36 deermice (out of the 155 that survived to breed) that made "innate" dispersals of 550 to 3300 feet into unfamiliar territory (Fig. 2), the following deductions have been drawn: 1) twenty-four (31%) of the males and 12 (15%) of the females made the extensive dispersals; 2) some individuals traveled through unfavorable habitats (woodland) even though their home range areas were largely surrounded by favorable grasslands; 3) only one animal (a male) was known to return to its birthplace a short time after dispersing (2050 feet); and 4) mice that dispersed the greatest distance did not have time to explore the areas adjacent to their birthplaces or the availability of the habitats through which they passed.

Discussion

If we assume that an innate dispersal trait exists, we can speculate on how the trait may be of value to a species: 1) it would bring about wide outbreeding; 2) it would help reduce the likelihood of too close inbreeding; 3) it would further the spread of new genes; 4) it would rapidly extend the range of the species as favorable habitats become available; 5) it would enable a species to reinvade areas depopulated by catastrophes, such as disease, fire, flood, abnormal weather, or man's activities, without having to start repopulating just at the edges and gradually overflow inward; 6) it would tend to reduce intraspecific conflict and bring about a more efficient utilization of habitat resources; and 7) it might spread a reservoir of characters of possible future value (not adaptive in the new situation yet not selected against, which might help explain why closely related species often differ by characters that are not adaptive).

Domestication probably favors the "homing" or environmental dispersal trait. If such is the case, domesticated animals introduced and released in foreign lands should spread their range more slowly than would be the case with their undomesticated relatives. There is some evidence that this is what happened with many of the introductions in New Zealand and with rabbits in Australia. Perhaps "wildness" is related to the adolescent wanderlust resulting from inheritance of the innate dispersal trait.

For the well-being of migratory species it is not only necessary that some individuals make extensive innate dispersal movements, but it is equally important that all other individuals return to the vicinity where raised. What would happen if most individuals of migratory populations did not tend to return to the home range? Chaos would develop, for every individual would then be expected to return to the more preferred habitats in the breeding range of the species. There would be so much time and energy spent in establishing individual breeding territories that only a fraction of the total population would be able to breed, if the local food supply were not completely exhausted first. As it is, when the entire breeding range of a migratory bird, fish, or mammal is taken into consideration, there is a far more orderly establishment of the individual breeding ranges and territories than could possibly happen by chance, and natural selection has probably been important in the evolution of this pattern. Territoriality and environmental dispersals are of local importance rather than of geographic significance.

Since such a high proportion of the individuals making up mass emigrations are juveniles (Lack, 1954), it seems highly possible that they may also be responding to the innate dispersal trait. The higher the population density, the greater the number of innate dispersers. Such animals often pass up areas of abundant food supply, and the movements sometimes actually begin before there is any shortage of food. It seems possible that some emigrations, such as an occasional influx of snowy owls into northeastern United States, are merely an increased number of innate dispersers, brought about by an increase in population density rather than by an extensive shift in the home range of older adults. The age of this type of disperser needs to be determined to see if there are any old adults included, *i.e.*, individuals that have bred at least once before making the extensive dispersal. We also need to know if the distance of dispersal is any greater when population density is high.

If animals freely disperse in response to population pressure, then localized epidemics, or "outbreaks," should not occur. On the other hand, even though the direction of dispersal is probably inherently random, the direction of actual dispersal movements are probably greatly modified by many environmental factors. Consequently, there would be a greater tendency — for environmental dispersal movements in particular — toward the more densely populated regions of a species' range, where the more preferred habitats are to be found.

Howell (1922) put many of the factors to be considered in the dispersal of life into chart form. It seems quite plausible that the trigger mechanism initiating dispersals in many vertebrates is associated with the maturing of sex organs, as metabolic processes are speeded up by the activity of reproductive hormones. Lashley (1938) said that the "evidence points to the conclusion that the neural mechanism is already laid down before the action of the hormone, and that the latter is only an activator, increasing the excitability of a mechanism already present." Slonaker (1924 and 1927) recorded the rate of activity of albino rats in a revolving wheel, and noted that there was a marked increase in voluntary activity for males at puberty and for females during each estrus. It would be interesting to see if individuals possessing the presumed innate dispersal trait had larger adrenals than the environmental dispersers.

Instinctive behavior does not necessarily have to be advantageous to the individual in a social species, but rather it may benefit the group by serving to maintain and spread the species (Tinbergen, 1951: 157). And, as pointed out by Tinbergen, homologous behavior elements may shift their position within the pattern and come to serve different functions in different species, or in some species become lost completely.

Many basic phenomena of animal behavior are not well understood, because they are difficult to investigate. It is particularly hard to analyze the inheritance of behavior patterns. The phenomenon of dispersal alone is a broad subject that will require the participation of many investigators. Much more information is needed about 1) the frequency with which different kinds of animals disperse various distances with population density at different levels; 2) the randomness of the distances of dispersal; 3) the sex ratio of the animals making extensive dispersals in relation to the current sex ratio of the local population; 4) the time relationships of the movements, and whether dispersals are always made at puberty; 5) the motivating and terminating forces, whether physiological or ecological; and 6) the genetic explanations of the variability observed in dispersal distances (field and laboratory experiments). Even if the innate concept should prove to be sound, much information will still be required to learn the various ways in which this behavior is expressed in different animal populations under different situations.

Summary

Dispersal is the movement an organism makes away from its point of origin to the place where it reproduces or would have reproduced if it had survived and found a mate. For the most part, the major dispersal movements are made by virgins about the time they attain puberty.

Possession of the innate dispersal trait implies that such an animal is predisposed at birth to leave home at puberty and make one dispersal into surroundings beyond the confines of its parental home range. Such density-independent individuals have inherited an urge to leave home voluntarily. They often pass up available and suitable niches and venture into unfavorable habitats.

Animals that make an innate dispersal movement are obsessed with a dispersal instinct. The "purposiveness" of the innate concept is not for the individual's welfare; rather, in spite of the high rate of mortality of innate dispersers, it has distinct survival value for the species. Innate dispersers are particularly important to a species because they, 1) increase the spread of new genes, 2) create wide outbreeding, 3) enable a species to spread its range rapidly as favorable habitats are created, 4) permit the species to have a discontinuous distribution, and 5) help the species quickly reinvade areas that may have been depopulated by catastrophes, such as floods, fires, or man's activities.

Points that appear more or less to favor the existence of an innate dispersal concept include: 1) the distances of dispersal are, at least sometimes, significantly not random; 2) some introduced species spread their range too rapidly to be the result of population pressure factors; 3) reinvasion of a depopulated area does not commence at the edge and gradually overflow inward, but, instead, the density of the species builds up almost simultaneously over all of the area that is within the maximum limits of innate dispersals; 4) the rate at which innate dispersals are made seems to be density-independent; 5) the movements are made instinctively without any prior experience or instructors to imitate; 6) innate dispersers frequently cross or attempt to cross regions of unfavorable habitat, regardless of the availability of adjacent suitable habitats; and 7) the stimulus is of short duration, apparently being expressed only once, when the animal becomes sexually active for the first time.

The presence of the environmental dispersal trait implies that the individual will remain where born or, by means of trial and error, eventually select a new home range usually within the confines of its parental home range. It will have a strong homing tendency and move only as far as forced by population pressure factors (intraspecific competition or density-dependent factors) such as parental ejection of young, voluntary avoidance of crowded areas, mating and territoriality, availability of food and homesites, or the presence of other organisms including predators. Minor shifts of homesites result in a dispersal, but these are all called environmental dispersals, even though a series of them by one individual might eventually result in a total dispersal distance that is quite extensive, even exceeding that of some innate dispersal movements. Environmental dispersal has only local significance, whereas innate dispersal is of geographic importance.

To verify or refute the existence of an innate dispersal trait, the assistance of other investigators is urgently solicited, for findings of many workers will be necessary before we can thoroughly understand the dispersal behaviorism and dispersal pattern in different species.

REFERENCES

Beer, J. R. and R. K. Meyer 1951. Seasonal changes in the endocrine organs and behavior patterns of the muskrat. *J. Mamm.*, **32**:173-191.

Blair, W. F. 1953. Population dynamics of rodents and other small mammals. *Adv. in Genetics*, **5**:1-41.

Burt, W. H. 1949. Territoriality. *J. Mamm.*, **30**:25-27.

Davenport, C. B. 1915. The feebly inhibited, 11. Nomadism or the wandering impulse, with special reference to heredity. *Proc. Nat. Acad. Sci.*, **1**:120-122.

Dice, L. R. and W. E. Howard 1951. Distance of dispersal by prairie deermice from birthplaces to breeding sites. *Univ. Mich. Contrib. Lab. Vert. Biol.*, **50**:1-15.

FITCH, H. S. 1948. Ecology of the California ground squirrel on grazing lands. *Amer. Midl. Nat.*, **39**:513-596.

HOWARD, W. E. 1949. Dispersal, amount of inbreeding, and longevity in a local population of prairie deermice on the George Reserve, Southern Michigan. *Univ. Mich. Contrib. Lab. Vert. Biol.*, **43**:1-52.

—— AND J. T. EMLEN, JR. 1942. Intercovey social relationships in the valley quail. *Wilson Bull.*, **54**:162-170.

HOWELL, A. B. 1922. Agencies which govern the distribution of life. *Amer. Nat.*, **56**:428-438.

HUNTINGTON, C. E. 1951. "Ortstreue" and subspecies formation in the pied flycatcher. *Ecol.*, **32**:352-355.

IMLER, R. H. 1945. Bullsnakes and their control on a Nebraska wildlife refuge. *J. Wildlife Mgt.* **9**:265-273.

JOHNSTON, R. F. 1956. Population structure in salt marsh song sparrows. Part 1. Environment and annual cycle. *Condor*, **5**:24-44.

LACK, DAVID 1954. *The natural regulation of animal numbers.* Oxford Univ. Press, New York, 343 pp.

LASHLEY, K. S. 1938. Experimental analysis of instinctive behaviour. *Psychol. Rev.*, **45**:445-471.

MOHR, ERNA 1933. The muskrat, *Ondatra zibethica* (Linnaeus), in Europe. *J. Mamm.*, **14**:58-63.

SLONAKER, J. R. 1924. The effect of pubescence, oestruation and menopause on the voluntary activity in the albino rat. *Amer. J. Physiol.*, **68**:294-315.

——. 1927. The effect of follicular hormone on old albino rats. *Amer. J. Physiol.*, **81**:325-335.

TINBERGEN, N. 1951. *The study of instinct.* Oxford Clarendon Press, London, 228 pp.

VON HAARTMANN, LARS 1951. Der Trauerfliegenschnapper. I. Ortstreue and Rassenbildung. *Soc. Fauna Flora Fennica, Acta zool. Fenn.*, **56**:1-104.

Copyright ©1961 by the Cooper Ornithological Society
Reprinted from *Condor* **63**:386–389 (1961)

POPULATION MOVEMENTS OF BIRDS

By Richard F. Johnston

This short paper is an exploratory discussion of three kinds of population movements of birds: dispersal, spacing, and migration. The last of these is of interest here only in its occurrence coincident with dispersal and spacing, and it is to the latter that chief attention is directed.

DISPERSAL refers only to movement, usually of young, from sites of birth to sites of breeding. "Effective distance of dispersal" refers to the least distance in an air line between site of birth and site (or sites) of subsequent breeding, and of course this is less in almost every instance than the distance actually traveled by an individual. The main significance of dispersal is ecological—by means of dispersal a species avails itself of suitable habitat within, peripheral to, or isolated from an occupied area. This general adaptation can be broken down into subcategories (Howard, 1960:157). There is, of course, important correlated function concerning movement of genetic elements through populations or between neighboring populations.

SPACING is the movement of birds that brings about territorial dispersion. Spacing is guided partly by configuration of their preferred vegetation and topography and partly by their territorial behavior. Territories of optimal quality tend to be occupied before those of suboptimal quality; consequently, a patchwork of occupied places can be established within a frame of generally suitable habitat. The routes that individuals follow in finding suitable places will reflect the fact that routes tend to be unknown beforehand, as are the locations of unoccupied places. Thus, the important elements of the environment in which spacing occurs are essentially distributed at random, as far as their influence on the direction and magnitude of spacing is concerned. For a population, therefore, spacing movements tend to cancel one another, that is, the vectorial sum of such movements approaches zero.

For a migratory bird, migration and dispersal (but not effective dispersal) more or less coincide in time. Additionally, terminal stages of migration and dispersal can merge with initial stages of spacing. Consequently, there is some reason for confounding these kinds of movement with one another; that we actually gain in distinguishing between them may be demonstrated in the following discussion.

DISPERSAL

There is general agreement that the chief means of dispersal in birds consists of movements by young individuals. Almost forty years ago Grinnell (1922) recognized that individuals that moved long distances (so-called distributional accidentals) were predominantly young-of-the-year; extralimital occurrence was almost always of young birds. Fisher (1955) adduced further evidence showing that such movement, which he called dispersal, was a function of young, mainly first-year birds. Howard (1960) discussed this kind of movement by young individuals of several kinds of vertebrates. Most significant is the evidence from studies on banded, sedentary bird populations (Nice, 1937; Erickson, 1938; Kluijver, 1951; Gibb, 1954; Johnston, 1956), which has demonstrated movement from sites of birth by young-of-the-year and remarkable constancy of adults in remaining at sites of breeding.

In sedentary bird populations, but not necessarily in non-sedentary ones, the difference in amount of movement recorded for adults *versus* first-year birds is so great that a qualitative difference in behavior seems to be involved. Moreover, there are data supporting the idea that certain first-year birds characteristically move long distances in the process of dispersal. Table 1 presents a summary of distances moved in the dispersal of Song Sparrows (*Melospiza melodia*) of two, distinct populations. Earlier (Johnston, *op. cit.*:42), the correspondence in distances moved in the two samples had been noted. Curves drawn from these data are bimodal and differ from curves characteristic of random dispersal mainly near the extremes of the frequency distributions. In both samples, more individuals move relatively short distances and relatively great distances than are expected to do so on the basis of chance alone. Therefore, dispersal can be considered to be an organized or directed characteristic in Song Sparrows.

TABLE 1

DISTANCES OF DISPERSAL IN TWO POPULATIONS OF SONG SPARROWS

Population	Distance of dispersal in meters							
	100	200	300	400	500	600	700	800+
California[1]	39[3]	30	12	8	0	0	8	3
Ohio[2]	12	27	30	8	6	8	3	6

[1] From Johnston, 1956:42.
[2] From Nice, 1937:83.
[3] Per cent of total instances.

Studies on other kinds of sedentary animals have shown their dispersal qualitatively to resemble that of Song Sparrows. Bateman (1950) noted several examples of bimodal curves in the dispersal of flying insects in his analysis of "dispersal" of genes in populations. He concluded that "many if not most methods of gene dispersal produce such distributions," although he had no examples from terrestrial vertebrates at that time and he did not limit his conclusion to strictly sedentary populations. Bimodal curves of dispersal have been demonstrated for a leafhopper, *Macrosteles divisus* (Frampton, Linn, and Hansing, 1942), a weevil, *Bruchus pisorum* (Wakeland, 1934), a lizard, *Sceloporus olivaceus* (Blair, 1960), the House Finch, *Carpodacus mexicanus* (Thompson, 1960), the House Sparrow, *Passer domesticus* (Wagner, 1959), and a mouse, *Peromyscus maniculatus* (Dice and Howard, 1951). All available evidence indicates that such bimodal distributions are reflections of species-specific behavioral tendencies in dispersal; the preponderant fraction of the dispersing element is characterized by movement of relatively short distance (the "homing tendency" of Howard, *op. cit.*), and the lesser fraction of the dispersing element by movement of relatively long distance.

Thus, the tendency by subadults to move is evident in any fraction of this age class. Were there actually no tendency to move on the part of some individuals, the curves of dispersal would show a greater frequency of individuals staying at birthsites. Yet, few occur exactly at birthsites, and modal distances of dispersal for Song Sparrows, one of the most sedentary of bird species, are 200 to 400 meters. If it is necessary to emphasize one thing in order to establish dispersal as an innate tendency, the emphasis is properly placed on those young having the capacity to disperse long distances. Howard's review (*op. cit.*) has led him to a similar conclusion.

It is appropriate to note at this point that the genetic background for dispersal is complex. That only a small fraction of young disperse long distances is of adaptive significance. Also, there are different selective values for long-distance (or short-distance) dispersal in accord with whether the phenotype is present in centrally-located populations or at the periphery of the distribution of the species. Such fluctuating selective value on phenotypes (which may be composed of several genotypes) can, as Alden H. Miller has noted (personal communication), result in behavioral polymorphism. In such an instance, central populations would show a greater, and peripheral populations a lesser, incidence of phenotypes dispersing long distances—conditions that could be demonstrated by appropriate intensive population studies.

SPACING

The fact that young individuals are the agents of dispersal provides one measure of the distinction between dispersal and spacing. Additionally, spacing, motivated in part by territorial behavior, results in the dispersion of breeding units in a population, and it would seem, to consider only two alternatives, that territorial aggression, rather than spacing itself, is the significant selective element in territorial behavior. It would also appear that spacing is not immediately related to dispersal.

Nevertheless, territorial spacing and dispersal have been associated causally (Fisher, 1955:440; Howard, 1960:152). The suggestions were that dispersal ("environmental dispersal" of Howard, *loc. cit.*) in sedentary (*ortstreuen*) bird species is attributable to the "territorial system." These suggestions were probably motivated by undue emphasis on the significance of territoriality. A tremendous amount of work in the past two decades and especially a paper by Pitelka (1959:253) serve clearly to point up the general conclusion that spacing of individuals or breeding units is the immediate adaptive advantage of territoriality; behavioral systems of territoriality result in ecologic systems of spacing. Indeed, the 31 functional categories listed by Carpenter (1958) chiefly have as their common denominator the phenomenon of spacing.

Territoriality in the northern hemisphere is a phenomenon chiefly of spring and summer; yet, by early spring (or autumn and winter for some species) dispersal has already been achieved. This is true almost without exception for sedentary, resident birds (Johnston, 1956:40), but it is less true for migratory species. In the latter, dispersal and migratory movement are identical, as has been noted previously; dispersal is thus nearly complete only when migration ceases. However, spacing functionally equivalent to dispersal occurs in early stages of territorial behavior. Effective distances of dispersal can consequently be increased or decreased depending on chance encounters of non-established birds with established birds. Such modification is the chief bearing of territorially-activated spacing movement on dispersal, and it should be clear that such bearing means little. The essential meaninglessness is emphasized particularly at the level of populations, in which as many individuals will be forced to move one way as another. In sedentary populations showing autumnal territoriality, the action of territorial aggression on juveniles that are dispersing would be evident in autumn, not spring, but the significance of territoriality in affecting meaningful modification of dispersal would be the same as that discussed for migratory species.

The term spacing is available and suitable for discussing movement resulting from territorial behavior, and it is here proposed that it be used as distinct from dispersal.

CONCLUSIONS

Dispersal, defined as movement from site of birth to site of breeding, is a mechanism that tends to ensure complete testing or investigation by a species of all suitable habitat within and beyond the area of established distribution of this species. This fragment of the population that disperses is the young; in birds such individuals are usually less than one year old. The genetic heritage of some of the individuals probably casts them in the roles of dispersers to long distances.

Spacing, resulting from territorial behavior of adults and responsible for dispersion of breeding units, does not effect, or in any meaningful way affect, dispersal. The capacities for these two types of movement exist independently of one another, in spite of the fact that some dispersal can occur coincident with spacing.

Migratory movement includes dispersal *in sensu lato*, but effective dispersal is independent of migration.

LITERATURE CITED

Bateman, A. J.
    1950. Is gene dispersal normal? Heredity, 4:353–363.

Blair, W. F.
    1960. The rusty lizard, a population study (Univ. Texas Press, Austin).

Carpenter, C. R.
    1958. Territoriality: a review of concepts and problems. *In* Behavior and evolution, edited by A. Roe and G. G. Simpson (Yale Univ. Press, New Haven), 224–250.

Dice, L. R., and Howard, W. E.
    1951. Distance of dispersal by prairie deermice from birthplaces to breeding sites. Contr. Lab. Vert. Biol., Univ. Mich., No. 50:1–15.

Erickson, M. M.
    1938. Territory, annual cycle, and numbers in a population of wren-tits (*Chamaea fasciata*). Univ. Calif. Publ. Zool., 42:247–334.

Fisher, J.
    1955. The dispersal mechanisms of some birds. Acta Congr. Internat. Ornith., 11:437–442.

Frampton, V. L., Linn, M. B., and Hansing, E. D.
    1942. The spread of virus diseases of the yellows type under field conditions. Phytopathology, 32:799–808.

Gibb, J.
    1954. Population changes of titmice, 1947–1951. Bird Study, 1:40–48.

Grinnell, J.
    1922. The role of the "accidental." Auk, 39:373–380.

Howard, W. E.
    1960. Innate and environmental dispersal of individual vertebrates. Amer. Midl. Nat., 63:152–161.

Johnston, R. F.
    1956. Population structure in salt marsh song sparrows. Part I. Environment and annual cycle. Condor, 58:24–44.

Kluijver, H. N.
    1951. The population ecology of the great tit, *Parus m. major* L. Ardea, 39:1–135.

Nice, M. M.
    1937. Studies in the life history of the song sparrow. I. A population study of the song sparrow. Trans. Linn. Soc. N.Y., 4:iv+1–247.

Pitelka, F. A.
    1959. Numbers, breeding schedule, and territoriality in pectoral sandpipers of northern Alaska. Condor, 61:233–264.

Thompson, W. L.
    1960. Agonistic behavior in the house finch. Part I: Annual cycle and display patterns. Part II: Factors in aggressiveness and sociality. Condor, 62:245–271; 378–402.

Wagner, H. O.
    1959. Die Einwanderung des Haussperlings in Mexiko. Zeits. f. Tierpsych., 16:584–592.

Wakeland, C.
    1934. The influence of forested areas on pea field populations of *Bruchus pisorum*, L. (Coleoptera, Bruchidae). Jour. Econ. Entom., 27:981–986.

Museum of Natural History, The University of Kansas, Lawrence, Kansas, January 17, 1961.

MIGRATION OF TERRESTRIAL ARTHROPODS IN RELATION TO HABITAT

By T. R. E. SOUTHWOOD

Imperial College of Science and Technology, London

CONTENTS

I. Introduction 171	(7) Thysanoptera 186
(1) The evolution of migration: an hypothesis 171	(8) Lepidoptera 186
	(9) Coleoptera 190
(2) Characteristics of migratory movement 172	(10) Diptera 194
	(11) Other taxa 196
(3) Assessment of the level of migratory movement . . 174	III. General aspects of migration . . 198
	(1) The interactions of environmental change, migratory movement and diapause . . 198
(4) Habitats: temporary and permanent 174	
II. Migratory movement in various taxa 175	(2) The maintenance of a variable level of migratory movement in a species 199
(1) Pseudoscorpionidea . . . 175	
(2) Araneida 176	IV. Conclusions 200
(3) Odonata 178	V. Summary 202
(4) Orthoptera 181	VI. References 203
(5) Hemiptera, Heteroptera . . 183	VII. Addendum 211
(6) Hemiptera, Homoptera . . 185	

I. INTRODUCTION

(1) The evolution of migration: an hypothesis

The evolutionary significance of migration has often been disputed. As Elton (1927, 1930) comments, it appears to be an expensive process for the species, since a large number of individuals must fail to find a suitable new habitat and so die without leaving progeny. It is, no doubt, such observations that have led to the suggestion that the evolutionary importance of migration lies in its acting as a safety valve, being a means of getting rid of excess population.

There seem to be two main objections to accepting this idea. First, as Williams (1930, 1958) and Lack & Lack (1951) have pointed out that on such an hypothesis migration would be a lethal character and it is difficult to see how it could have been selected. Secondly, many workers have found that migratory movement is independent of the immediate environment, for example, in spiders (Duffey, 1956), in *Thrips imagines* (Davidson & Andrewartha, 1948), in certain Lepidoptera (Allan, 1942; Roer, 1959; Urquhart, 1960), in various trypetid flies (Christenson & Foote, 1960) and in Coccinellidae (Dobrzhanski, 1922). C. G. Johnson (1960a) comments that 'I see migration as an evolved adaption rather than as a current reaction to adversity'. Such observations lead Andrewartha & Birch (1954) to propose an 'innate tendency for dispersal'; their discussion of Sheppard's (1951) work on the moth *Panaxia dominula* makes it apparent, however, that they were not clearly separating movements within a population from those away from a population and only the latter are considered as migratory in this article.

It is thus not only to the start but also to the end of migratory movement that we must look for its *raison d'être*. At the end of a successful migratory movement the animal has reached a new habitat. These new habitats are continually arising and if an insect species is to occupy the majority of the available niches at any given time, then it must have a level of migratory movement geared to the rate of change of its habitat. Looking at the problem from another point of view, that of habitat classification, Elton & Miller (1954) concluded that those animals that depend upon temporary habitats must have remarkably well-developed senses adapted to finding each new unit as it is produced.

The following hypothesis is therefore proposed: that the prime evolutionary advantage of migratory movement lies in its enabling a species to keep pace with the changes in the locations of its habitats. If this is correct, then within a taxon one should find a higher level of migratory movement in those species associated with temporary habitats than in those with more permanent ones.

In this article the evidence on the level of migratory movements in various groups will be considered in the light of this hypothesis. First, however, I must define what I mean by migratory movement, indicate how it may be recognized and assessed, and describe the types of habitat considered as permanent and temporary.

(2) *Characteristics of migratory movement*

Locomotory movement is a characteristic of most animals; it may be active walking, flying and swimming, or it may be passive, as ballooning and phoresy. From many points of view these movements can be classified into two types. Pearson & Blakeman (1906) divided flights into 'flights' from habitat to habitat and 'flitters', i.e. 'mere to and fro motion associated with the quest for food or mate in the neighbourhood of the habitat'. More recently Nielsen & Nielsen (1950), working on a butterfly, and Provost (1952) on mosquitoes distinguished between what they called appetitive movements and migratory movements, whilst Moericke (1955) recognized different flight 'moods' in aphids. At the International Entomological Congress in Vienna Dr J. S. Kennedy and myself both stressed the view that insect movements are of two types: these are here called trivial movements (following Heape, 1931) and migratory movements, and in fact correspond to the division of Pearson & Blakeman, Nielsen & Nielsen, Provost and Moericke.

Migratory movements take an animal away from its population territory or habitat and frequently result in an increase in the mean distance between the individuals of the original population. Trivial movements are restricted to the animal's habitat and hence can lead only to a limited increase in the mean distance between individuals.

C. G. Johnson (1960 a, b, c) has pointed out that migratory movement occurs early in the life of the adult insect. The first post-teneral flight is frequently migratory, whilst insects that are sexually mature seldom migrate. The discovery of this relationship between age and migration points the way to an experimental approach to the underlying physiological, possibly hormonal, mechanism of migration.

Migratory movements also have certain behavioural characteristics. The take-off usually follows a definite pattern; a diurnal filter mounts to the top of the vegetation, faces the sun and flies upwards. Such a take-off usually ensures that the insect will quickly pass through the boundary layer (Taylor, 1958, 1960), in which it can control its flight direction completely, and into the region beyond where its flight speed is exceeded by that of the wind.

During a migratory movement animals are often positively phototactic (C. G. Johnson, 1960a) and changes of direction are less frequent than in trivial movement. All migratory movements have a fixed minimum period during which there is 'persistent locomotor activity' (Kennedy, 1951) or persistent holding on to the carrier (in phoretic animals). A migrating animal will not stop on perceiving a mate, food or shelter (vegetative stimuli). In contrast, the duration of a trivial movement is completely variable and it may be terminated at almost any time on perceiving a vegetative stimulus.

These behavioural distinctions are exemplified by Nielsen & Nielsen's (1950) observations on the butterfly *Ascia monuste* and have been interpreted by these authors in the terms of the Lorenz–Tinbergen theory of instinctive behaviour. Migratory movement is non-appetential, whilst trivial movement is appetential with an urge for food, a mate or shelter, which is the goal. Alternatively, we can distinguish the two types of movement in the light of the Sherrington–Kennedy theory (Kennedy, 1958, 1960). The thresholds for vegetative stimuli (mate, food, shelter) are high during migratory movement; in contrast, the thresholds for vegetative stimuli are low during trivial movement.

It must be noted that a migratory movement may, after some time, become a trivial movement (e.g. with certain aphids (Moericke, 1955) and the sexuales of ants and termites) and that it is unlikely that there is a hard and fast division into these two types, but rather a complete spectrum.

My concept of migratory movement differs from that of many other workers (e.g. Amanshauser, 1955; Williams, 1958, 1959; Urquart, 1960) but largely agrees with that of C. G. Johnson (1960a, b, c) and Kennedy (1961) who discuss many of the controversial points. Two must be touched on here.

First, the distinction between active and passive migration is not made, for it is apparent rather than real; thus Rainey (1960) shows that strongly flying locusts are virtually as dependent on winds as feeble aphids (see also Kennedy, 1951, 1961; Taylor, 1958; C. G. Johnson 1960a, b, c). It is true, however, that occasionally an individual engaging in a trivial movement may be swept up and carried away by a sudden gust of wind or transported passively in ships, planes or otherwise by man. Such individuals are dispersed passively and are referred to as 'vagrants' in the present article; their transport may have the same ecological effects as migratory movements, but behaviourally are quite distinct.

Secondly, many workers regard migration and dispersal as distinct phenomena. Dispersal means a scattering, an increase in the mean distance between individuals, and may be used, as by Andrewartha & Birch (1954), for movements within the population territory, as well as for those away from it. Only the latter type of dispersive movements are considered here (and in my Congress paper (Southwood, 1960b)); all such dispersive movements, except those of vagrants, are seen as migratory, and conversely, the conspicuous phenomenon of migration is considered to lead commonly to dispersal. Both phenomena take the animals away from the area where they developed and although whole populations appear to migrate, in fact some individuals are left behind, at the start and all along the route, so that the end result of a migration is the scattering of the original population over a wide area, that is to say, dispersal. Some movements into diapausing sites are the only exceptions to this (see p. 198).

Summarizing, the basic feature of migratory movement, as defined here, is that the animal leaves the population territory or habitat in which it has previously lived and whilst engaged in such movement does not respond normally to any vegetative stimulus such as food, a mate or shelter, but undertakes so-called 'movement for the sake of movement.' Such a definition is close to the *Oxford Dictionary* definition of 'migrate', namely 'to move from one place of abode to another'.

(3) *Assessment of the level of migratory movement*

The level of migratory movement is a combination of its frequency within a given-sized population and its duration. It is not yet possible to find a universal method of measuring this level in all terrestrial arthropods. With large conspicuous insects, those with a high level could be recorded in the literature as migrants. With smaller ones a better quantitative measure might be a comparison of the proportions of different species in the aerial plankton to that of their actual populations on the ground or vegetation from which the sampled air had come. The degree of development of the flight organs in allied species gives another measure. Also it is important to distinguish between those insects in which flight is the normal means of locomotion (*vide* Southwood & Johnson, 1957)—such as Diptera and Lepidoptera, where a flight may be trivial or migratory—and those in which walking is the normal means of locomotion —such as Coleoptera and most Hemiptera, in which most, and sometimes all, trivial movement will be expressed as walking, allowing a measure of flight activity to be equated with migratory movement (as with aquatic beetles and bugs).

It is by no means always possible to distinguish, with certainty, migratory from trivial movement in published accounts. The ideal evidence of a migratory movement is that while engaged in it the animal does not respond to food, a mate or habitat, and moves from the actual territory where it has developed into an inhospitable terrain; such movement is normally at the start of adult life.

(4) *Habitats: temporary and permanent*

The terms temporary and permanent are imprecise and relative. It is, however, easy to recognize the extremes: a dung pat lasts but a short while, allowing only one or two generations to be passed in that location, whilst a large river may remain unchanged in its position for thousands of years and countless generations can live in the same location as their forebears. Furthermore, fresh dung pats are continually being produced, whereas a new river is a rare phenomenon.

When a new permanent habitat arises, a long period of time is available for its colonization and this period is utilized; for example, McDonough (1939) found that it took 4 years for a psychid moth to colonize young beech trees. In contrast, flea beetles (Halticini) discover and attack annual crucifers when the seedlings have just emerged from the soil. From the aspect of migration, I consider that the frequencies both of the passing away of the old habitats and of the birth of new ones are of importance; fortunately these are linked: for any habitat they are like the two sides of a coin.

Permanent habitats are rivers, lakes, perennial plants including trees of climax vegetation such as woodlands, salt marshes, heath lands and marshes fringing lakes and rivers.

Temporary habitats are dung, carrion, fungi, plant debris (i.e. logs, straw, hot-beds), and annual and perennial plants of seral communities (e.g. wastelands, fields). Such habitats, being early stages in the biological succession, are only in one locality for a relatively short time. Some ponds are of a very temporary nature, soon drying out, others, notably the bog pools of heathlands and brackish ponds of saltings are more permanent.

In the arid and semi-arid regions of the world the irregular climate causes the vegetation to vary from season to season and from year to year (Buxton, 1923; Wiltshire, 1946; Boyko, 1949; Vesey-Fitzgerald, 1957). Many insects feed on the mass of ephemeral plants or on the growths from perennials that develop in the wake of the erratic rains; the locations of such habitats are always shifting and thus they must be considered as temporary.

Another type of temporary habitat is found when the animal lacks a resting stage and can only utilize a given host plant for a limited period in any season; many flower- and fruit-feeding insects fall into this category and pass from plant to plant as these come into season.

Occasionally the impermanence is on the part of the insect whose habitat requirements change during the life cycle; for example, acridids need a different habitat for oviposition and for feeding. These are also cases of temporary habitats, but they are clearly distinguishable from impermanence due to environmental change (p. 198).

[*Editors' Note:* Material has been omitted at this point.]

(9) *Coleoptera*

The primary methods of locomotion in beetles are walking and swimming, thus in contrast to, say Diptera, the great majority of flights will be migratory in nature. This is especially true of aquatic beetles and Jackson (1950, 1952, 1956a, b, 1958) has made a series of studies on the ability for flight, based on an examination of the wing structure and muscles of ninety-three species. From her work it is possible to divide these species into those in which all individuals are able to fly, those in which some can fly and others in which all are flightless. The habitats of these species are given by Balfour-Browne (1940, 1950, 1953, 1958). Table 4 shows the relation between ability to fly and habitat; any one species may be recorded for several habitats. It is seen that in general the more permanent habitats on the left have a higher proportion of entirely or partly flightless species than the more temporary habitats on the right.

The relation holds much more strongly if the entire habitat range for each species is considered; all the species of ponds and temporary pools are able to fly and all but three of the running water species are flightless. The exceptions are *Hydroporus lepidus*, *Agabus didymus* and *A. biguttatus*. However, *H. lepidus* occurs in clear pools in other parts of its range (Balfour-Browne, 1940), *A. biguttatus* lives in springs, some of which dry up in the summer months (Balfour-Browne, 1950) and *A. didymus* lives in small streams. In some of the variable species Jackson gives an indication of the proportion of flightless individuals. For example, most *Agabus arcticus* are flightless, whilst the reverse condition holds in the closely allied *A. sturmi*. The former occupies more permanent habitats in mountain tarns and pools than the latter in stagnant ponds. That the fully winged pond species frequently engage in migratory movement is shown by the observations of Greensted (1939) and Fernando (1958) on the colonization of small temporary pools.

Table 4. *The relation of flight ability to habitat in water-beetles; the figures give numbers of species (after Jackson 1952, 1956a, b and Balfour-Browne 1940, 1950, 1958).*

	rivers, springs	lakes, tarns, canals	brackish water, peat pools, bog ponds	ponds, ditches	artificial ponds, gravel pits, cattle troughs
able to fly	3	10	30	27	24
variable	6	13	14	15	8
unable to fly	13	7	7	2	0

Wing polymorphism in Carbidae has been extensively studied by both Darlington (1943) and Lindroth (1945, 1949); the latter also made observations on the occurrence of flight in fully winged individuals. He found that flight is a 'daily function' only in *Cicindela* species and certain *Bembidion* species; the other 'flying forms use their wings only at certain seasons', e.g. after hibernation; females caught flying in the field were always found to be immature. This evidence can be expressed as follows: only *Cicindela* and *Bembidion* species engage in trivial flight (in Sweden); in the fully winged individuals of other species trivial movement is expressed as walking and all flight is migratory, occurring typically before sexual maturity.

Flightless carabids are found in especially large numbers on 'old' islands and mountain tops, whether wind-swept or wooded; in all these situations the vegetation has reached its climate and new habitats are not likely to arise, Lindroth (1949) discusses how the gene for the fully winged state would be selected against in such environments; his arguments could be applied with equal force to all dwellers in permanent habitats. He also shows that with a dimorphic species fully winged individuals are most numerous at the edge of the range; that is, the region where the amount of colonizable habitat will fluctuate, with climate, from year to year. Lindroth concludes that stability and isolation of habitat favours flightlessness and variability while 'moderate splitting up of the habitat' favours the fully winged conditions; the main function of flight in carabids is, he says, 'to facilitate a change in quarter'.

Another measure of the levels of migratory flight in Coleoptera from different habitats is obtained by analysing the compositions of catches of flying beetles made in areas where permanent and temporary habits are both present in some numbers. Freeman (1939, 1945) provides a set of such data, he trapped with aerial nets over an agricultural area with few trees and therefore the fact that almost all the beetles he caught had temporary habitats might be attributed to population, not activity differences. Hardy & Milne (1938) collected insects from the air in nets flown from kites over the coast in Suffolk and Kent, when off shore winds were blowing from the undulating wold or downland hinterlands, and also over a playing field at Hull. They collected thirty-seven Coleoptera, representing fourteen species; all had temporary habitats.

Larger numbers of Coleoptera were taken in a suction trap run for 5 days in early May (Southwood & Johnson, 1957). The majority of the catch were Staphylinidae (5838 individuals), whilst Ptiliidae were also numerous (110); the members of both these families are mainly inhabitants of temporary habitats, i.e. compost, haystack bottoms. The remaining species were for the greater part identified down to genus and commonly to species. If these are divided up according to habitats it is found that only four species (or species groups), accounting for six individuals, have permanent habitats (trees and perennial herbs). In contrast seventy-two species, representing 2039 individuals have temporary habitats (ponds, pond margins, annual herbs, compost, dung carrion, logs, dying trees and flowers). There was no reason to suppose that temporary habitat species were much more abundant in the area than permanent ones and thus this evidence indicates that Coleoptera from temporary habitats have a higher level of migratory flight than those from permanent ones.

The inclusion of logs and pond margins in the temporary habitat category should perhaps be justified. Dying trees and fallen logs are suitable environments for most subcortical insects, such as Scolytidae, for a comparatively short period (Webb & Jones, 1956). In Scolytidae, after feeding and sometimes after pairing, either both sexes or the females alone leave the galleries (Lavabre 1958–59) and it has been found by Chapman (1956) in *Trypodendron lineatum* that periods of feeding and breeding in which the flight muscles are reduced, alternate throughout the adult life with short periods of, presumably migratory, flight. Various carabids and heterocerids live on the exposed mud at the sides of ponds and ditches, as Lindroth (1949) observes these habitats are always changing, becoming too dry or too wet.

The last and most extensive type of evidence on migratory movement in beetles is provided by records of mass flights or studies on the migration of individual species. A large number of observers have recorded mass flights of various ladybirds (Coccinellidae); the most common evidence is the occurrence of large numbers on beaches (Moncreaff, 1868; Tutt, 1901; Marriner, 1939; Oliver, 1943; Burton, 1950; Crawford, 1953; Riggall, 1953). Such swarms always consist of aphidophagous forms, mainly *Coccinella* and *Adalia*, which prey especially on the summer colonies of aphids, such as *Aphis fabae*, found on annual plants, crops and grasses. Their highly mobile prey (p. 185), could be regarded as their habitat and indeed other aphid predators, namely Syrphidae (p. 194) are accepted migrants. Dobrzhanski (1922) showed that these coccinellid movements were not due to a shortage of food or other immediate cause, but were part of the beetle's life cycle. These movements may carry the coccinellid into a hibernation site, often on mountain tops (Piper, 1897; Camerano, 1914; Hawkes, 1926; Radzievskaya, 1939; Chapman, Romer & Stark, 1955; Hodek, 1960) or be typical migratory movement from one territory to another. Bodenheimer (1943) made a study of these movements in the Middle East; he found that only a small part of the population actually hibernates gregariously on mountain tops, the majority move seasonally from the mountains to the lowlands, in much the same way as the Lepidoptera in the same area (Wiltshire, 1946).

Swarms, including beetles other than Coccinellidae, have been observed (Bond, 1868; Scott, 1926, 1950; Nijveldt, 1950; Green, 1952; Riggall, 1953; Callan, 1954), the species involved always having temporary habitats, notably annual plants and decaying vegetation.

Detailed studies have been made on the migrations of a few species of beetles. Extensive flight, often pre- or post-hibernation, has been recorded in *Sitona* (Jackson, 1928, 1933; Hans, 1959), *Apion* (Shcerbinouskii, 1939), various Halticini (Moreton, 1945; Watzl, 1950) and *Meligethes* (Fritzche, 1957). *Meligethes* are flower feeders, whilst the others feed on various Cruciferae and Papilionaceae, now occurring mainly as crops or weeds, but originally plants of serial or semi-arid communities. All these, then, have temporary habitats and further evidence of their high level of flight activity is provided by the records of Glick (1939), Freeman (1945) and Southwood & Johnson (1957) who found them well represented in catches of flying Coleoptera.

Two chrysomelids also come into this category of species with temporary habitat host plants and a high level of migratory movement; they are the Colorado beetle, *Leptinotarsa decemlineata*, and the spotted cucumber beetle, *Diabrotica undecimpunctata howardi*. The former overwinters as an adult and has extensive pre- and post-hibernation flights; these can be considered as the cause of its rapid spread and considerable success as a pest species. They can be correlated with the original host plant of *L. decemlineata*, *Solanum rostratum* being a scattered plant of disturbed areas. Its burrs may have been carried by bison from one drinking place to another (Tower, 1906). *Diabrotica* feeds on a wide range of weeds and crop plants. Its migration in the Mississippi valley was studied by Smith & Allen (1932) who found that 'migratory flights' were unique and differed in several aspects from ordinary flights about the field. Such migratory flights occurred either just after emerging or after mating, the beetles ascending to the tips of various plants and then taking off into the air. Smith & Allen considered that by limiting such take-off to southerly winds in the spring and northerly ones in the autumn, the beetles regularly migrated north in the spring and south in the autumn. As the beetles do not fly when the temperature is below $15°C$, and in the spring northerly winds would surely be cold, the spring movement, at least, does not demand any recognition of wind direction by the beetles. It may well be that in autumn dispersal takes place in all directions, only some beetles returning south.

In a different category come the limited migrations of three beetles with more permanent habitats. The hispid leaf-miner of oil palms, *Coelaenomendera elaeidis*, was found by Cotterell (1925) to migrate down wind, the outbreak centres moving comparatively short distances—a few miles in several seasons. The cockchafer, *Melolontha melolontha*, breeds in grass swards adjacent to woodlands, its populations remaining in the same location for a long period. After emergence males and females fly from the sward to the woodlands for feeding and copulation; the females make the journey twice more, back to the sward for oviposition and then a final return to the trees. There is a fixed orientation for all these flights, which are along more or less the same path and there is, it seems, virtually no true migratory movement, (Schneider, 1952; Couturier & Robert, 1958).

The garden chafer, *Phyllopertha horticola*, breeds in grass lands; only a few females migrate, the remainder of the population remaining within the territory (Milne, 1958, 1959 a, b). This migratory movement described as 'bee-lining' occurs, atypically, in the second phase of the adult life of certain females, those that did not pair a sufficient number of times in the first phase (Milne, 1960). The level of 'bee-lining' therefore is determined by the success of the males in finding females. Leaving aside the differential effect of climate on male and female activity, a female has the greatest chance of being found by a male in a dense population. The level of migration is highest then when the population density is lowest: this is a curious reversal of the normal situation, but one that is irrelevant to my hypothesis.

The occurrence of fully winged and flightless individuals within a species has already been discussed; this results of course in a polymorphism for migratory movement. It seems that such a polymorphism also exists in certain bruchids where all the individuals are fully winged. *Callosobruchus maculatus* has an active form which differs in certain morphological characters and is apparently genetically controlled (Caswell, 1960), although earlier work suggested that it might be the effect of crowding (Utida, 1954, 1956). In *Acanthoscelides obtectus* some individuals usually remain in a collection of beans, after the majority have migrated away (Larson & Fisher, 1938).

[*Editors' Note:* Material has been omitted at this point.]

III. GENERAL ASPECTS OF MIGRATION

(1) *The interactions of environmental change, migratory movement and diapause*

The evidence given in the foregoing sections shows that in a stable habitat the migratory tendency is minimal, but in the most temporary habitats it reaches full expression. There are, however, various kinds of impermanence, each producing a different response according to its nature. Figure 1 is an attempt to systematize these various degrees of impermanence and the animals' response.

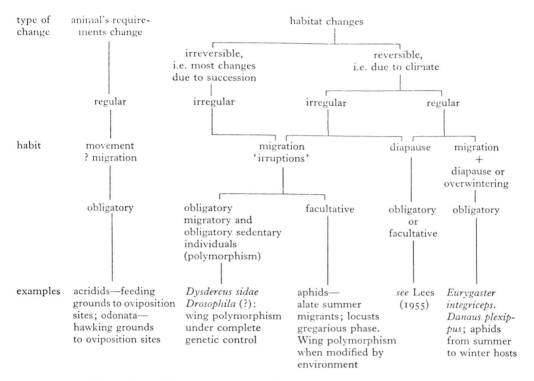

Fig. 1. The relation of movement and diapause to environmental change.

In some arthropods a movement is associated with a change in requirements rather than a change in habitat, all such movements being obligatory. In other cases, however, it is the habitat that changes. If considered over a period of time, habitats described as temporary show their impermanence in two ways: (1) change irreversible—most changes associated with ecological successions, e.g. the ageing of a dung pat; (2) change reversible—changes due to the effects of climate. Changes of habitat due to climate may be regular, as for example the alteration of summer and winter seasons in temperature zones, or irregular, as the rainfall of semi-arid and arid areas.

The fundamental differences in the type of habitat change have been of profound importance in determining the evolution of methods of surviving such environmental change by the Arthropoda.

It has been suggested by many other authors (e.g. Eliot, 1943; Wiltshire, 1946; Andrewartha & Birch, 1954) that migrations and diapause are the two possible methods of surviving an unfavourable period. When the change in the habitat is reversible, diapause, remaining *in situ* until the return of favourable conditions, appears to be the safest method, especially when the change is regular. The diapause site is unlikely to be identical with the feeding site and a movement will take place from one to the other. The extent of this movement will depend on the relative risks associated with losing touch with the feeding site on one hand and finding a protected site for diapause on the other. This being so, extensive pre- and post-diapause movements would be expected in arthropods whose feeding sites are ephemeral, changing their position from season to season, or very exposed. The outstanding examples of such movements, as *Eurygaster integriceps*, many coccinellids, *Danaus plexippus* and *Agrotis infusa*, fulfil these requirements. For example, Urquart (1960) showed that *Danaus* was unable to withstand the cold of winter in Canada.

Movements are also made by insects with annual hosts that would need to be rediscovered the next year, for example, halticine beetles (p. 193) and the bug *Lygus rugulipennis*. Whilst, in contrast, in the case of the boll weevil, *Anthonomus grandis*, living on the somewhat more permanent cotton bush in America, almost all individuals leave the cotton field, but few hibernate more than 75 m. away from its periphery (Beckham, 1957; Fye, McMillian, Walker & Hopkins, 1959). Movements into or away from the diapause site are always migratory at the start, but towards the end are probably trivial; as such changes in the environment are the effects of regular climatic cycles, the diapause and its associated movement are usually obligatory.

The changes of habitat may, however, be irregular, as in the areas of the warm temperate or subtropical zones. Coupled with this irregularity in time is an irregularity in space, and quite frequently a biotype with different climate, and hence suitable conditions, will occur within travelling distance. It is therefore not surprising that whilst some arthropods of habitats with an irregular climate undergo diapause *in situ*, others have developed a tendency for migration. The level of migratory movement in these species has been finely balanced by natural selection to allow for the frequency of change of the habitat. This means that the level may vary from one individual to another in the same species; this variability also occurs in species whose habitats change as the result of ecological succession and possibly in those whose range extends over a variety of climatic zones.

(2) *The maintenance of a variable level of migratory movement in a species*

A variable level of migratory movement can be maintained in two ways. First, by the species being polymorphic, so that in any given population a proportion will be obligatory migrants and the remainder non-migrants. The proportions in which these two forms occur will be geared, by natural selection, to the frequency of change of the habitat: the more temporary the habitat, the more obligatory migratory individuals and vice versa. All cases of completely genetically controlled wing polymorphism are examples of this (e.g. in beetles, Jackson, 1928; Lindroth, 1949), whilst there is evidence suggestive of a polymorphism for movement in fully winged forms of

Dysdercus sidae (Ballard & Evans, 1927) and of certain bruchids (Larson & Fisher, 1938; Caswell, 1960) and possibly in *Drosophila* (Dobzhansky & Wright, 1947). In the grasshopper *Melanoplus bilituratus* Paul & Putnam (1960) found that only certain individuals left the 'outbreak area' and that their elytron and femur lengths were significantly greater than those of the individuals which did not migrate. It is noteworthy that Lack (1954, pp. 244–6) gives instances of polymorphism for migration in birds.

I propose to call the second way 'facultative migration'. This is perhaps the highest evolutionary adaption for the exploitation of irregularly changing environments and is characterized by the migration being triggered off by some factor that heralds the advance of unsatisfactory conditions. The production of summer migrant alates in aphids (in which migration is obligatory) should be regarded as a facultative response by the species to the early signs—crowding or water shortage—of the deterioration of the environment, and likewise with the gregarious phase of locusts (other than *Schistocerca*). The production of the fully winged morph in many Heteroptera is probably a facultative response to environmental conditions (Southwood, 1961).

The last type of dispersal, which may be facultative, occurs when the trivial movement of an individual becomes dispersive, because the insect is caught up in winds of greater velocity than its flight speed. Such individuals have been referred to as vagrants in this article and should not be confused with migrants; they lack the characteristic dispersive-movement behaviour pattern, i.e. ascending to the tips of vegetation, launching themselves into the air, flying upwards, which has been described in the Araneida, Orthoptera, Heteroptera, Aphididae and Diptera, and undoubtedly occurs in other groups. It is postulated that this vagrant type of dispersal may become significant in a permanent habitat when a large population has been built up and the consequent jostling or excessive searching for unoccupied oviposition sites increases the frequency of trivial movements to such a level that if only a fraction of a percent of these result in dispersal away from the habitat, a considerable number of individuals will be involved. The 'migrations' of the larvae and probably of the adults of the spruce budworm, *Choristoneura fumiferana* (p. 190), appear to be an example of this type of dispersal.

IV. CONCLUSIONS

It can be concluded that the level of migratory movement (by definition, away from the habitat) is positively correlated with the degree of impermanence of the habitat, and that the prime evolutionary value of migration lies in the colonization of changing or temporary habitats. A general definition of a temporary habitat is that after a short time it ceases to meet the environmental requirements of the animal—time being measured in terms of generations. Such a definition covers all types of change—changes of requirements by the animals and changes in the habitat—due to ecological succession or climate (see Fig. 1). This view of the value of migration is in accord with Lack's (1954) conclusions on bird migration, although with birds a distinction is still drawn between typical migration, associated with regular changes in the environment and irruptions associated with irregular changes. As indicated already, such a distinction does not seem justified in the case of terrestrial arthropods, where the movements associated with both types of change seem identical, both in terms of behaviour and of chronological position.

This article has been mainly concerned with climatic (physical) and botanical (biotic) change in the habitat. Andrewartha & Birch (1954) stress that the physical environment should not be considered apart from the biotic. Excesses of animals, of the same sort (i.e. crowding) or of a different sort (including predators), are other biotic factors that are likely to cause the environment to change so that it no longer meets the species requirements. Crowding is often considered as the primary cause of migration and Andrewartha & Birch (1954) point out the value of dispersal (*sensu lato*) as a means of protection from predators. What part have such factors played in the evolution of migratory movement?

The correlation found between the climatic or botanical impermanence of the habitat and the level of migratory movement, taken together with other arguments advanced in the Introduction, shows that crowding and predator effects must have played a secondary role in the evolution of migration; were this not so there would surely be more instances of animals of permanent habitats with high levels of migratory movement. If this is accepted, then crowding and a low level of predation are the effects of migratory movement; although they are extremely important in the population dynamics of the species concerned, they would tend to cancel each other out.

When an animal first colonizes a new habitat there is a strong probability that its normal predators (including parasites) will be absent and during this first period after colonization the population will, if the reproductive rate is sufficiently high, build up quickly (Gause, 1934). During this period of build-up intraspecific competition, i.e. overcrowding, may occur (as in *Lucilia*, Nicholson, 1950). Now denizens of temporary habitats are frequently (by definition) occupying new habitats and therefore might be considered to be recurrently exposed to the risk of overcrowding for this cause. But the very habit that has led to this situation, i.e. the high level of migratory movement, also causes it to be alleviated by numbers of individuals moving out of the habitat. Theoretically this would also allow the maintenance of a high rate of reproduction, a desirable feature for an animal that only has a very limited time in which to utilize any habitat. The comparative freedom from predator control is, of course, only a feature of any population early in its life, subsequently the effects of predators may become important and even lead to the extinction of the colony, as observed with one colony of the migratory butterfly *Ascia* by Nielson & Nielsen (1950).

In contrast to the above, those movements within the habitat that result in an increase in the mean distance between the individuals of the population are likely to be closely related to the phenomena of overcrowding and predator avoidance. The movements of some newly hatched larvae away from the site of the egg batch are of this type and are outside the scope of this article, but they are discussed by Henson (1959) who terms them 'secondary dispersive processes'. It may be, of course, that it is from such movements that migration has been evolved.

It is suggested that the correlation between overcrowding and migration, sometimes false, sometimes true, accepted by so many authors (e.g. Abbott, 1951) has arisen in four ways. First, the mass migrations of dense populations are far more likely to be noticed than those of the few individuals of sparser populations. Secondly, one-way movements observed in one locality are often taken as evidence that the migratory individuals are lost and neither they, nor their progeny, contribute to future generations; but the return may be in smaller, unnoticed numbers (Williams, 1951) or along another route, as with some birds (Lack & Lack, 1951) or locusts (Waloff, 1960). Thirdly, in temporary-habitat species crowding may frequently be the trigger for facultative migration and, lastly and more rarely, in a permanent habitat intense overcrowding could increase the general level of trivial movement so much that large numbers may become vagrants.

V. SUMMARY

It is suggested that animal movements fall basically into two types: trivial and migratory. Trivial movements are normally confined to the territory or habitat of the population to which the animal belongs, migratory movements carry the animal away from this area. Although there is undoubtedly no sharp line but a gradation between these two types, they can be distinguished by various ecological, physiological and behavioural characteristics. It is pointed out how the evolutionary value of migration has often been disputed and, with many animals, workers have failed to find an association between migration and a concurrent deterioration of the habitat, and have concluded that migration is a definite phase in the life cycle of these animals.

The following hypothesis is therefore proposed: that the prime evolutionary advantage of migratory movement lies in its enabling a species to keep pace with the changes in location of its habitats. If this is true, then the level of migratory movement in any species should be geared to the rate of change of its habitat and be highest in those species with the most temporary habitats.

The evidence on migratory movement in the major taxa of terrestrial Arthropoda is considered in the light of this hypothesis and it is shown that the level of migratory movement is positively correlated with impermanence of the habitat. A habitat may cease to meet the environmental requirements of the insects because of a change in these requirements or because of changes in the habitat itself. Changes in the habitat may be regular or irregular, reversible or irreversible, and due to climate or to ecological succession. Migration and diapause are the two methods animals have evolved to meet such changes. It is suggested that, as with diapause, migration can be considered as obligatory and facultative.

Obligatory migration occurs completely independently of any factor in the immediate environment; but there is evidence that in some species this is geared to the frequency of the appearance of new habitats and the disappearance of the old ones, by the species being polymorphic in respect of migratory movement. Perhaps the most highly evolved condition occurs in those animals that possess facultative migration; this is not a response to immediate adversity, but is triggered off by some factor that heralds the advance of unsatisfactory conditions. In some cases species respond, facultatively, to such factors by producing forms for whose individuals migration is obligatory (e.g. alate *Aphis fabae* on summer hosts).

It is concluded that the changing pattern of the environment has been the primary factor in the evolution of migratory movement. Nevertheless, migration also plays an important part in the population dynamics of denizens of temporary habitats, being on the one hand a defence against predators and on the other a means of reducing density and preventing overcrowding.

I have been greatly helped and encouraged in the preparation of this article by the advice of Drs R. D. Alexander, M. I. Crichton, H. H. Crowell and V. F. Eastop, Lt.-Col. F. C. Fraser, Mr R. A. French. Drs B. R. Laurence, T. Lewis, A. C. Neville and L. F. Steiner ('USDA, unpublished'), Dr F. Schneider, Mr H. L. G. Stroyan, Sir Boris Uvarov, Dr N. Waloff and Mr M. J. Way, who have not only drawn my attention to relevant data and facts in their respective groups, but have in many instances generously allowed me to quote their unpublished work. Drs C. G. Johnson and J. S. Kennedy, Professor O. W. Richards and Mr L. R. Taylor have kindly read the whole or part of the manuscript and it is a pleasure to acknowledge my indebtedness to them for numerous constructive criticisms and invaluable discussions.

VI. REFERENCES

ABBOTT, C. H. (1950). Twenty-five years of migration of the Painted Lady butterfly, *Vanessa cardui*, in southern California. *Pan-Pacif. Ent.* **26**, 161–72.
ABBOTT, C. H. (1951). A quantitative study of the migration of the Painted Lady butterfly, *Vanessa cardui* L. *Ecology*, **32**, 155–71.
ALEXANDER, R. D. (1957). The taxonomy of the Field Crickets of the Eastern United States (Orthoptera: Gryllidae: Acheta). *Ann. ent. Soc. Amer.* **50**, 584–602.
ALEXANDER, R. D. (1961). Aggressiveness, territoriality and sexual behaviour in field crickets. *Behaviour* (in press).
ALEXANDER, R. D. & BIGELOW, R. S. (1960). Allochronic speciation in field crickets, and a new species, *Acheta veletis*. *Evolution*, **14**, 334–46.
ALEXANDER, R. D. & THOMAS, E. S. (1959). Systematic and behavioural studies on the crickets of the *Nemobius fasciatus* group (Orthoptera: Gryllidae: Nemobiinae). *Ann. ent. Soc. Amer.* **52**, 591–605.
ALLAN, P. B. M. (1942). Travelling butterflies. *Entomologist*, **75**, 147–50.
ALLEN, H. W. (1958). Orchard studies on the effect of organic insecticides on parasitism of the oriental fruit moth. *J. econ. Ent.* **51**, 82–7.
AMANSHAUSER, H. (1955). Was ist ein Wanderfalter? *Z. wien. Ent. Ges.* **40**, 273–6.
ANDREWARTHA, H. G. & BIRCH, L. C. (1954). *The distribution and abundance of animals.* Chicago.
AYROZA GALVÃO, A. L. (1948). Active and passive dispersion of anopheline species. *Proc. 4th Cong. trop. Med. Malar,* **1**, 656–71.
BALCH, R. E. (1939). The outbreak of the European spruce sawfly in Canada and some important features of its bionomics. *J. econ. Ent.* **32**, 414–18.
BALFOUR-BROWNE, F. (1940, 1950, 1958). *British water beetles,* **1, 2, 3**. London.
BALFOUR-BROWNE, F. (1953). Coleoptera. Hydradephaga. *Handbooks for the identification of British insects,* **4** (3), 1–33.
BALLARD, E. & EVANS, M. G. (1927). *Dysdercus sidae.* Montr., in Queensland. *Bull. ent. Res.* **18**, 405–32.
BARNES, M. M. (1959). Radiotracer labeling of a natural tephritid population and flight range of the walnut husk fly. *Ann. ent. Soc. Amer.* **52**, 90–2.
BEALL, G. (1941*a*). The monarch butterfly, *Danaus archippus* Fab. I. General observations in Southern Ontario. *Canad. Fld Nat.* **55**, 177–84.
BEALL, G. (1941*b*). The monarch butterfly, *Danaus archippus* Fab. II. The movement in southern Ontario. *Canad. Fld Nat.* **55**, 133–7.
BEALL, G. (1942). Mass movement of the wasp, *Polistes fuscatus* var. pallipes LeP. *Canad. Fld Nat.* **56**, 64–7.
BECKHAM, C. M. (1957). Hibernation sites of the bollweevil in relation to a small, Georgia Piedmont, cotton field. *J. econ. Ent.* **50**, 833–4.
BEEBE, W. (1949). Insect migration at Rancho Grande in north-central Venezuala. General account. *Zoologica*, **34**, 107–10.
BENSON, R. B. (1946). A mass-movement of the sawfly *Athalia cordata* Lep. in north Devonshire (Hym, Tenthredinidae). *Ent. mon. Mag.* **82**, 87.

BENSON, R. B. (1950). An introduction to the natural history of British sawflies (Hymenoptera Symphyta). *Trans. Soc. Brit. Ent.* **10**, 45–142.

BIRD, R. D. & DENNING, D. (1936). A migration of the dragonfly, *Libellula quadrimaculata* L. *Canad. Ent.* **68**, 283.

BISHOPP, F. C. & LAAKE, E. W. (1921). Dispersion of flies by flight. *J. agric. Res.* **21**, 729–66.

BLACKWALL, J. (1827). Observations and experiments, made with a view to ascertain the means by which the spiders that produce gossamer effect their aerial excursions. *Trans. Linn. Soc. Lond.* **15**, 449–59.

BLAIS, J. R. (1953). Effects of the destruction of the current year's foliage of balsam fir on the fecundity and habits of flight of the spruce budworm. *Canad. Ent.* **85**, 446–8.

BLUNCK, H. (1954). Beobachtungen über Wanderflüge von *Pieris brassicae* L. *Beitr. Ent.* **4**, 485–528.

BODENHEIMER, F. S. (1943). Studies on the life-history and ecology of Coccinellidae: 1. The life-history of *Coccinella septempunctata* L. in four different zoo-geographical regions (Coleoptera: Coccinellidae) *Bull. Soc. Fouad Ent.* **27**, 1–28.

BOND, F. (1868). (Remarks at meeting.) *Proc. ent. Soc. Lond.* 1868, 41.

BONNEMAISON, L. (1951). Contribution à l'étude des facteurs provoquant l'apparition des formes ailées et sexuées chez les Aphidinae. *Ann. Épiphyt.* **2**, 1–380.

BOYKO, H. (1949). On climatic extremes as decisive factors for plant distribution. *Palestine J. Bot. Rehovot*, **7**, 41–52.

BRAENDEGAARD, J. (1937). Observations on spiders starting off on 'ballooning excursions'. *Vidensk. Medd. dansk. naturh. Foren. Kbh.* **101**, 115–17.

BRINKHURST, R. O. (1958). Alary polymorphism in Gerroidea (Hemiptera–Heteroptera). *Nature, Lond.*, **182**, 1461–2.

BRINKHURST, R. O. (1959). Alary polymorphism in Gerroidea (Hemiptera–Heteroptera). *J. anim. Ecol.* **28**, 211–30.

BRISTOWE, W. S. (1939). *The comity of spiders*, 1. London.

BROWN, C. E. (1958). Dispersal of the pine needle scale, *Phenacaspis pinifoliae* (Fitch), (Diaspididae: Homoptera). *Canad. Ent.* **90**, 685–90.

BROWN, E. S. (1951). The relation between migration-rate and type of habitat in aquatic insects, with special reference to certain species of Corixidae. *Proc. zool. Soc. Lond.* **121**, 539–45.

BROWN, E. S. (1954). Report on Corixidae (Hemiptera) taken in light traps at Rothamsted Experimental Station. *Proc. R. ent. Soc. Lond.* (A), **29**, 17–22.

BURLA, H., BRITO DA CUNHA, A., CAVALCANTI, A. G. L., DOBZHANSKY, T. & PAVAN, C. (1950). Population density and dispersal rates in Brazilian *Drosophila willistoni*. *Ecology*, **31**, 393–404.

BURMANN, K. (1955). Nordtiroler Wanderfalterbeobachtungen 1954. *Z. wien. ent. Ges.* **40**, 241–4.

BURTON, J. F. (1950). Observations on a migration of insects over the Thames estuary. *Entomologist*, **83**, 203–6.

BUXTON, P. A. (1923). *Animal life in deserts*. London.

CALLAN, E. McC. (1954). An association of Hemiptera and Coleoptera in Trinidad. *Ent. mon. Mag.* **90**, 102.

CAMERANO, L. (1914). Le riunioni delle Coccinelle. *Z. wiss. InsktBiol.* **10**, 187–9.

CAMPBELL, K. G. (1960). Preliminary studies in population estimation of two species of stick insects (Phasmatidae Phasmatodea) occurring in plague numbers in Highland Forest areas of south-eastern Australia. *Proc. Linn. Soc. N.S.W.* **85**, 121–37.

CASWELL, G. H. (1960). Observations on an abnormal form of *Callosobruchus maculatus* (F.). *Bull. ent. Res.* **50**, 671–80.

CHAPMAN, J. A. (1956). Flight muscle changes during adult life in a Scolytid beetle. *Nature, Lond.*, **177**, 1183.

CHAPMAN, J. A., ROMER, J. I. & STARK, J. (1955). Ladybird beetles and army cutworm adults as food for Grizzly Bears in Montana. *Ecology*, **36**, 156–8.

CHNÉOUR, A. (1953). Migrations des papillons dans la zone méditerranéenne, principalement en ce qui concerne la Tunisie. *Bull. Soc. Sci. nat. Tunisie*, **6**, 63–70.

CHRISTENSON, L. R. & FOOTE, R. H. (1960). Biology of fruit flies. *Ann. Rev. Ent.* **5**, 171–92.

CLARK, L. R. (1947a). An ecological study of the Australian Plague Locust (*Chortoicetes terminifera* Walk.) in the Bogan–Macquarie outbreak area, N.S.W. *Bull. Coun. sci. industr. Res. Aust.* **226**, 1–71.

CLARK, L. R. (1947b). Ecological observations on the small Plague Grasshopper, *Austroicetes cruciata* (Sauss.) in the Trangie District, Central Western New South Wales. *Bull. Coun. sci. industr. Res. Aust.* **228**, 1–26.

CLARK, L. R. (1949). Habitats and community relations of the acrididae of the Trangie District, N.S.W. *Bull. sci. ind. Res. Org., Melbourne*, **250**, 1–20.

CLOUDSLEY-THOMPSON, J. L. (1949). The significance of migration in myriapods. *Ann. Mag. nat. Hist.* (12), **2**, 947–62.

CLOUDSLEY-THOMPSON, J. L. (1951). Supplementary notes on Myriapoda. *Naturalist*, 1951, 16–17.
CLOUDSLEY-THOMPSON, J. L. (1958). *Spiders, scorpions, centipedes and mites*. London.
COMMON, I. F. B. (1954). A study of the ecology of the adult Bogong moth, *Agrotis infusa* (Boisd.) (Lepidoptera: Noctuidae), with special reference to its behaviour during migration and aestivation. *Aust. J. Zool.* **2**, 223–63.
CORBET, P. S. (1957). The life-history of the emperor dragonfly, *Anax imperator* Leach (Odonata: Aeshnidae). *J. Anim. Ecol.* **26**, 1–69.
CORBET, P. S. (1958). Temperature in relation to seasonal development of British dragonflies (Odonata). *10th int. Congr. Ent.* (2), 755–7.
CORNWELL, P. B. (1958). Movements of the vectors of virus diseases of cacao in Ghana. I. Canopy movement in and between trees. *Bull. ent. Res.* **49**, 613–30.
CORNWELL, P. B. (1960). Movements of the vectors of virus diseases of cacao in Ghana. II. Wind movements and aerial dispersal. *Bull. ent. Res.* **51**, 175–201.
COTTERELL, G. S. (1925). The hispid leaf miner (*Coelaenomenodera elaeidis*) of oil palms on the Gold Coast. *Bull. ent. Res.* **16**, 77–83.
COUTURIER, A. & ROBERT, P. (1958). Recherches sur les migrations du Hanneton commun (*Melolontha melolontha* L.). *Ann. Épiphyt.* **9**, 257–328.
CRAWFORD, G. I. (1953). Swarms of Coccinellidae (Col.) on Norfolk coast. *Entomologist*, **86**, 24.
CRICHTON, M. I. (1961). Observations on the longevity and dispersal of adult Limnephilidae (Trichoptera). *11th int. Cong. Ent.* **1**, 366–71.
CURTIS, J. (1834). *Brit. Entomology*, **11**, 509.
DANNREUTHER, T. (1946). Migration records 1945. *Entomologist*, **79**, 97–110.
DANNREUTHER, T. (1947). Migration records 1946. *Entomologist*, **80**, 107–12, 137–44.
DARLINGTON, P. J. (1943). Carabidae of mountains and islands: data on the evolution of isolated faunas, and on atrophy of wings. *Ecol. Monogr.* **13**, 37–61.
DARLOW, H. M. (1951). Observations on migrant Lepidoptera in the Mediterranean. *Entomologist*, **84**, 246–54.
DAVEY, J. T. (1953). Possibility of movements of the African migratory locust in the solitary phase and the dynamics of its outbreaks. *Nature, Lond.*, **172**, 720–4.
DAVEY, J. T. (1959). The African migratory locust (*Locusta migratoria migratorioides* Rel. & Frm., Orth.) in the Central Niger Delta. II. The ecology of *Locusta* in the semi-arid lands and seasonal movements of populations. *Locusta*, **7**, 1–180.
DAVIDSON, J. & ANDREWARTHA, H. G. (1948). The influence of rainfall, evaporation and atmospheric temperature on fluctuations in the size of a natural population of *Thrips imaginis* (Thysanoptera). *J. Anim. Ecol.* **17**, 200–22.
DICKSON, R. C. & LAIRD, E. F. (1959). California desert and coastal populations of flying aphids and the spread of lettuce-mosaic virus. *J. econ. Ent.* **52**, 440–3.
Добржанский Ф. Г. (DOBRZHANSKI, F. G.) (1922). Скопления и перелеты у божьих коровок (Coccinellidae). *Известия отдела прикладной энтомологии*. *Rep. Bur. appl. Ent., Leningr.*, **2**, 103–24. (Abstract in *Rev. appl. Ent.* (A), **11**, 305.)
DOBZHANSKY, T. & WRIGHT, S. (1943). Genetics of natural populations. X. Dispersion rates in *Drosophila pseudo-obscura*. *Genetics*, **28**, 304–40.
DOBZHANSKY, T. & WRIGHT, S. (1947). Genetics of natural populations. XV. Rate of diffusion of a mutant gene through a population of *Drosophila pseudoobscura*. *Genetics*, **32**, 303–24.
DORST, H. E. & DAVIS, E. W. (1937). Tracing long-distance movements of beet leafhopper in the desert. *J. econ. Ent.* **30**, 948–54.
DUBININ, N. P. & TINIAKOV, C. G. (1946). Inversion gradients and natural selection in ecological races of *Drosophila funebris*. *Genetics*, **31**, 537–45.
DUFFEY, E. (1956). Aerial dispersal in a known spider population. *J. Anim. Ecol.* **25**, 85–111.
DUNNING, J. W. (1869). (Remarks at meeting.) *Proc. ent. Soc. Lond.* 1869, 25.
EIMER, T. (1882). Eine Dipteren- und Libellen-wanderung beobachtet in September 1880. *Jb. ver. Vaterl. Naturk. Württemb., Stuttgart*, **38**, 105–13.
ELIOT, N. (1943). Migration v. hibernation. *Entomologist*, **76**, 193–8.
ELTON, C. S. (1925). The dispersal of insects to Spitsbergen. *Trans. ent. Soc. Lond.* 1925, 289–99.
ELTON, C. S. (1927). *Animal ecology*. London.
ELTON, C. S. (1930). *Animal ecology and evolution*. Oxford.
ELTON, C. S. & MILLER, R. S. (1954). The ecological survey of animal communities: with a practical system of classifying habitats by structural characters. *J. Ecol.* **42**, 460–96.
Федотов Д. М., ред. (FEDOTOV, D. M., ed.) (1947, 1955). *Вредная черепашка*. Eurygaster integriceps Put. 1–3. Москва.
FERNANDO, C. H. (1958). The colonization of small freshwater habitats by aquatic insects. 1. General discussion, methods and colonization in the aquatic Coleoptera. *Ceylon J. Sci. (Bio Sci.)*, **1**, 117–54.

FERNANDO, C. H. (1959). The colonization of small freshwater habitats by aquatic insects. 2. Hemiptera (the water-bugs). *Ceylon J. Sci. (Biol. Sci.)*, **2**, 5–32.
FRAENKEL, G. (1932). Die Wanderungen der Insekten. *Ergeb. Biol.* **9**, 1–238.
FRASER, F. C. (1924). A survey of the Odonata (dragonfly) fauna of Western India with special remarks on the genra *Macromia* and *Idionyx* and descriptions of thirty new species. *Rec. Indian Mus.* **26**, 423–522.
FRASER, F. C. (1941). A note on the 1941 immigration of *Sympetrum fonscolombii* (Selys) (Odon.). *J. Soc. Brit. Ent.* **2**, 133–6.
FRASER, F. C. (1949). Odonata. *Handbooks for the identification of British insects*, **1** (10), 1–48.
FREEMAN, J. A. (1939). A contribution to the study of wind-borne insects with special reference to the vertical distribution and dispersal. Ph.D. Thesis, London University.
FREEMAN, J. A. (1945). Studies in the distribution of insects by aerial currents. The insect population of the air from ground level to 300 feet. *J. Anim. Ecol.* **14**, 128–54.
FRENCH, R. A. (1959). Migration records 1958. *Entomologist*, **92**, 164–76.
FRITZCHE, R. (1957). Zur Biologie und Ökologie der Rapsschädlinge aus der Gattung *Meligethes*. *Z. angew. Ent.* **40**, 222–80.
FYE, R. E., MCMILLAN, W. W., WALKER, R. L. & HOPKINS, A. R. (1959). The distance into woods along a cotton field at which the boll weevil hibernates. *J. econ. Ent.* **52**, 310–12.
GAINES, J. C. & EWING, K. P. (1938). The relation of wind currents, as indicated by balloon drifts, to cotton flea hopper dispersal. *J. econ. Ent.* **31**, 674–7.
GAMBLES, R. M. (1951). A dragonfly migration at Vom. *Niger. Fld*, **16**, 135–8.
GARDNER, A. E. (1954). A key to the larvae of the British Odonata. *Ent. Gaz.* **5**, 157–71, 193–213.
GARDNER, A. E. (1955). The biology of dragonflies. *Proc. S. Lond. ent. nat. Hist. Soc.* 1954–1955, 109–34.
GAUSE, G. F. (1934). *The struggle for existence.* Baltimore.
GHESQUIÈRE, J. (1932). Les migrations de Papillons en Afrique centrale. *Rev. Zool. Bot. afr.* **22** (suppl.), 20–38.
GILMOUR, D., WATERHOUSE, D. F. & MCINTYRE, G. A. (1946). An account of experiments undertaken to determine the natural population density of sheep blowfly, *Lucilia cuprina* Wied. *Bull Coun. sci. industr. Res. Aust.* **195**, 1–39.
GLICK, P. A. (1939). The distribution of insects, spiders and mites in the air. *Tech. Bull. U.S. Dep. Agric.* **373**, 1–150.
GREATHEAD, D. J. (1959). The biology of *Stormorhina lunata* (Fabricus) (Diptera: Calliphoridae), predator of the eggs of Acrididae. Ph.D. Thesis, London University.
GREEN, J. (1952). A swarm of beetles in Carmarthenshire. *Ent. mon. Mag.* **88**, 143.
GREENBANK, D. O. (1957). The role of climate and dispersal in the initiation of outbreaks of the spruce budworm in New Brunswick. II. The role of dispersal. *Canad. J. Zool.* **35**, 385–403.
GREENSTED, L. W. (1939). Colonization of new areas by water-beetles. *Ent. mon. Mag.* **75**, 174–5.
HAINE, E. (1955). Aphid take-off in controlled wind speeds. *Nature, Lond.*, **175**, 474–5.
HANS, H. (1959). Beiträge zur Biologie von *Sitona lineatus* L. *Z. angew. Ent.* **44**, 341–86.
HARDY, A. C. & MILNE, P. S. (1938). Studies in the distribution of insects by aerial currents. Experiments in aerial tow-netting from kites. *J. Anim. Ecol.* **7**, 199–229.
HAWKES, O. A. M. (1926). On the massing of the ladybird *Hippodamia convergens* (Coleoptera) in the Yosemite Valley. *Proc. zool. Soc. Lond.* 1926, 693–705.
HAYWARD, K. J. (1928). Migration of insects in north-eastern Argentina. *Entomologist*, **61**, 210–12.
HAYWARD, K. J. (1931). Some further notes on insect migration in Argentina. *Entomologist*, **64**, 40–1.
HAYWARD, K. J. (1958). Migration of butterflies in Argentina (summer 1956–57). *Entomologist*, **91**, 163–4.
HEAPE, W. (1931). *Emigration, migration and nomadism.* Cambridge.
HEED, W. B. (1956). Apuntes sobre la ecologia y la dispersion de los Drosophilidae (Diptera) de El Salvador. *Commun. Inst. trop. El Salvador*, **5**, 59–74.
HENSON, W. R. (1951). Mass flights of the spruce budworm. *Canad. Ent.* **83**, 240.
HENSON, W. R. (1959). Some effects of secondary dispersive processes on distribution. *Amer. Nat.* **93**, 315–20.
HOCKING, B. (1960). Migration of flying Diptera. *11th int. Congr. Ent.* (in press).
HODEK, I. (1960). Migration to the hibernation-quarters in Coccinellidae. *11th int. Congr. Ent.* (in press).
JACKSON, D. J. (1928). The inheritance of long and short wings in the weevil *Sitona hispidula*, with a discussion of wing reduction among beetles. *Trans. roy. Soc. Edinb.* **55**, 655–735.
JACKSON, D. J. (1933). Observations on the flight and muscles of *Sitona* weevils. *Ann. appl. Biol.* **20**, 731–70.

JACKSON, D. J. (1950). *Noterus clavicornis* Degeer and *N. capricornis* Herbst. (Col. Dytiscidae) in Fife. *Ent. mon. Mag.* **86**, 39–43.
JACKSON, D. J. (1952). Observations on the capacity for flight of water beetles. *Proc. R. ent. Soc. Lond.* (A), **27**, 57–70.
JACKSON, D. J. (1956a). The capacity for flight of certain water beetles and its bearing on their origin in the Western Scottish Isles. *Proc. Linn. Soc. Lond.* **167**, 76–96.
JACKSON, D. J. (1956b). Observations on flying and flightless water beetles. *J. Linn. Soc. (Zool.)*, **43**, 18–42.
JACKSON, D. J. (1958). Observations on *Hydroporus ferrugineus* Steph. (Col. Dytiscidae), and some further evidence indicating incapacity for flight. *Ent. Gaz.* **9**, 55–9.
JANNONE, G. (1948). Migrazioni periodiche di Lepidotteri in Eritrea e loro riflessi sull' agricultura (seconda nota). *Ann. Mus. Stor. nat. Genova*, **93**, 142–67.
JOHNSON, B. (1953). Flight muscle autolysis and reproduction in aphids. *Nature, Lond.*, **172**, 813.
JOHNSON, B. (1954). Effect of flight on behaviour of *Aphis fabae* Scop. *Nature, Lond.*, **173**, 831.
JOHNSON, B. (1957). Studies on the degeneration of the flight muscles of alate aphids. I. A comparative study of the occurrence of muscle breakdown in relation to reproduction in several species. *J. ins. Physiol.* **1**, 248–56.
JOHNSON, B. (1959). Studies on the degeneration of the flight muscles of alate aphids. II. Histology and control of muscle breakdown. *J. ins. Physiol.* **3**, 367–77.
JOHNSON, C. G. (1954). Aphid migration in relation to weather. *Biol. Rev.* **29**, 87–118.
JOHNSON, C. G. (1960a). A basis for a general system of insect migration and dispersal by flight. *Nature, Lond.*, **186**, 348–50.
JOHNSON, C. G. (1960b). The present position in the study of insect dispersal. *Rep. 7th Commonw. ent. Confr.* 140–5.
JOHNSON, C. G. (1960c). A functional approach to insect migration and dispersal and its bearing on future study. *11th int. Congr. Ent.* (in press).
JOHNSON, C. G. & SOUTHWOOD, T. R. E. (1949). Seasonal records in 1947 and 1948 of flying Hemiptera–Heteroptera, particularly *Lygus pratensis* L., caught in nets 50 ft. to 3,000 ft. above ground. *Proc. R. ent. Soc. Lond.* (A), **24**, 128–30.
JOHNSON, C. G., TAYLOR, L. R. & HAINE, E. (1957). The analysis and reconstruction of diurnal flight curves in alienicolae of *Aphis fabae* Scop. *Ann. appl. Biol.* **45**, 682–701.
JONES, S. C. & WALLACE, L. (1955). Cherry fruit fly dispersion studies. *J. econ. Ent.* **48**, 616–17.
JOYCE, R. J. V. (1952). The ecology of grasshoppers in east central Sudan. *Anti-Locust Bull.* **11**, 1–99.
KEILHOLZ, (1925). Massenzüge von Libellen. *Ent. Z.* **39**, 38–9.
KENNEDY, J. S. (1951). The migration of the desert locust (*Schistocerca gregaria* Forsk.) *Phil. Trans.* B, **235**, 163–290.
KENNEDY, J. S. (1956). Phase transformation in locust biology. *Biol. Rev.* **31**, 349–70.
KENNEDY, J. S. (1958). The experimental analysis of aphid behaviour and its bearing on current theories of instinct. *10th int. Congr. Ent.* (2), 397–404.
KENNEDY, J. S. (1961). A turning point in the study of insect migration. *Nature, Lond.* **189**, 785–91.
KENNEDY, J. S., BOOTH, C. O. & KERSHAW, J. (1961). Host finding by aphids in the field. III. Visual attraction. *Ann. appl. Biol.* **49**, 1–21.
KENNEDY, J. S., LAMB, K. P. & BOOTH, C. O. (1958). Responses of *Aphis fabae* Scop. to water shortage in host plants in pots. *Ent. exp. appl.* **1**, 274–91.
KENNEDY, J. S. & STROYAN, H. L. G. (1959). Biology of aphids. *Ann. Rev. Ent.* **4**, 139–60.
KEY, K. H. L. (1945). The general ecological characteristics of the outbreak areas and outbreak years of the Australian plague locust (*Chortoicetes terminifera* Walk.). *Bull. Coun. sci. industr. Res. Aust.* **186**.
KÖHLER, K. (1927). Der Libellenzug durch Schlesien im Jahre 1925. *Mitt. naturw. Ver. Troppau*, **33**, 26–30.
KÖHLER, K. (1935). Massenflüge von Libellen in Schlesien (C.S.R.). *Mitt. naturw. Ver. Troppau*, **40**, 27–31.
LACK, D. (1954). *The natural regulation of animal numbers.* Oxford.
LACK, D. & LACK, E. (1951). Migration of insects and birds through a Pyrenean pass. *J. Anim. Ecol.* **20**, 63–7.
LANE, C. (1955). Insect migration on the north coast of France. *Ent. mon. Mag.* **91**, 301–6.
LARSON, A. O. & FISHER, C. K. (1938). The bean weevil and the southern cowpen weevil in California. *Tech. Bull. U.S. Dep. Agric.* **593**, 1–70.
LAVABRE, E. M. (1958–59). Le Scolyte des branchettes du caféier robuste, *Xyleborus morstatti* Haged. *Café, cacao, Thé*, **2**, 119–30; **3**, 21–33.
LEES, A. D. (1955). *The physiology of diapause in arthropods.* Cambridge.
LEMPKE, B. J. (1950). The migrating Macrolepidoptera of Holland in comparison with those of Great Britain. *Proc. S. Lond. ent. nat. Hist. Soc.* 1948-9, 148–58.

LEMPKE, B. J. (1957). *De nederlandse trekvlinders*. Zutphen.
LESTAGE, J. A. (1935). Note sur un Psoque hôte multiple des dunes coxydoises (*Lachesilla pedicularia* L.). *Bull. Soc. ent. Belg.* **75**, 344–5.
LESTON, D. (1953). Corixidae (Hem.) at ultra-violet light in Middlesex with remarks on migrations. *Ent. mon. Mag.* **89**, 291.
LESTON, D. (1954). Corixidae (Hem.) at ultra-violet light: additional data. *Ent. mon. Mag.* **90**, 166.
LESTON, D. & GARDNER, A. E. (1953). Corixidae (Hemiptera) at mercury-vapour light and some records from Surrey, England. *Ent. Gaz.* **4**, 269–72.
LEWIS, T. (1961). Records of Thysanoptera at Silwood Park, Berks., with notes on their biology. *Proc. R. ent. Soc.* (A), **36**, 89–95.
LICHTENSTEIN, J. L. & GRASSE, P. (1922). Une migration d'Odonates. *Bull. Soc. ent. Fr.*, 1922, 160–3.
LINDQUIST, A. W., YATES, W. W. & HOFFMAN, R. A. (1951). Studies of the flight habits of three species of flies tagged with radioactive phosphorus. *J. econ. Ent.* **44**, 397–400.
LINDROTH, C. H. (1945, 1949). Die Fennoskandischen Carabidae. Eine tiergeographische Studie. *Medd. Göteborgs mus. zool. Avd.* **109**, 1–707; **110**, 1–277; **122**, 1–911.
LOCKET, G. H. & MILLIDGE, A. F. (1951, 1953). *British spiders* **1, 2**. London.
LONGFIELD, C. (1937). *The dragonflies of the British Isles*. London.
LONGFIELD, C. (1948). A vast migration of dragonflies into the south coast of Co. Cork. *Irish. Nat. J.* **9**, 133–41.
MACAN, T. T. (1939). Notes on the migration of some aquatic insects. *J. Soc. Brit. Ent.* **2**, 1–6.
MACLEOD, J. (1956). A preliminary experiment on the local distribution of blowflies. *J. Anim. Ecol.* **25**, 303–18.
MACLEOD, J. & DONNELLY, J. (1957). Some ecological relationships of natural populations of calliphorine blowflies. *J. Anim. Ecol.* **26**, 135–70.
MACLEOD, J. & DONNELLY, J. (1958). Local distribution and dispersal paths of blowflies in hill country. *J. Anim. Ecol.* **27**, 349–74.
MACLEOD, J. & DONNELLY, J. (1960). Natural features and blowfly movement. *J. Anim. Ecol.* **29**, 85–93.
MACY, R. W. (1949). On a migration of *Tarnetrum corruptum* (Hagen) (Odonata) in western Oregon. *Canad. Ent.* **81**, 50–1.
MARRINER, T. F. (1939). Movements of Coccinellidae. *Ent. Rec.* **51**, 104–6.
MARSHALL, W. (1783). Account of the black canker caterpillar which destroys the turnips in Norfolk. *Phil. Trans.* 1783, 217–22.
MCDONOGH, R. S. (1939). The habitat, distribution and dispersal of the psychid moth, *Luffia ferchaultella*, in England and Wales. *J. Anim. Ecol.* **8**, 10–28.
MCLACHLAN, R. (1896). Oceanic migration of a nearly cosmopolitan dragonfly (*Pantala flavescens* F.). *Ent. mon. Mag.* **32**, 254.
MCLACHLAN, R.(1900). Abstract of an article by Mons. A. Lancaster, Migration of *Libellula quadrimaculata* in Belgium in June 1900. *Ent. mon. Mag.* **36**, 222–6.
MILNE, A. (1958, 1959a, b, 1960). Biology and ecology of the garden chafer, *Phyllopertha horticola* (L.); (1958), IV. The flight season: introduction and general aspects. *Bull. ent. Res.* **49**, 685–99; (1959a), V. The flight reason: sex proportions. *Bull. ent. Res.* **50**, 39–52; (1959b), VI. The flight season: reproductive state of females. *Bull. ent. Res.* **50**, 467–86; (1960), VII. The flight season: male and female behaviour and concluding discussion. *Bull. ent. Res.* **51**, 353–78.
MITCHELL, D. F. & EPLING, C. (1951). The diurnal periodicity of *Drosophila pseudoobscura* in Southern California. *Ecology*, **32**, 696–708.
MOERICKE, V. (1955). Über die Lebensgewohnheiten der geflügelten Blattläuse (Aphidine) unter besonderer Berücksichtigung des Verhaltens beim Landen. *Z. angew. Ent.* **37**, 29–91.
MONCREAFF, H. (1868). Notes from Southsea: swarms of Coleoptera. *Entomologist*, **4**, 142–3.
MORETON, B. D. (1945). On the migration of flea beetles (*Phyllotreta* spp.) (Col., Chrysomelidae) attacking Brassica crops. *Ent. mon. Mag.* **81**, 59–60.
MORLEY, C. (1942). Machaon come astray. *Trans. Suffolk nat. Hist. Soc.* **5**, 46.
MUNDT, A. H. (1882). Migration of dragon-flies (*Aeschna heros* Fabr.). *Canad. Ent.* **14**, 56–7.
NEWMAN, L. H. (1955). *Nymphalis antiopa*: migrant or stowaway. *Entomologist*, **88**, 25–7.
NICHOLSON, A. J. (1950). Competition for food amongst *Lucilia cuprina* larvae. *8th int. Congr. Ent.* 277–81.
NIELSEN, A. T. & NIELSEN, E. T. (1952). Migrations of the pieride butterfly *Ascia monuste* L. in Florida *Ent. Medd.* **26**, 386–91.
NIELSEN, E. (1932). *The biology of spiders* **1, 2**. Copenhagen.
NIELSEN, E. T. (1958). The initial stage of migration in salt-marsh mosquitos. *Bull. ent. Res.* **49** 305–13.

NIELSEN, E. T. & NIELSEN, A. T. (1950). Contributions towards the knowledge of the migration of butterflies. *Amer. Mus. Novit.* **1471**, 1–29.

NIJVELDT, W. (1950). Een massale vlucht van *Bledius tricornis* Herbst (Col.). *Ent. Ber., Amst.*, **13**, 156.

NORDMAN, A. F. (1935). Über Wanderungen der *Libellula quadrimaculata* L. bei der Zoologischen Station Tvärminne in S. Finland im Juni 1932 und 1933. *Notul. ent., Helsingf.*, **15**, 1–8.

NORDMAN, A. F. (1937). Further observations on the migrations of *Libellula quadrimaculata* at the Zoological Station of Tvärminne, S. Finland in June 1936. *Notul. ent., Helsingf.*, **17**, 24–8.

OLIVER, F. W. (1943). A swarm of ladybirds (Coleoptera) on the Libyan Desert coast of Egypt between Hamman and Abusir. *Proc. R. ent. Soc. Lond.* (A), **18**, 87–8.

OSTEN-SACKEN, C. R. (1886). Some new facts concerning *Eristalis tenax*. *Ent. mon. Mag.* **23**, 97.

OWEN, D. F. (1956). A migration of insects at Spurn Point, Yorkshire. *Ent. mon. Mag.* **92**, 43–4.

OWEN, D. F. (1958). Dragonfly migration in south-west Portugal, autumn 1957. *Entomologist*, **91**, 91–5.

PARSHLEY, H. M. (1922). A note on the migration of certain water-striders (Hemiptera). *Bull. Brooklyn Ent. Soc.* **17**, 136–7.

PATON, C. I. (1929). Migration of dragonflies and uraniid moth in British Guiana. *Entomologist*, **62**, 212.

PAUL, L. C. & PUTNAM, L. G. (1960). Morphometrics, parasites and predators, of migrant *Melanoplus bilituratus* (Wlk.) (Orthoptera: Acrididae) in Saskatchewan in 1940. *Canad. Ent.* **92**, 488–93.

PEARSON, E. O. (1958). *The insect pests of cotton in tropical Africa.* London.

PEARSON, K. & BLAKEMAN, J. (1906). Mathematical contributions to the theory of evolution. XV. A mathematical theory of random migration. *Drap. Co. Mem. biom. ser.* **3**, 1–54.

PIMENTEL, D. & FAY, R. W. (1955). Dispersion of radioactively tagged *Drosophila* from pit privies. *J. econ. Ent.* **48**, 19–22.

PIPER, C. V. (1897). A remarkable sembling habit of *Coccinella transverosguttata*. *Ent. News*, **8**, 49–51.

POPHAM, E. J. (1951). A study of the changes of the water-bug fauna of North Surrey from 1946 to 1950 with special reference to the migration of Corixids. *J. Soc. Brit. Ent.* **3**, 268–73.

PROVOST, M. W. (1952). The dispersal of *Aedes taeniorhynchus* I. Preliminary studies. *Mosquito News*, **12**, 174–90.

PROVOST, M. W. (1953). Motives behind mosquito flights. *Mosquito News*, **13**, 106–9.

PROVOST, M. W. (1957). The dispersal of *Aedes taeniorhynchus* II. The second experiment. *Mosquito News*, **17**, 233–47.

QUARTERMAN, K. D., KILPATRICK, J. W. & MATHIS, W. (1954). Fly dispersal in a rural area near Savannah, Georgia. *J. econ. Ent.* **47**, 413–19.

QUARTERMAN, K. D., MATHIS, W. & KILPATRICK, J. W. (1954). Urban fly dispersal in the area of Savannah, Georgia. *J. econ. Ent.* **47**, 405–12.

Радзиевская С.Б. (RADZIEVSKAYA, S. B.) (1939). К вопросу о зимовках божьих коровок и борьбе с хлопковыми тлями. *Вопросы экологии и биоценологии, Москва-Ленинград.* (*Probl. Ecol. Biocenol., Moscow*, **4**, 268–75 (abs. *Rev. appl. Ent.* (A), **29**, 486).

RAINEY, R. C. (1951). Weather and the movements of locust swarms: A new hypothesis. *Nature, Lond.*, **168**, 1057–60.

RAINEY, R. C. (1958a). Atmospheric movements and the biology of the desert locust (*Schistocerca gregaria* Forskål). *Proc. Linn. Soc. Lond.* **169**, 73–4.

RAINEY, R. C. (1958b). Some observations on flying locusts and atmospheric turbulence in eastern Africa. *Quart. J. R. met. Soc.* **84**, 334–54.

RAINEY, R. C. (1960). The mechanisms of desert locust swarm movements and the migration of insects. 11*th int. Congr. Ent.* (in press).

RAINEY, R. C. & WALOFF, Z. (1948). Desert locust migrations and synoptic meterology in the Gulf of Aden area. *J. Anim. Ecol.* **17**, 101–12.

RAINEY, R. C. & WALOFF, Z. (1951). Flying locusts and convective currents. *Anti-Locust Bull.* **9**, 51–70.

RAO, Y. R. (1936). A study of migration among the solitaries of the desert locust (*Schistocerca gregaria* Forsk.). 4*th int. Locust Confr.* (10), 14 pp.

RICHARD, G. (1958). Contribution à l'étude des vols migratoires des Corixidae (Insectes hétéroptères): les vols de l'été. *Vie et Milieu*, **9**, 179–99.

RICHARDS, O. W. & WALOFF, N. (1954). Studies on the biology and population dynamics of British grasshoppers. *Anti-Locust. Bull.* **17**, 1–184.

RIGGALL, E. C. (1953). Mass movements of Coleoptera on the Lincolnshire coast. *Ent. mon. Mag.* **89**, 130–1.

ROER, H. (1959). Über Flug- und Wandergewohnheiten von *Pieris brassicae* L. *Z. angew. Ent.* **44**, 272–309.

ROOT, F. M. (1912). Dragonflies collected at Point Pelee and Pelee Island, Ontario, in the summer of 1910 and 1911. *Canad. Ent.* **44**, 208–9.
ROTH, P. (1928). Les Ammophiles de l'Afrique du Nord. *Ann. Soc. ent. Fr.* **97**, 153–240.
ROTHSCHILD, M. (1956). Notes on insect migration on the north coast of France. *Ent. mon. Mag.* **92**, 375–6.
SAKAI, K-I., NARISE, T., HIRAIZUMI, Y. & IYAMI, S. (1958). Studies on competition in plants and animals. IX. Experimental studies on migration in *Drosophilia melanogaster*. *Evolution*, **12**, 93–101.
SAUNT, J. W. (1945). Migration of Syrphidae (Diptera). *Ent. mon. Mag.* **81**, 131.
SAVORY, T. H. (1945). *The spiders and allied orders of the British Isles*, second ed. London.
SCHNEIDER, F. (1952). Untersuchungen über die optische Orientierung der Maikäfer (*Melolontha vulgaris* F. und *M. hippocastani* F.) sowie über die Entstehung von Schwärmbahnen und Beifallskonzentrationen. *Mitt. Schweiz. ent. Ges.* **25**, 269–340.
SCHOOF, H. F. & MAIL, G. A. (1953). Dispersal habits of *Phormia regina* in Charleston, West Virginia. *J. econ. Ent.* **46**, 258–62.
SCHOOF, H. F., SIVERLY, R. E. & JENSEN, J. A. (1952). Housefly dispersion studies in metropolitan areas. *J. econ. Ent.* **45**, 675–83.
SCOTT, H. (1926). Coleoptera in the sea. *Ent. mon. Mag.* **62**, 115.
SCOTT, H. (1950). *Paederus fuscipes* Curtis (Col., Staphylinidae) and other migrant insects attracted to a ship's lights in the Red Sea. *Ent. mon. Mag.* **86**, 217–18.
SHANNON, H. J. (1926). A preliminary report on the seasonal migrations of insects. *J. N.Y. ent. Soc.* **34**, 199–206.
SHEPPARD, P. M. (1951). A quantitative study of two populations of the moth, *Panaxia dominula* (L.). *Heredity*, **5**, 349–78.
Щербиновский Н.С. (SCHERBINOVSKII, N. S.) (1939). Сезонные миграции и поведение жуков клеверных семеедов. *Защита растений*. *Plant Prot.* **18**, 136–41 (abs. *Rev. appl. Ent.* A, **27**, 681–2).
SMITH, C. E. & ALLEN, N. (1932). The migratory habit of the spotted cucumber beetle. *J. econ. Ent.* **25**, 53–7.
SMITH, K. G. (1951). A dragonfly migration at Lagos. *Niger. Fd*, **16**, 138–9.
SMITH, W. W. (1890). Notes on *Eristalis tenax* in New Zealand. *Ent. mon. Mag.* **26**, 240–2.
SNOW, D. W. & ROSS, K. F. A. (1952). Insect migration in Pyrennes. *Ent. mon. Mag.* **88**, 1–6.
SOUTHWOOD, T. R. E. (1960a). The flight activity of Heteroptera. *Trans. R. ent. Soc. Lond.* **112**, 173–220.
SOUTHWOOD, T. R. E. (1960b). Migration—an evolutionary necessity for denizens of temporary habitats. *11th int. Congr. Ent.* (in press).
SOUTHWOOD, T. R. E. (1961). A hormonal theory of the mechanism of wing polymorphism in Heteroptera. *Proc. R. ent. Soc. Lond.* (A), **36**, 63–6.
SOUTHWOOD, T. R. E., JEPSON, W. F. & VAN EMDEN, H. F. (1961). Studies on the behaviour of *Oscinella frit* L. (Diptera) adults of the panicle generation. I. Observations in the field. *Ent. exp. appl.* **4**, 196–210.
SOUTHWOOD, T. R. E. & JOHNSON, C. G. (1957). Some records of insect flight activity in May 1954, with particular reference to the massed flights of Coleoptera and Heteroptera from concealing habitats. *Ent. mon. Mag.* **93**, 121–6.
STOKOE, W. J. & STOVIN, G. H. T. (1944). *The caterpillars of the British butterflies*. London.
STOKOE, W. J. & STOVIN, G. H. T. (1948). *The caterpillars of the British moths*. London.
STRICKLAND, A. H. (1950). The dispersal of Pseudococcidae (Hemiptera-Homoptera) by air currents in the Gold Coast. *Proc. R. ent. Soc. Lond.* **25**, 1–9.
SWEENEY, R. C. H. (1960). Cotton insect pest investigations in the federation of Rhodesia and Nyasaland. Part II. Cotton stainer investigations. *Emp. Cott. Gr. Rev.* **37**, 32–44.
TAKADA, H. (1958). Drosophila survey of Hokkaido. X. Drosophilidae from several localities of Hokkaido. *J. Fac. Sci. Hokkaido Univ. Zool.* (6) **14**, 120–7.
TAYLOR, L. R. (1958). Aphid dispersal and diurnal periodicity. *Proc. Linn. Soc. Lond.* **169**, 67–73.
TAYLOR, L. R. (1960). The distribution of insects at low levels in the air. *J. Anim. Ecol.* **29**, 45–63.
TAYLOR, L. R. & KALMUS, H. (1954). Dawn and dusk flight of *Drosophila subobscura* Collin. *Nature, Lond.*, **174**, 221–3.
TEMPLE, V. (1949). The courtship flight of butterflies as the means, of extending the range of certain species. *Entomologist*, **82**, 145–7.
THEOBOLD, F. T. (1927). *The plant lice or Aphididae of Great Britain*, **2**. London.
THOMPSON, A. T. (1960). (Remarks at meeting.) *Proc. R. ent. Soc. Lond.* (C), **25**, 26–7.
TIMOFEEFF-RESSOVSKY, N. H. & TIMOFEEFF-RESSOVSKY, E. A. (1940). Populations-genetische Versuche an *Drosophila*. *Z. indukt. Abstamm.- u. VererbLehre*, **79**, 28–49.

TISCHLER, W. (1937). Untersuchungen über Wanzen an Getreide. *Arb. physiol. angew. Ent.* **4**, 193-231.
TOWER, W. L. (1906). *An investigation of evolution in the chrysomelid beetles of the genus* Leptinotarsa. Carnegie Inst., Washington.
TULLOCH, J. B. G. (1929). Dragonfly migration. *Entomologist*, **62**, 213.
TUTT, J. W. (1898-1902). Migration and dispersal of insects. *Ent. Rec.* **10**, 209-13, 233-8; **11**, 14-18, 43-5, 64-7, 89-93, 117-21, 153-5, 181-3, 213-15, 319-24; **12**, 13-16, 69-72, 127-8, 173-86, 206-9, 236-8, 253-7; **13**, 97-102, 124-5, 145-7, 233-7, 255-6, 281-4, 317-20, 353-8; **14**, 73-5, 207-14, 232-3, 262-5, 292-5, 315-19.
URQUHART, F. A. (1960). *The Monarch butterfly.* Toronto.
UTIDA, S. (1954). 'Phase' dimorphism observed in the laboratory population of the cowpea weevil, *Callosobruchus quadrimaculatus.* (In Japanese.) *Oyo-Dobuts. Zasshi* **18**, 161-8 (abs. *Rev. appl. Ent.* (A), **47**, 389).
UTIDA, S. (1956). 'Phase' dimorphism observed in the laboratory population of the cowpea weevil, *Callosobruchus quadrimaculatus.* 2nd Report. Differential effects of temperature, humidity and population density upon some ecological characters of the two phases. *Res. Popul. Ecol.* **3**, 93-104.
UVAROV, B. P. (1957). The aridity factor in the ecology of locusts and grasshoppers of the Old World. *Arid Zone Research, UNESCO Rev. Res.* **8**, 164-98.
VACHON, M. (1947). Nouvelles remarques à propos de la phorésie des Pseudoscorpions. *Bull. Mus. Hist. nat. Paris,* **19**, 84-7.
VALLE, K. J. (1946). Zur Invasion von *Libellula depressa* L. (Odon., Libellulidae) nach Finland. *Ann. ent. fenn.* **12**, 45-51.
VERRIER, M. L. (1954). Rassemblements et migrations chez les Ephémères. *Bull. biol.* **88**, 68-89.
VESEY-FITZGERALD, D. F. (1957). The vegetation of central and eastern Arabia. *J. Ecol.* **45**, 779-98.
VODJDANI, S. (1954). Contribution à l'étude des Punaises de céréales et en particulier d'*Eurygaster integriceps* Put. (Hemiptera, Pentatomidae, Scutellerinae). *Ann. Éphiphyt.* **5**, 105-60.
WAINWRIGHT, C. J. (1944). Migratory Diptera. *Ent. mon. Mag.* **80**, 225-6.
WALOFF, Z. V. (1940). The distribution and migrations of *Locusta* in Europe. *Bull. ent. Res.* **43**, 211-46.
WALOFF, Z. V. (1960). The fluctuating distributions of the desert locust in relation to the strategy of control. *Rep. 7th Commonw. ent. Confr.* 132-40.
WALOFF, Z. V. & RAINEY, R. C. (1951). Field studies on factors affecting the displacements of desert locust swarms in eastern Africa. *Anti-Locust Bull.* **9**, 1-50.
WATZL, O. (1950). Zur Lebensweise und Bekämpfung des Rübenerdflohs. *PflSchBer.* **4**, 129-49.
WAY, M. J. & BANKS, C. J. (1961). Population studies on *Aphis fabae* Scop. (Hemiptera: Aphidiae). *Proc. R. ent. Soc. Lond. C.* **26**, 21, 27-8.
WEBB, W. E. & JONES, T. (1956). A study of the biology and control of Ambrosia beetles (Scolytoidea) attacking timber in West Africa. *10th int. Congr. Ent.* **4**, 381-4.
WILLIAMS, C. B. (1929). Some records of dragonfly migration. *Entomologist*, **62**, 145-8.
WILLIAMS, C. B. (1930). *The migration of butterflies.* Edinburgh.
WILLIAMS, C. B. (1951). Seasonal changes in the flight direction of migrant butterflies in the British Isles. *J. Anim. Ecol.* **20**, 180-90.
WILLIAMS, C. B. (1958). Insect migration. *Ann. Rev. Ent.* **2**, 163-80.
WILLIAMS, C. B. (1959). *Insect migration.* London.
WILLIAMS, C. B., COMMON, I. F. B., FRENCH, R. A., MUSPRATT, V. & WILLIAMS, M. C. (1956). Observations on the migration of insects in the Pyrenees in the autumn of 1953. *Trans. R. ent. Soc. Lond.* **108**, 385-407.
WILTSHIRE, E. P. (1940). Some notes on migrant Lepidoptera in Syria, Iraq and Iran. *Entomologist*, **73**, 231-4.
WILTSHIRE, E. P. (1945). Studies in the geography of Lepidoptera. II. Swallow-tails in desertic S.W. Asia. *Proc. R. ent. Soc. Lond.* (A), **20**, 16-25.
WILTSHIRE, E. P. (1946). Studies in the geography of Lepidoptera. III. Some Middle East migrants, their phenology and ecology. *Trans. R. ent. Soc. Lond.* **96**, 163-86.

[Editors' Note: The addendum has been omitted.]

ARTIFICIAL SELECTION FOR DISPERSAL IN FLOUR BEETLES (TENEBRIONIDAE: *TRIBOLIUM*)[1]

John C. Ogden[2]

Department of Biology, Stanford University, Stanford, California 94305

In flour beetles of the genus *Tribolium*, there are certain behavioral traits which can be easily studied and are open to modification by artificial selection. Dawson (1964) observed that individuals of certain highly inbred lines of *T. confusum* spent a greater proportion of time on the surface of the flour medium than the normal wild type. Sokoloff (1966) extended this observation to certain mutant strains of *T. castaneum* and showed that the response was maintained independently of population density in the culture container. Other workers have compared the general activity level of wild type and certain body color mutants (McDonald and Fitting 1965). The mutant types were usually more active. Lerner and Inouye (1968) were able to show rapid responses to selection for speed of running through a hierarchical T-maze using both *T. castaneum* and *T. confusum*. In addition, Lerner (pers. commun.) has been able to select for flying in *T. castaneum* with similarly rapid responses.

The present preliminary investigation was carried out as part of a larger study of dispersal behavior in *Tribolium* (Ogden 1968, 1969). Ogden (1969) showed some of the olfactory cues which mediate dispersal behavior in *T. confusum*. Here, an attempt was made to determine if dispersal behavior, defined as emigratory movement by walking in a laboratory apparatus, would respond to directional selection.

Materials and Methods

The *Tribolium* material used in this study was obtained from the Department of Genetics in the University of California, Berkeley. The *T. confusum* (CF) and *T. castaneum* (CS) strains were the same as used by Ogden (1969). *Tribolium* were raised in a mixture by weight of 95% stoneground whole wheat flour and 5% powdered brewer's yeast. Standard handling methods were used. All experiments were done in the dark in a Jamesway poultry incubator which was held at 29°C and 70% ($\pm 5\%$) relative humidity.

Following original work by Prus (1963), an apparatus was constructed in which beetles could move from one container to another (Fig. 1). Two 8-dram shell vials (I and A vials) were connected with a piece of 4-mm I.D. Tygon tubing 15 cm long (U tube). Aluminum bacteriological caps were used to hold the tubing in place and cover the tops of the vials. A thread, tied at one end of the tubing, was run doubled through it, forming a loop in the other vial. Beetles could climb up this loop out of the I vial, crawl through the U tube on the thread, and drop into the A vial. The apparatus was set so that only one-way

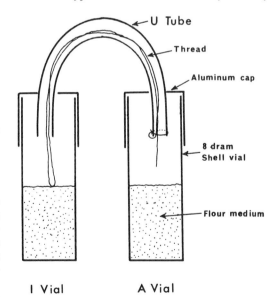

Fig. 1. Dispersal apparatus.

[1] Received May 9, 1969; accepted October 20, 1969
[2] Present address: Smithsonian Tropical Research Institute, P.O. Box 2072, Balboa, Canal Zone.

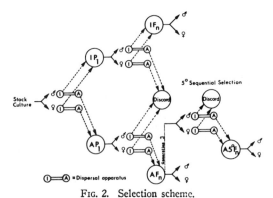

Fig. 2. Selection scheme.

movement, from the I vial to the A vial, was possible. Eight grams of medium were placed in all tubes and 5 replicates were employed for each run of a selected strain with 50 beetles in each replicate.

Figure 2 shows the selection scheme employed in these experiments. Lines were established from foundation populations of CS and CF by first separating males and females in the pupal stage. Fifty virgin beetles about 2 weeks old were then placed in the I vial of each replicate and the run proceeded for 1 week. Separate runs were performed for each sex. At the end of this period, the beetles remaining in the I vials were mated and formed the IP_1 generation; beetles in the A vials were mated and formed the AP_1 generation. After mass mating for 3 days, large numbers of eggs were collected from each of the parental lines and allowed to develop to pupae. Sexes were separated at the pupal stage, emerged adults were aged at least 2 weeks, and the selection runs were repeated. In the I lines, those selected for dispersal *inactivity*, runs lasted for 1 week for both sexes. At the end of this period, animals in the A vials were discarded and those remaining in the I vials were mass mated and formed the IF_n generation.

In the A lines, those selected for dispersal *activity*, the selection scheme was modified several times. For the first two generations, 50 virgin beetles were introduced into the I vial of each replicate and the run proceeded for 1 hour. Those beetles in the A vials were then mated as the AF_n generation and those in the I vials were discarded. In generation 3, a new active line was established in order to increase the intensity of selection. Beetles moving to the A vials in the first hour were saved as the AF_n generation. The run was continued and those beetles entering the A vials in the next 24 hours were put through 5 successive runs of 24 hours. At the end of each 24-hour run, beetles in the A vials were returned to fresh medium in the I vials and another run was allowed to proceed. Beetles remaining in the I vials were discarded after each run. At the end of the fifth 24-hour run, those beetles moving into the A vials of the apparatus were removed and mass mated to form the $A5°F_n$ generation. In generation 4, this line was continued with five successive runs, but instead of 24-hour runs, 1-hour runs were employed.

Complete controls were not used in these experiments. In generations 3 to 5, beetles of the same age as selected beetles were taken from the foundation populations of CS and CF and run in the dispersal apparatus. The sexes were not separated in control runs. The results of assays of both control and selected lines were then compared.

Except for generation 1, assays of dispersal activity were done in each generation at the same time that beetles of that generation were being selected. The number of beetles moving to the A vial in 24 hours constituted the assay. The assay included those beetles remaining in the U tube of the apparatus, but usually by this time all beetles had dropped into the A vial.

The selection experiment described here proceeded for only five generations. The laboratory sustained a massive infection of what was probably the coccidian parasite *Adelina* and the selected lines had to be destroyed in order to save other stocks and experiments. This relatively short term of selection, however, has established some trends, the importance of which will be discussed later.

Results

The results of assays on selected lines of CS are presented in Table 1. The data from the two sexes were combined in the table as the response to selection in both males and females was qualitatively the same. Assays for generation 1 do not appear as the data were taken in the wrong way. Since the $A5°$ lines were not started until generation 3, assays for this line do not appear until generation 4.

By inspection of Table 1, it can be seen that there is a great deal of variability in the response to selection from generation to generation. Furthermore it appears that much change in the lines took place before the first assays were performed in generation 2. For these reasons the data were difficult to analyze. There are too few points to fit a regression curve. The analysis of variance (ANOVA) was used in a one-factor design on the transformed data of the fifth generation of selection including the control lines. The F test showed that the selected lines are a highly significant source of variation ($P < .005$). The four

TABLE 1. Assay of selected lines in *T. castaneum*. Per cent beetles moving in 24 hours (Mean of 5 replicates, 50 beetles/replicate)

Strain	Generation			
	2	3	4	5
CS I	41.2	67.2	28.0	34.0
CS A	53.6	81.0	53.8	74.5[a]
CS A 5°	—	—	85.4	76.3[a]
Control	—	54.4	50.0	46.8

[a] $P < .05$ Significantly different from control

TABLE 2. Assay of selected lines in *T. confusum*. Per cent beetles moving in 24 hours (Mean of 5 replicates, 50 beetles/replicate)

Strain	Generation			
	2	3	4	5
CF I	67.0	73.0	80.0	78.6
CF A	88.5	84.0	88.4	89.6
CF A 5°	—	—	88.2	95.0[a]
Control	—	89.2	85.2	83.2

[a] $P < .05$ Significantly different from control

means were compared using the studentized range test (Goldstein 1964). Both CS A and CS A5° lines are significantly different ($P < .05$) from control level. The CS I line approaches a significant difference. The A5° line shows a trend to be more active than the A line. As determined by this analysis, it seems as if a degree of separation has been achieved in five generations.

The results of assays on selected lines of CF are presented in Table 2. A one-factor ANOVA was performed on the transformed data of the fifth generation. This analysis shows that strains are a highly significant source of variation ($P < .005$). Using the studentized range test, the means were compared. The CF A5° line is significantly different ($P < .05$) from control level. The CF A line approaches a significant difference. The CF I line, however, while lower in activity than the controls, does not differ significantly. The results here are very similar to the CS results with some degree of separation achieved by generation 5.

DISCUSSION

Other work (Naylor 1959, 1961; Ghent 1963; Ogden 1969) has indicated that the odor of flour medium condition by the products of a contained population of beetles plays an important role in dispersal behavior in *Tribolium*. In the fourth generation, beetles of all selected lines were tested for rate of dispersal from homotypic conditioned medium. In both species the relationship between the selected lines was preserved, but in CS the rate of dispersal of all lines was elevated, compared to dispersal from fresh medium, while in CF the rate of dispersal of all lines was depressed. It appears as if selection has not altered the basic attraction and repulsion responses of CF and CS to conditioned medium.

The fact that some divergence between selected lines was achieved after five generations indicates that this sort of dispersal behavior in *Tribolium* has a genetic component. The extent of this selection in natural populations, and its population dynamic consequences are unknown, but preliminary work in this area (Wellington 1957, 1964; Southwood 1960, 1962; Lidicker 1962) has yielded some interesting results and speculations.

ACKNOWLEDGMENTS

I wish to thank Paul R. Ehrlich for advice and stimulating discussion through the course of this work. Alexander Sokoloff provided the *Tribolium* material and offered many helpful suggestions. William Z. Lidicker, Everett Dempster, and Graham A. E. Gall were kind enough to read and criticize the manuscript. This research was supported in part by N.I.H. Training Grant in Population Biology No. 5T1GM36503 (Paul R. Ehrlich, Director). This paper was prepared while the author was supported by USPHS Grant No. GM367 (Everett Dempster, Director).

LITERATURE CITED

Dawson, P. S. 1964. An interesting behavioral phenomenon in *Tribolium confusum*. Tribolium Inform. Bull. 7: 50–52.

Ghent, A. W. 1963. Studies of behavior of *Tribolium* flour beetles. Contrasting responses of *T. castaneum* and *T. confusum* to fresh and conditioned flours. Ecology 44: 269–283.

Goldstein, A. 1964. Biostatistics, an introductory text. Macmillan, New York, 272 p.

Lidicker, W. Z., Jr. 1962. Emigration as a possible mechanism permitting the regulation of population density below carrying capacity. Amer. Naturalist 96: 29–33.

Lerner, I. M., and N. Inouye. 1968. Behavior genetics, with special reference to maze-running of *Tribolium*. In Dronamraju K. R. (ed.), 1968. Haldane and modern biology. Johns Hopkins Press.

McDonald, D. J., and L. Fitting. 1965. Activity differences in black and wild type strains of *Tribolium confusum*. Tribolium Inform. Bull. 8: 124–125.

Naylor, A. F. 1959. An experimental analysis of dispersal in the flour beetle *Tribolium confusum*. Ecology 40: 453–465.

———. 1961. Dispersal in the red flour beetle, *Tribolium castaneum*. Ecology 42: 231–237.

Ogden, J. C. 1968. An experimental and genetic analysis of dispersal behavior in *Tribolium*. Unpublished doctoral dissertation, Stanford University, Stanford, Calif., 91 p.

———. 1969. Effect of components of conditioned medium on behavior in *Tribolium confusum*. Physiol. Zool. **42**: 266–274.
Prus, T. 1963. Search for methods to investigate momobility in Tribolium. Ecology **44**: 801–803.
Sokoloff, A. 1966. A behavioral difference of two mutant strains in *Tribolium castaneum*. Tribolium Inform. Bull. **9**: 109–110.
Southwood, T. R. E. 1962. Migration—an evolutionary necessity for denizens of temporary habitats. Intl. Cong. Entomol. Trans., 11th Cong. **3**: 54–58.
———. 1962. Migration of terrestrial arthropods in relation to habitat. Biol. Rev. **37**: 171–214.
Wellington, W. G. 1957. Individual differences as a factor in population dynamics: the development of a problem. Canad. J. Zool. **35**: 293–323.
———. 1964. Qualitative changes in populations in unstable environments. Canad. Entomol. **96**: 436–451.

HERITABILITY OF FLIGHT DURATION IN THE MILKWEED BUG LYGAEUS KALMII

Roy L. Caldwell
Joseph P. Hegmann
Department of Zoology
University of Iowa
Iowa City 52240

MEASUREMENT of tethered flight provides an index of flight behaviour in migratory insects[1]. Although tethered flight probably overestimates field performance[2], the frequency distributions are similar for flight duration in *Lygaeus kalmii* caught in the field (Iowa City, Iowa populations) and reared in the laboratory (Fig. 1). Thus it is possible to obtain from populations reared in the laboratory data which index behaviour of ecological significance and to examine in detail individual differences in flight behaviour. These differences may reflect effects of both genetic and environmental variation.

Although numerous studies have suggested the importance of environmental factors influencing flight in various species[3,4], there is a paucity of information about behavioural variation caused by genetic differences. So far most evidence of genetic variation for flight has been derived from the presence of polymorphisms with respect to morphometric characters such as wing structure. Flight is a continuous variable, probably dependent on allelic differences at many loci, and is easily measured: it is amenable to quantitative-genetic analysis. To assess the relative importance of genetic and environmental differences affecting flight, we have estimated the heritability of the behaviour in a population using regression of offspring scores on parent values.

Heritability is a population-specific parameter describing the ratio of the additive genetic variance to the total variance for a trait. It also has predictive value, for progress by selection is a function of the selection intensity, the phenotypic standard deviation of the trait in the population, and the heritability of the trait. The use of estimates of heritability in analysis of behavioural differences has been discussed already[5,6].

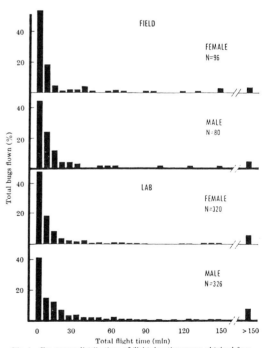

Fig. 1. Frequency distributions of flight duration scores obtained from male and female *L. kalmii* captured in the field during the time of spring post-hibernacular flights, and from virgin male and female *L. kalmii* reared from eggs in laboratory conditions. All bugs were tested while tethered in still air and the times for five consecutive flights were summed. Intervals are 6 min.

Table 1. MEANS AND VARIANCES OF SQUARE ROOT TRANSFORMED FLIGHT SCORES (MIN) FOR MALE AND FEMALE *L. kalmii* OF THE PARENT AND OFFSPRING GENERATIONS

	Parents		Offspring	
	Males	Females	Males	Females
$\bar{X}$	4.37	4.03	4.30	3.96
V	20.38	21.42	19.33	23.17
N	33	33	326	320

We used the offspring of *L. kalmii* caught in the field and their offspring to assess the relative importance of genotype and environment in determining individual differences in flight. All bugs were flight tested at 25° C according to Dingle's method[1]. *L. kalmii* reared in the laboratory were maintained in constant crowding conditions and their flight tested while they were virgins, 6 to 8 days after the imaginal moult. This is the age of maximum flight activity in the conditions maintained. Thirty-three male and thirty-three female offspring of the *L. kalmii* caught in the field were selected at random to serve as the parent generation, and their flight was tested. They were mated at random and ten males and ten females were randomly selected from the offspring of each pair and their flight was tested. When less than ten offspring of either sex matured, all available offspring of that sex were tested. Flight scores were obtained from sums over five consecutive trials. Because of high mean-variance correlations within these groups of offspring and the skewed distributions of values for both offspring and parents, flight scores were square-root transformed before further analysis.

Means and variances of the transformed flight scores for males and females of parent and offspring generations are shown in Table 1. Our estimates of heritability for flight in *L. kalmii* calculated from regression of offspring scores (pooled across sex) on male parent and on female parent were 0.20 ± 0.06 and 0.41 ± 0.05, respectively. Differences in estimates of heritability derived from male and from female parents may indicate a maternal effect on flight and merit further investigation. Our estimates, however, are consistent with heritabilities obtained for other behavioural traits[5]. A comparison of the mean scores for males and females within the generations provides no evidence for a sex difference for flight ($t < 1$, df = 64; $t < 1$, df = 644). Although males and females of migrant insects often differ in duration and timing of flight[4], Southwood[7] concluded from a review of several sampling studies that flight is similar in male and female Lygaeids. Our data for *L. kalmii*, if tethered flight is used as an index, agree with his conclusions.

Southwood[8] has repeatedly emphasized the importance of flight in insects, such as *L. kalmii*, which inhabit changing environments because it enables them to change their habitats. Furthermore, Kennedy[9] has argued that maximum utilization of the habitat in such species might be made possible by the maintenance of a high degree of variability with respect to flight. Although studies on flight behaviour have often stressed the importance of environmental factors producing such variance, evidence has recently been reported of genetic differences affecting flight behaviour in another milkweed bug, *Oncopeltus fasciatus*[3]. We found that a substantial proportion of the variance of flight performance is attributable to differences in additive genetic values. Flight behaviour in this population of *L. kalmii* should therefore respond to mass selection. In view of these results, and current emphasis on the importance of behavioural differences in migration[6], quantitative genetic analysis of such differences may be useful in discerning strategies of migration and dispersion.

This work was supported by a pre-doctoral fellowship and grants from the US National Science Foundation.

Received March 31, 1969.

[1] Dingle, H., *J. Exp. Biol.*, **42**, 269 (1965).
[2] Kennedy, J. S., and Booth, C. O., *J. Exp. Biol.*, **40**, 67 (1963).
[3] Dingle, H., *J. Exp. Biol.*, **48**, 175 (1968).
[4] Johnson, C. G., *Nature*, **198**, 423 (1963).
[5] DeFries, J. C., in *Behavior-Genetics Analysis* (edit. by Hirsch, J.) (McGraw-Hill, New York, 1967).
[6] Ewing, A. W., and Manning, A., *Ann. Rev. Entomol.*, **12**, 471 (1967).
[7] Southwood, T. R. E., *Trans. Roy. Entomol. Soc.*, **112**, 173 (1960).
[8] Southwood, T. R. E., *Biol. Rev.*, **37**, 171 (1962).
[9] Kennedy, J. S., *Nature*, **189**, 785 (1961).

9

Copyright ©1977 by The Genetics Society of Canada
Reprinted by permission from Can. J. Genet. Cytol. **19**:717-722 (1977)

THE GENETIC BASIS OF DISPERSAL BEHAVIOR IN THE FLOUR BEETLE *TRIBOLIUM CASTANEUM*

UZI RITTE AND BATIA LAVIE

Department of Genetics, The Hebrew University of Jerusalem, Jerusalem, Israel

Flour beetles of the genus *Tribolium* are suitable for the study of dispersal. In an attempt to gain an understanding of the genetic basis of this trait in *T. castaneum*, a two-way selection experiment was carried out, for 11 successive generations, in a stock derived from an Israeli wild population. When dispersants were defined as beetles that disperse more than once in several attempts, separation between the lines was achieved after one generation of selection. The difference between the lines remained more or less constant in subsequent generations of selection, and did not disappear when selection was relaxed. Reciprocal crosses between the lines suggest that the tendency to disperse is determined by the genotype at a single, sex-linked locus. The allele for dispersal is dominant. Males tend to disperse more readily than females of equivalent genotypes.

Introduction

Dispersal plays an important role in the life history of many species. Contrary to migration, which is defined as a directed mass transfer of populations from place to place, often seasonally and temporarily (Johnson, 1969), dispersal refers to less regular and directed movements, involving individuals that leave their former population, sometimes to join other populations or to establish new populations in suitable, empty habitats. Yet, in spite of the great ecological and evolutionary significance that may be associated with dispersal (see, for example, Karlin, 1976 and Lidicker, 1975), relatively little is known about the genetic basis of dispersal ability, and the mechanisms responsible for its maintenance in natural populations.

Flour beetles of the genus *Tribolium*, which have become a model organism for many aspects of population biology (King and Dawson, 1972), are also suitable for the study of dispersal (Prus, 1963). In laboratory stocks, which had been maintained in closed containers for many generations, dispersal activity was shown to depend on the species, genetic strain, sex, density, age and degree of conditioning of the flour (Prus, 1966; Zygromska-Rudzka, 1966; Ogden, 1970a; Ziegler, 1976).

In a natural population of *T. castaneum* studied by Ritte and Agur (1977), a variability for the tendency to disperse was found between families and members of the same family. The present work describes a two-way selection experiment for dispersal activity in this population. The results of this experiment, which also included crosses between the selection lines, suggest not only that the differences in dispersal ability among individuals are genetic, but that the genetic basis of this trait is much simpler than had been anticipated.

Materials and Methods

Dispersal Apparatus and Experimental Conditions

Dispersal was studied in an apparatus similar to the one designed by Prus (1963). In this apparatus, beetles are scored as dispersants if they move on a string, through a 30 cm piece of polyvinyl tubing, from one vial (A), with flour, into another, empty vial (B). All experiments were carried out in a dark Controlled Environments incubator, with a temperature of 29°C and 70% relative humidity. The beetles were kept in whole wheat baking flour, supplemented, for every 100 g of flour, with 5 g of brewers yeast. In an attempt to obtain dispersal rates which are independent of the density in vial A, initial densities in these vials were four beetles per 1 g of flour. This value is much below the lowest density that was shown by Zyromska-Rudzka (1966) to affect dispersal rates.

Source Population and Selection Lines

The stock, from which the selection lines were derived, was established from a sample collected in a wild population of *Tribolium castaneum* inhabiting a grain storage silo in Rehovot, Israel. Ritte and Agur (1977) studied dispersal behavior at three ages (10, 20 and 30 days) in 19 groups of full sibs, members of the second generation of nonselected progeny of the original sample. For the present study beetles from the three families with the overall highest rates of dispersal were selected as parents for the High Dispersal (HD) lines, while beetles from the three families with the lowest overall rates were selected for the Low Dispersal (LD) lines.

Selection Program

Selection was carried out in two separate replicas, each replica including a HD line and an LD line. Each generation, each line in each replica was composed of several full-sib families. When beetles were mated as parents of these families, care was taken to avoid sib matings. The number of families in each replica (see Table I) varied according to the number of successful matings that were obtained. Twelve randomly drawn progeny from each family (16 in generation S_1), all of equal age, were tested as a group. Family size was chosen to fit the number of beetles of equal age which could be obtained in all families. In generations S_1, S_4 and S_5 control groups, similar in age and number of beetles to the families in the selection lines, were tested together with the selection lines. They were drawn at random from the unselected stock obtained from the original sample. After generation S_5 the test of control groups was discontinued, as it was difficult to assume that the unselected stock, from which the departure of dispersants was prevented, continued to represent, in terms of average dispersal acticity, the original wild population.

Criterion of Selection

Generation S_1, in addition to being the first generation of selection, served also for the establishment of the criteria for selection. In particular we wanted to find out whether there was a difference between defining each beetle found in vial B as a dispersant, as compared to defining as dispersants only beetles that tend to move to vial B repeatedly. Generation S_1 included, in each replica, 3 HD (High Dispersal) families, 4 LD (Low Dispersal) families and 4 control groups. As stated above, the parents of the HD families were drawn at random from the three families of Ritte and Agur (1977) showing the highest rates of dispersal, while the parents of the LD families came from the three families with the lowest rates. The beetles were tested in groups of 16. They were first introduced into vials A at the age of 14 days. The dispersal tubes leading to vials B were attached 24 hours later. After another 24 h, and for three additional days, the beetles in each vial B were counted, marked with a dot of quick drying nitrocellulose paint on the elytrae, and returned to their respective vial A. For each family we calculated two values: (1) the total number of beetles that were found in vial B at least once during the 4-day period; (2) the number of beetles that moved to vial B at least twice in that period.

An analysis of variance, using the angular transformations of the proportion of dispersing beetles, showed no statistically significant difference between the lines using the first criterion ($F_{1/5} = 6.28$, $p > .05$), and a significant difference between the lines using the second criterion ($F_{1/5} = 20.40$, $p < .01$). Defining dispersants as beetles that entered vial B at least three times showed no further improvement compared to the definition of dispersants as beetles that moved to vial B twice.

As a result we decided to define as dispersants beetles that move to vial B more than once. To continue the HD lines, individuals from the S_1 generation that had moved at least three times were selected from the HD families, and for the LD line, individuals that had not moved at all were selected from the LD families.

The same criterion was used in all subsequent generations of selection. In generations S_2 and S_3, in which selection was accompanied by another experiment, all beetles were given six opportunities to disperse. In the other generations the beetles had only two opportunities to disperse, each test being preceded by one day of "rest", in which the beetles that were returned to vial A could "settle down". In the two test experiments the procedure was as follows:

Day 1. 10 day old beetles are introduced into vial A.
Day 2. The thread leading to vial B is connected.
Day 3. The thread is disconnected from vial A, and the beetles found in vial B are scored, marked and returned to vial A.
Day 4. The thread leading to vial B is connected.
Day 5. The beetles found in vial B are scored.

Beetles that had moved to vial B twice (at least twice in S_2 and S_3) were scored as dispersants. Only dispersants from the HD families were selected as parents in the HD line, and only beetles from the LD families that had not moved at all were selected as parents in the LD line.

Relaxation of Selection

At generation S_8 all the beetles that were not selected as parents for the next generation served to form the nuclei for relaxed selection lines. All beetles from both HD lines served to establish a single relaxed selection line of High Dispersal (RHD), and all beetles from both LD lines formed similarly a single RLD line. The relaxed lines were transferred to fresh flour every 35 days without selection, for a total of six transfers (generations). In the same period of time the selection lines themselves were reproduced at intervals of 2-3 months, for a total of three generations. The relaxed lines were thus tested together with generation S_{11} of the selection lines.

Crosses Between the Selection Lines

At generations S_5 and S_7, beetles that met the selection criteria were also used in reciprocal single pair crosses between the two selection lines (HD females × LD males and LD females × HD males). The progeny of these crosses were tested together with the selection lines. At S_8 all tested groups had a 1:1 sex ratio, with the two sexes marked differently and scored separately.

Results

Selection Experiment

Each generation the two replicas, although not tested simultaneously, usually gave similar results. Analysis of variance indicated that the replicas were not significantly

TABLE I

Rates of dispersal (frequency of beetles that disperse at least twice) of the two selection lines, in 11 generations of selection (weighted means and standard errors of the two replicate lines in each direction). In parentheses, the average number of families tested in each replica[1]

Generation	High dispersal line	Low dispersal line	Controls
1	.34±.12(3)	.08±.02(4)	.23±.05(4)
2	.73±.11(4)	.22±.09(8)	
3	.81±.16(6)	.18±.05(21)	
4	.44±.06(11)	.04±.01(18)	.07±.02(7)
5	.50±.10(17)	.06±.04(13)	.14±.04(10)
6	.29±.08(11)	.01±.01(8)	
7	.43±.09(10)	.04±.01(16)	
8	.48±.22(10)	.02±.01(6)	
9	.74±.17(12)	.08±.03(11)	
10	.56±.06(17)	.02±.01(32)	
11	.78±.07(9)	.05±.03(9)	

[1]The number of families that were established each generation varied according to the number of beetles meeting the criterion for selection, while for a family to enter the testing program it also had to provide a sufficient number of 10-day old progeny.

different from one another, so that the average proportion of dispersants, pooled over both replicas, was representative of the overall response to selection. The pooled results, for eleven generations of selection, are given in Table I. A significant difference ($p < 0.05$) was achieved after only one generation of selection, and this difference was maintained, and even increased, in subsequent generations ($p < .001$ in all generations after S_1). By generation S_4 no overlap existed between the various families of the HD and LD lines.

The rather wide and irregular fluctuations of generation means are probably due to a variety of environmental factors, not all of which can be identified at present. It should be remembered that in S_2 and S_3 the beetles were given six opportunities to disperse, a fact which may be responsible for the higher frequencies of dispersants (defined as beetles that disperse at least twice) in these generations. Other causes for the differences between generations may be due to the fact that dispersal periods varied somewhat around the average of 24 hours, and the exact time of day in which each test was started and terminated could not be kept constant.

The average rates of dispersal of the control groups (generations S_1, S_4 and S_5) were intermediate to those of the two selection lines. In S_4 and S_5 the controls deviated significantly from the HD lines, while the differences from the LD lines were not significant.

Relaxation of Selection

In generation S_{11}, six groups of 12 beetles each, drawn from each of the relaxed selection lines (RHD and RLD), were tested together with the selection lines themselves. The average rate of dispersal among the RHD groups was $.78 \pm .30$, while that of the RLD groups was $.10 \pm .06$. These values were practically identical to the corresponding rates of the HD and LD lines (see Table I).

Crosses Between the Lines and the Genetic Model

The results of the crosses between the lines (together with a more detailed representation of the results of generations S_6 and S_8) are given in Tables II and III.

In S_6 (Table II) no distinction was made between males and females, and the important feature of this experiment is the significant differences between the results of the reciprocal crosses. The HD × LD progeny showed dispersal rates which were much higher than the mid-point between the parental lines (in replica 2 even higher than the higher parent), while in the LD × HD cross the values were much lower.

This raised the possibility of a sex-linked major gene affecting dispersal. The differences between the reciprocal crosses can be explained by assuming that one sex-linked locus with two alleles — D for dispersal and d for non-dispersal — is involved. D is dominant over d, but not completely. On this hypothesis the females and males of the HD line would be DD and DY, respectively, while the LD females and males would be dd and dY.

TABLE II

Rates of dispersal (means and standard errors) of the progeny of the reciprocal crosses between the lines, together with the values of the selection lines themselves (generation S_6). In parentheses – the number of families tested

Type of mating (female × male)	Replica 1	Replica 2
HD × HD	$.27 \pm .02(11)$	$.30 \pm .05(10)$
HD × LD	$.16 \pm .03(14)$	$.34 \pm .05(9)$
LD × HD	$.03 \pm .01(8)$	$.08 \pm .01(11)$
LD × LD	$.01 \pm .01(9)$	$0(7)$

TABLE III

Rates of dispersal (means and standard errors), separately for male and female progeny of the reciprocal crosses between the lines, together with similar values of the selection lines themselves (generation S_8)

Type of mating (female × male)	Postulated genotype ♀♀	♂♂	Replica 1 ♀♀	♂♂	Number of families tested	Replica 2 ♀♀	♂♂	Number of families tested
HD × HD	DD	D	.50±.15	.53±.19	16	.46±.04	.67±.10	4
HD × LD	Dd	D	.30±.05	.44±.06	24	.42±.05	.50±.10	9
LD × HD	Dd	d	.16±.04	.13±.07	13	.40±.05	.35±.06	20
LD × LD	dd	d	.02±.02	.12±.07	7	0	.07±.04	5

In order to verify the model the crosses were repeated in generation S_8, but this time with a 1:1 sex ratio in all tested groups, and with dispersal rates calculated separately for females and males. The results (Table III) agreed with our genetic model, except that the three crosses in which, according to the model, there should be no difference in dispersal rates between males and females (HD × HD, HD × LD and LD × LD), males dispersed more readily than females (significantly so, however, only in the HD × LD cross, replica 1).

It seems, therefore, that the tendency to disperse is determined mainly by the combination of the genotype at the sex-linked locus and the sex itself. This fact of males dispersing more than females of the same genotype renders even more striking the reversal of sex rank, predicted by the model and observed in the LD × HD cross.

Two points in Table III, although agreeing with the general trend of the predictions of our model, have values that seem to be a bit out of line (dispersal rate of female progeny of the LD × HD cross in replica 1 lower than expected, and rate of male progeny of the LD × HD cross in replica 2 higher than expected). As in each case the complementary group in the other replica is in good accordance with our original expectation, we assume that these exceptional points are due to sampling errors or uncontrolled environmental factors. The possibility that the difference in dispersal rates between the Dd females in the two reciprocal crosses of replica 1 is due to maternal effect has to await the accumulation of additional data.

Discussion

In our work beetles were defined as dispersants if they dispersed more than once, while nondispersants were beetles that did not disperse at all during the course of the test. Ziegler (1976) has recently shown dispersal to involve all members of experimental, laboratory populations of *T. castaneum* in which the opportunity to disperse lasted for 15 consecutive days, suggesting that the distinction between these morphs should be quantitative, based on when the beetles leave, rather than qualitative. It may be that our nondispersants too are actually late dispersants, but this should not affect the meaning of the distinction between dispersants and nondispersants. Due to the relatively high fecundity of young adults, the differences between dispersants and nondispersants in their contribution to the population should be significant even if the former disperse during the first days of adulthood and the latter stay longer for only a few more days.

The interpretation of the results of the selection experiment, on the basis of the proposed genetic model, is rather straightforward: Separation between the selection lines was achieved in one generation, as is expected from a character in which the genetic variance is determined by a very small number of genes. A quick response to selection for dispersal activity was also obtained by Ogden (1970b) and Schurr and Bolduan (1967). Homozygosis is also expected to be quickly achieved, and after that selection ought not to lead to further progress. Conversely, relaxation of selection should not reduce the difference between the lines. Both these expectations are supported by our results.

How variability for dispersal ability is maintained is still a matter for speculation. According to Van Valen (1971), such variability may result from the interplay between two opposing selection forces — selection *within* populations against dispersal, caused by the departure of dispersants, and selection *between* local populations, which are initially founded by dispersants, in favor of dispersal. *Tribolium* beetles have long been associated with grain and flour storage facilities, and the instability of at least some of these habitats must have played a major role in the evolution of the ecological characteristics of the beetles (Dawson, 1977). Dispersal can thus be interpreted as the mechanism by which *Tribolium* avoids total extinction, in spite of the almost complete certainty of local extinctions. Natural selection in a changing environment should favor local populations which are variable with respect to the tendency of their members to disperse. Some disperse spontaneously and colonize new habitats, while others stay behind for as long as possible.

Roff (1975) has suggested several genetic models which are consistent with the existence of variability for dispersal. One of his models includes heterozygosity of dispersants, as is also suggested by our model. If the dispersants that found new populations indeed include heterozygous females, nondispersants can appear in the population, by segregation, soon after its establishment. As the dispersants start leaving the population, the frequency of nondispersants will increase with time. The relative frequency of the two alleles can serve as a rough estimate for the age of the population. The fact that our controls were closer to the LD lines suggests that in our population nondispersants were rather abundant, which implies that the population was not of very recent origin. The relative paucity of dispersants in the original sample may also account for the rapid response to selection for dispersal, in spite of the fact that dispersal is controlled by a dominant allele.

Acknowledgments

We thank Avigdor Beiles and especially Morris Soller for helpful discussions. This work was supported by the Leslie Lavi Research Fund.

References

Dawson, P. S. 1977. Life history strategy and evolutionary history of *Tribolium* flour beetles. Evolution, **31**: 226-229.

Johnson, C. G. 1969. Migration and dispersal of insects by flight. Methuen and Co., London.

Karlin, S. 1976. Population subdivision and selection migration interaction. *In* Population genetics and ecology. *Edited by* S. Karlin and E. Nevo. Academic Press, N.Y. pp. 617-657.

King, C. E. and Dawson, P.S. 1972. Population biology and the *Tribolium* model. Evol. Biol. **5**: 133-227.

Lidicker, W. Z., Jr. 1975. The role of dispersal in the demography of small mammals. *In* Small mammals: their productivity and population dynamics. *Edited by* F. B. Golley, K. Petrusewicz and L. Ryszkowski. Cambridge University Press. pp. 103-128.

Ogden, J. C. 1970a. Aspects of disperal in *Tribolium* flour beetles. Physiol. Zool. **43**: 124-131.

Ogden, J. C. 1970b. Artificial selection for dispersal in flour beetles (Tenebrionidae: *Tribolium*). Ecology, **51**: 130-133.

Prus, T. 1963. Search for methods to investigate mobility in *Tribolium*. Ecology, **44**: 801-803.

Prus, T. 1966. Emigrational ability and surface numbers of adult beetles in 12 strains of *Tribolium confusum* Duval and *T. castaneum* Herbst (*Coleoptera, Tenebrionidae*). Ekol. Pol. Ser. A, **14**: 547-588.

Ritte, U. and Agur, Z. 1977. Variability for dispersal behavior in a wild population in *Tribolium castaneum*. Tribolium Inform. Bull., in press.

Roff, D. A. 1975. Population stability and the evolution of dispersal in heterogeneous environment. Oecologia (Berlin), **19**: 217-237.

Schurr, K. and Bolduan, J. 1967. Genetic selection for disperal in *Tribolium castaneum*. Proc. North Cent. Branch Entomol. Soc. Am. **22**: 76-79.

Van Valen, L. 1971. Group selection and the evolution of dispersal. Evolution, **25**: 591-598.

Ziegler, J. R. 1976. Evolution of migration response: Emigration by *Tribolium* and the influence of age. Evolution, **30**: 579-592.

Zyromska-Rudzka, H. 1966. Abundance and emigrations of *Tribolium* in a laboratory model. Ekol. Pol. Ser. A, **14**: 491-518.

SIMILARITIES IN DISPERSAL TENDENCY AMONG SIBLINGS IN FOUR SPECIES OF VOLES (*MICROTUS*)[1]

Ray Hilborn

Institute of Animal Resource Ecology, University of British Columbia, Vancouver 8, British Columbia, Canada

Introduction

Microtine rodent populations fluctuate periodically, and Chitty (1967) suggested that genetic changes in social behavior are the driving mechanisms for these cycles. A modification of the original Chitty hypothesis was put forward recently by Krebs et al. (1973). In this hypothesis dispersal is stressed as a process leading to genetic change within the population. During the phase of population increase, density-intolerant voles leave the population. Dispersal has been shown to account for a high proportion of the loss from populations during the increase phase (Myers and Krebs 1971). The density tolerant, nondispersing individuals remain in the expanding population, and at peak densities the population is composed of those individuals that are also characterized by their aggressive behavior (Krebs 1970). The high level of aggression in peak populations reduces both reproductive success and survival, causing the population to decrease in numbers. Once the population has reached relatively low numbers, density-intolerant individuals can immigrate from refuge populations, and the population can begin to increase. This hypothesis considers the tendency to disperse (density intolerance) and aggressiveness to be characteristics which have a genetic basis. Therefore, selection through dispersal will alter the genetic composition of the population as the population increases.

What empirical predictions does this new hypothesis make? If dispersal tendency is under genetic control, then it can be predicted that members of a single sibship will be more similar in their dispersal behavior than individuals from different sibships. Under this hypothesis, there should be more variability in dispersal tendency between sibships than within sibships when the population is increasing. When the population reaches the peak phase or the decline phase, dispersal is unimportant and survival of individuals will occur at random with respect to families. This paper discusses a test of the null hypothesis that between-sibship variability in dispersal tendency is equal to within-sibship variability in dispersal tendency.

Methods

I have attempted to test this hypothesis with field data on four species of *Microtus* collected by Krebs (1966), Krebs et al. (1969), and recent unpublished data collected at the University of British Columbia (Table 1). There are two steps in the analysis: individuals are classified into litters, and the dispersal tendency for individuals is measured. Present field methods make it difficult to measure dispersal directly, but Myers and Krebs (1971) have done so for a limited number of individuals. They showed that in increasing populations most of the loss from populations could be accounted for by dispersal, but during peak and declining populations this was not the case and losses were believed to represent deaths in situ. Since dispersal of individuals can rarely be observed, the best measure of dispersal available is the lifetime of an individual on a study area. Recent work (Hilborn 1974) using radioactive markers has shown that in situ death rarely occurs during expanding populations of *Microtus townsendi*. Individuals that disperse have a short observed lifetime because they move off the study area. Those that do not disperse continue to be trapped on the study area until they die. The data of Myers and Krebs (1971) and Hilborn (1974) suggest that the observed lifetime of individuals would be a relatively good measure of dispersal during increasing populations and a relatively poor one during peak and declining populations when loss is more often due to death than dispersal. I have calculated between- and within-litter variability in observed lifetime.

[1] Received 21 June 1974; accepted 26 March 1975.

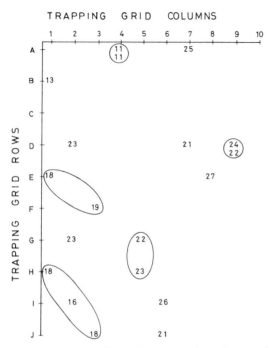

FIG. 1. Body weights (in grams) of newly tagged mice from one trapping period. Five groups of mice with similar body weights at adjacent trapping sites are encircled.

All censusing of *Microtus* populations was done by the method of Krebs (1966). Each population was trapped with Longworth live traps for 2 nights biweekly throughout the year. A typical study area consists of a grid, 68.6-m square, with 10 rows of 10 traps placed at 7.6-m intervals. During periods of high numbers additional traps were added to the grids to make sure that there were always sufficient empty traps to accommodate all mice. At every trapping period, tag number, grid location, weight, and sexual information were recorded for each individual. Field workers have noticed that groups of young, untagged mice with similar body weights were frequently caught very close to each other on an area in a trapping period. Figure 1 illustrates the body weights of new mice captured during one trapping period on a single area with individuals classified as littermates circled in this analysis. Such groups usually consist of two or three individuals captured within 7.6 m of each other and weighing within 1 or 2 g of each other. There will frequently be several such groups on a grid during a trapping period of moderate recruitment. This set of data was chosen because it shows the range of similarity in body weight and location found in the field. The two 11-g mice at A4 are almost surely littermates, but it is questionable if the 13-g mouse at B1 was

also part of the same litter. I will assume that these groups represent sibships which have all been weaned at the same time. This is the critical assumption of the analysis to follow: the probability that the new recruits are littermates will be related to the similarity between two individuals in first capture location and body weight.

Two techniques were used to classify individuals into litters. First, a subjective appraisal of data, such as that in Fig. 1, was used to determine litters. I could not get complete agreement on litter classification among (five) rodent trappers at the University of British Columbia and therefore decided to use a more objective method. A form of cluster analysis described by Gillie and Peto (1969) was used to classify individuals into litters. This was essential because a rigorous quantitative method was needed to separate groups. The maximum ranges of weight and distance between littermates at first capture were arbitrarily set based on subjective judgments of people familiar with *Microtus*. Litters were defined in a computer program to include as many individuals as possible within the constraints of the weight and distance criteria. In order to reduce the possibility of misclassification, the criteria used for litter definition were quite strict. To be included in a single sibship, two mice could be (at most) three units different where one unit was defined as 1 g of weight or 7.6 m apart at first capture. Only mice < 35 g at first capture were considered; most individuals of this size would be < 2-mo-old. The program considered all newly tagged mice for each trapping period and classified them as being members of a litter or unattached. The unattached mice were eliminated from further consideration; 50%–70% of individuals were unattached.

It is difficult to determine if litters classified in this way correspond to actual litters. Most litters classified consisted of only two individuals, which is much lower than the litter size at birth in these species. It seems likely that we are looking at a subsample of true litters. One reason for this is the strict criteria used in the litter classification; also, individuals in the litters may have failed to survive weaning, may have grown at different rates, moved away, or failed to enter traps. I tried marking litters in utero using radioactive techniques but was not successful. This technique would eliminate many of the problems associated with my indirect method. If we are looking at a subsample of each litter and if the sample of littermates is not random, the true distribution of life expectancy among littermates may be quite different. With my present techniques it is not possible to determine if my sample is random.

The lifetime for each mouse is defined as the

TABLE 1. Sources of data for experimental classifications. Weeks correspond to sampling periods in Krebs (1966), Krebs et al. (1969), and recent unpublished data collected at the University of British Columbia

Research site		Species	Weeks (from–to)
Unfenced increasing populations			
Berkeley	Tilden Control	M. californicus	1–34
Indiana	Grid I	M. pennsylvanicus	91–138
Indiana	Grid I	M. pennsylvanicus	218–230
Indiana	Grid A	M. pennsylvanicus	115–129
Indiana	Grid A	M. pennsylvanicus	219–225
Indiana	Grid A	M. ochrogaster	1–73
Indiana	Grid H	M. ochrogaster	109–125
Indiana	Grid H	M. ochrogaster	141–179
Vancouver	Grid C	M. townsendi	18–66
Vancouver	Grid I	M. townsendi	12–35
Unfenced peak populations			
Berkeley	Tilden Control	M. californicus	36–46
Indiana	Grid I	M. pennsylvanicus	140–192
Indiana	Grid I	M. pennsylvanicus	232–256
Indiana	Grid A	M. pennsylvanicus	39–89
Indiana	Grid A	M. pennsylvanicus	131–191
Vancouver	Grid C	M. townsendi	68–104
Vancouver	Grid I	M. townsendi	35–83
Unfenced declining populations			
Berkeley	Tilden Control	M. californicus	48–90
Indiana	Grid I	M. pennsylvanicus	194–218
Indiana	Grid A	M. pennsylvanicus	91–113
Indiana	Grid A	M. pennsylvanicus	193–217
Indiana	Grid A	M. ochrogaster	77–255
Vancouver	Grid C	M. townsendi	2–16
Fenced increasing populations			
Indiana	Grid B	M. ochrogaster	1–27
Indiana	Grid B	M. pennsylvanicus	1–53
Indiana	Grid D	M. ochrogaster	38–68
Fenced peak and declining populations			
Indiana	Grid B	M. ochrogaster	29–57
Indiana	Grid B	M. pennsylvanicus	55–71
Indiana	Grid D	M. ochrogaster	70–86

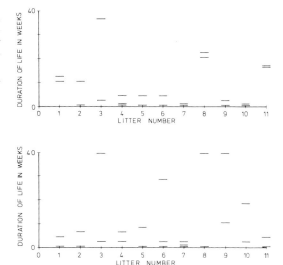

FIG. 2. Distribution of duration of life (lifetime) within litters in an increasing population (top graph) and a peak population (bottom graph). Note that in the top graph all the litters with the exception of litter 3 are quite closely grouped while in the bottom graph there are many litters with very large differences within the litter (3, 6, 8, 9, and 10). Also note that in the bottom graph only litter 7 is very tightly grouped, while in the top graph over half of the litters are quite tight. F in the top graph is 3.00. F in the bottom graph is 0.72. A perfectly random distribution of duration of life (lifetime) would produce an F of 1.00.

number of weeks between its first and last capture. Any mice that were accidentally killed or removed were eliminated from the analysis along with their entire litter. Each litter within a given week was assigned a number, and the grid, week, and litter numbers were recorded for each individual along with its lifetime. Each animal was put into one of five experimental classes depending upon whether the population was increasing ($r > 0.03$/wk), declining ($r < -0.02$), or at peak levels ($-0.02 < r < 0.03$), when the animal was born, and whether enclosed or not. These classifications were used by Krebs et al. (1973) to calculate population parameters for increasing, peak, and declining populations. To test more directly the relationship between dispersal and lifetime, similar analyses were made on populations from which dispersal could not occur. In Indiana, several of the populations were enclosed by mouse-proof fences (Krebs et al. 1969). Because data from declining populations on these enclosed grids were so limited, they were included with those from peak populations of the same grids. We pre-dicted that differences in the variability of the observed lifetime in open populations from which dispersal was occurring would not be observed in these enclosed populations. This design provides the five experimental classes shown in Table 1.

To test the hypothesis, the within- and between-litter variability in lifetime was measured. An analysis of variance was chosen because it facilitates the separation of variation in lifetime due to species, area, and season, from within- and between-litter variability. The analysis of variance also allows the combination of large sets of data to get accurate estimates of within-litter variability. A nested analysis of variance was performed on each of the five experimental classes. There were three sources of variation in addition to the within-litter variability, which is also known as the error term. The first was the grid-species complex listed in Table 1. If there was more than one increase, peak, or decline on a given grid for a species, each was treated separately. The second source of variation separated was that of week of first capture. This eliminates any variability resulting from animals living longer at the beginning of a peak phase than at the end or from a similar source of variations. The third effect was that resulting from between-litter differences. The null hypothesis

TABLE 2. Analysis of variance tables for five experimental classifications. The probabilities given are those for getting F-ratios as high or higher if the null hypothesis is true

Source	df	Sum of squares	Mean square	F	p
Unfenced increasing populations					
Grid-species	10	5,420.6	542.0	5.24	0.0000
Weeks	88	18,093.1	205.6	1.99	0.0000
Litters	124	19,323.2	155.8	1.51	0.0030
Error	269	27,812.0	103.3		
Total	491	70,648.9			
Unfenced peak populations					
Grid-species	6	3,990.4	665.1	4.55	0.0003
Weeks	62	13,527.7	218.1	1.49	0.0213
Litters	85	11,055.2	130.0	0.89	0.7238
Error	185	27,004.0	145.9		
Total	338	55,577.5			
Unfenced declining populations					
Grid-species	5	3,771.7	754.3	9.40	0.0000
Weeks	27	1,242.7	46.0	0.57	0.9472
Litters	31	990.4	31.9	0.39	0.9973
Error	80	6,415.8	80.1		
Total	143	12,420.6			
Fenced increasing populations					
Grid-species	1	29.9	29.9	0.18	0.6717
Weeks	26	3,887.2	149.5	0.91	0.5828
Litters	93	15,679.3	168.6	1.03	0.4206
Error	145	23,603.0	162.7		
Total	265	43,199.4			
Fenced peak and declining populations					
Grid-species	1	805.2	805.2	10.94	0.0015
Weeks	13	1,982.9	152.5	2.07	0.0237
Litters	56	5,518.9	98.5	1.33	0.1091
Error	87	6,399.6	73.5		
Total	157	14,706.6			

is that the ratio of the litter mean square to the error mean square is equal to one. The heritability can be calculated from these analyses of variance, but the test of significance is the F-ratio (Woolf 1968:119–121).

RESULTS AND DISCUSSION

Two principal conclusions emerged from the analysis of variance (Table 2). First, as shown in Fig. 2, in a population in the increase phase, the members of a litter tend to have similar lifespans, whereas during the peak phase of population growth there is more intralitter variation. The null hypothesis that the within- and between-litter variabilities are equal, is clearly rejected for unfenced, increasing populations ($p = 0.003$). Both fenced increase and fenced peak and decline populations show no differences between litters. Unfenced peaks also show no differences but unfenced declines show a significantly low value of F. Let us first consider the difference between the unfenced increase and the fenced increase.

Krebs et al. (1969) and Myers and Krebs (1971) show that dispersal is prevented by fencing populations, and that most losses during increasing populations are due to dispersal. Lifetime, as measured in live trapping studies, is thus a good measure of dispersal tendency in increasing populations. The highly significant value of F for unfenced increasing populations provides evidence that dispersal is non-random within sibships during normal periods of increase. This supports the hypothesis that there is genetic variation in dispersal tendency in the population, and during the population increase the dispersal-oriented sibships selectively leave the population. The fact that F was not significant in the fenced increasing populations supports the contention that differences in lifetime are due to differences in dispersal.

Reduced dispersal rates observed by Myers and Krebs (1971) during peak populations suggest that during these periods the lifetime is not a particularly good estimate of dispersal tendency, and we cannot say much about dispersal during those periods. The peak populations, as hypothesized by Krebs, should consist of aggressive, nondispersing mice. Our finding is consistent with this in that there was little differentiation among litters. It could be argued that within a study area there are good and poor habitats, and the significant value of F in the increase was because the litters born in the better areas stayed longer. We can perhaps reject this argument because of the nonsignificant value of F of both the unfenced peak populations and of the fenced populations, but it should not be ruled out completely. If the individuals classified as littermates are not true siblings, the change observed between the increase and the peak in unfenced populations would be difficult to explain, especially since there is no relationship between weight at first capture and lifetime.

There are two weaknesses in using an analysis of variance on these data. Since lifetime has a highly truncated distribution, the variances within litters are not likely to be homogeneous as assumed by the analysis of variance model (ANOVA). I have used several transformations of lifetime, and due to the robustness of ANOVA, the conclusions from these analyses have been identical to those presented in Table 2. The second potential problem with ANOVA is that since the mean lifetime changes between increase phase and peak phase, the power of ANOVA might possibly be reduced in the test on the peak populations. The data in Table 2 show, however, that the within-litter mean square is higher in the peak populations than in the increasing pop-

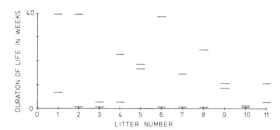

FIG. 3. Distribution of duration of life (lifetime) in a declining population. This figure illustrates that a high proportion of litters in declining populations have large differences in duration of life (lifetime) within the litter. F is 0.43.

ulations, but the between-litter mean square is higher in the increasing populations than in the peak populations. Thus, the failure to detect a significant effect in peak populations is certainly not caused by a weakness in the statistical test used.

The low value of F during the unfenced declines suggests that from each litter one individual lives a long time and the others live a short time. Figure 3 illustrates a typical set of data from declining populations. Assuming that little dispersal takes place during these declines, there may be some form of intralitter interaction occurring which causes only a few individuals from a litter to do well. Krebs' hypothesis proposes that by the time the population is declining it consists almost entirely of highly aggressive individuals. If there is any sort of territoriality or dominance which affects survival, we would expect a low value of F during the decline, but this finding is still difficult to explain.

The two other sources of variation presented in Table 2 are the differences between grid species complexes and the differences between weeks within a specific increase, peak, or decline phase. The grid species complex is almost always significantly different. This merely says that *Microtus californicus* from Berkeley have different survival rates from *Microtus pennsylvanicus* from Indiana. This is not surprising. The second source of variation was the effect of week of first capture. There are differences in lifetime between weeks in some populations but not in others. Since this does not relate to the general topic of this paper, I will not discuss the implications of this result.

Although the comparisons between the increasing, peak, and declining populations are completely consistent with Krebs' hypothesis, the differences between sibships do not have to be genetic but could be maternal. The test for heritability of dispersal during the population increases can be considered strong supporting evidence. It is apparent that significant behavioral changes which are heritable are occurring in these populations, although further experiments are required to establish the cause and effect relationships. The fact that the data are drawn from four species and three widely different areas demonstrates the widespread nature of these phenomena. Hopefully, better techniques for identifying sibships can be found which will replicate these results.

Acknowledgments

I thank A. Birdsall, C. Krebs, J. Myers, C. Wehrhan, R. Yorque, and two anonymous reviewers for comments on the manuscript and help in this work. C. Krebs and J. LeDuc provided the raw field data. The University of British Columbia provided generous computer time.

Literature Cited

Chitty, D. 1967. The natural selection of self-regulatory behaviour in animal populations. Proc. Ecol. Soc. Aust. **2**:51–78.

Gillie, O. J., and R. Peto. 1969. The detection of complementation map clusters by computer analysis. Genetics **63**:329–347.

Hilborn, R. 1974. Fates of disappearing individuals in fluctuating populations of *Microtus townsendi*. Ph.D. Thesis. Univ. British Columbia, Vancouver.

Krebs, C. J. 1966. Demographic changes in fluctuating populations of *Microtus californicus*. Ecol. Monogr. **36**:239–273.

Krebs, C. J., B. L. Keller, and R. H. Tamarin. 1969. *Microtus* population biology: Demographic changes in fluctuating populations of *M. ochrogaster* and *M. pennsylvanicus* in southern Indiana. Ecology **50**:587–607.

Krebs, C. J. 1970. *Microtus* population biology: Behavioral changes associated with the population cycle in *M. ochrogaster* and *M. pennsylvanicus*. Ecology **51**:34–52.

Krebs, C. J., M. S. Gains, B. L. Keller, J. H. Myers, and R. H. Tamarin. 1973. Population cycles in small rodents. Science **179**:35–41.

Myers, J. H., and C. J. Krebs. 1971. The role of dispersal in the regulation of fluctuating vole populations. Ecol. Monogr. **41**:53–78.

Woolf, C. M. 1968. Principles of biometry. C. Van Nostrand Co. Inc., Princeton, New Jersey. 359 p.

Part II

WHAT CONDITIONS MOTIVATE ORGANISMS TO MOVE?

Editors' Comments
on Papers 11 Through 16

11 LIDICKER
The Role of Dispersal in the Demography of Small Mammals

12 GRANT
Dispersal in Relation to Carrying Capacity

13 JOHNSON
Physiological Factors in Insect Migration by Flight

14 EVANS
Excerpt from *Studies of a Small Mammal Population in Bagley Wood, Berkshire*

15 NARISE
The Mode of Migration of Drosophila ananassae *Under Competitive Conditions*

16 HERRNKIND
Queuing Behavior of Spiny Lobsters

DISPERSAL

A "decision" to disperse will rationally consider both those motivating forces that are encouraging such action *and* some assessment of the potential advantages and disadvantages that will accrue to a disperser. At any given time and place and for any specific individual there is an imaginary balance sheet of pros and cons that will determine the merits of a dispersal decision. Of course, even accurate calculation of the dispersal balance sheet will not guarantee evolutionary success for the artful practitioner because of the large stochastic element (see Lidicker, 1981) influencing evolutionary success. Natural selection, however, favors those individuals genetically programmed or sufficiently rational to read the signs more correctly and make the appropriate decision more often than their fellows. Some of the elements that make up this dispersal balance sheet are shown in Table 1 (see also Lidicker, Paper 30).

Causes of dispersal are notoriously difficult to determine and

Table 1. Dispersal balance sheet

Types of factors	Potential advantages	Potential disadvantages
Environmental	Escape from unfavorable conditions (economic, physical, social)	Uncertainties of finding food, shelter, appropriate social milieu
	Reduce exposure to predators, competitors, diseases, parasites	Greater exposure to predators or competitors
		Increased energetic costs and physical deterioration
Genetic quantitative	May find uninhabited or low density area of suitable habitat	May not find any empty habitat
	Promiscuity	Uncertainty of finding a mate
	Frequency-dep. selection may favor rare phenotypes	Strange phenotypes may be avoided
qualitative	Heterosis and avoidance of inbreeding	Less viable offspring may be produced (breakdown of co-adapted system; disadvantageous recombinations)
	Greater chance for new and advantageous recombinations occurring	
		Founder effects

provide a challenging area of research in behavioral ecology. A dispersal event can be observed, but the specific and proximal motivating factor or factors are not necessarily apparent. We can, however, describe the general conditions prevailing when dispersal occurs, and this description should provide important clues to what the motivating factors really are. One thing is immediately evident—disperal occurs under an impressively large variety of conditions, which are not easy to classify meaningfully. Moreover, a dispersal decision must be based not only on the environmental circumstances but on the properties of the individual as well. The age, sex, physiological state, and genetic composition of the individual are all potentially important variables. Environmental conditions causing one individual to leave will be perceived differently by another. Some dispersal is a collective or social affair but mostly it is individualized. We are dealing therefore with a typical organism-environment interactive system, an understanding of which requires our appreciation of all three components: organism, environment, interactions. Because of this complexity it is often difficult to clearly establish sufficient motivating conditions. Some cases may be attributable wholly to environmental causes, some entirely to inherent organismal motivations; most will involve some mixture of the two. An early review outlining various environmental motivating factors is that of Stańczykowska and Wasilewski (1963). They show incisive insight in emphasizing the interconnected nature of the environmental complex. The importance of individual variation is emphasized by the work of Lavie and Ritte (1980) who report that in the flour beetle, *Tribolium castaneum,* individuals vary in their sensitivity to conditioned medium. The variation, which is genetically determined, affects both dispersal and the responsiveness of egg production to conditioned flour.

Any classification of dispersal based primarily on the environmental component of the organism-environment system will necessarily be less satisfactory for those cases in which the intrinsic motivations predominate. Given these difficulties, two major categories as proposed by Lidicker (Paper 11) provide a suitable beginning for the classification of dispersal. This paper (pp.103-108) defines "saturation" and "pre-saturation" types of dispersal, and provides examples from the literature on small mammals. These two types of dispersal are based on the demographic conditions prevailing in a population when dispersal events take place, the most important criterion being whether or not the population has reached its carrying capacity (saturation of essential resources). This classification is biologically meaningful because the specific motivating factors involved and the quality of the dispersers are very different between the two types. Moreover, the evolutionary backgrounds and implications of saturation and pre-saturation dispersal are quite distinct, and the ecological consequences are likely to be profoundly different. These two aspects are discussed in Parts IV and V.

Saturation dispersal is coincident with the traditional view that individuals leave home because their habitat is supporting all of or even more of the individuals than it can. Economic or physical conditions are limiting. This situation can occur even in relatively sedentary animals such as *Hydra* (Łomnicki and Slobodkin, 1966). Pre-saturation dispersal, on the other hand, is of special interest; in addition to the examples given in Paper 11, many other cases can be gleaned from the literature. Those cases of dispersal motivated by age, sex, or reproductive condition generally will fall into this category. A particularly fascinating example is provided by Birdsell (1957, pp. 49-54), who documents pre-saturation dispersal (at 30-68% of carrying capacity) in several human groups (from Tristan da Cunha,

Pitcairn Island, Bass Straight Islands, and the Australian mainland). An excellent study of dispersal in relation to carrying capacity has been reported by Peter Grant (University of Michigan), and is reprinted here as Paper 12. He offers a possible explanation for pre-saturation dispersal in herbivores involving nutrient limitation and selective feeding. Many examples from nonmammalian groups could also be provided; we will mention only a few.

Grinnell (1904) in his classic paper on the Chestnut-backed Chickadee, *Parus rufescens,* comments that ". . . the extremist intracompetition does not ensue until after further dissemination is impossible" (p. 373). He is saying that carrying capacity is not reached until all the dispersal sinks (in the sense of Lidicker, Paper 11) are filled. Kluyver and Tinbergen (1953) also describe pre-saturation dispersal in the Great Tit, *Parus major.* Insect examples also tend to emphasize the same categories of proximate factors in motivating pre-saturation dispersal. For example, in the greater milkweed bug, *Oncopeltus fasciatus,* dispersal has been shown to be affected by age, sexual activity, food quality, temperature, and photoperiod in addition to genetic differences among individuals. Adults are more likely to make long flights if exposed to low temperatures, short days, and lack of milkweed seed. Mating, long days, high temperature, and milkweed seed all promote reproduction that in turn inhibits flight (Dingle, 1966, 1968; Caldwell, 1974). Other insect examples emphasizing the importance of such variables in dispersal are provided by Green (1962) and Foster (1978). This point is well made by Johnson in Paper 13 in which he developed the idea of insect dispersal being controlled by an "eco-physiological system" interacting with differing genetic substrates to determine dispersal potential.

Within the large category of pre-saturation dispersal, we can group the array of environmental motivating forces into three major categories for convenience in thinking about them: economic (resources), tactical, and social. Economic forces include decreasing efficiency of utilizing resources such as food, decreasing quality of resources, or other clues to future shortages (Paper 11, p. 107; Paper 12). In other words, it may not be necessary to have resources become limiting before some individuals, capable of perceiving approaching limitations, move out. The frequency of emigrations by army ants can be indirectly correlated with the level of feeding experienced by the colonies (Topoff and Mirenda, 1980).

Tactical motivations for dispersal include the possibilities of avoiding predators, which are attracted to concentrations of prey, and avoiding the devastating effects of population crashes. Individuals involved in crashes not only risk predator influxes (Pearson, 1966) but also deleterious physiological and parasitological effects, as well as possible damage to their habitat. Little is known about these interesting possibilities, but two examples will be offered. Tschinkel and van Belle (1976) describe how the large larvae of the tenebrionid *Zophobus* disperse before pupating, presumably so as not to attract predators to a concentration of high quality food. In an underappreciated paper, Evans (1942) reports the significant observation that those individuals of the Bank Vole, *Clethrionomys glareolus,* who have the best chance of surviving a population crash are those that move to marginal habitats before peak densities are reached. These individuals then reoccupy the empty favorable sites. The last four pages of this paper contain Evans's information on habitat utilization, and are reprinted here (Paper 14).

The third class of motivations is that of social factors. Interactions among conspecifics (social) probably contribute to a great many "decisions" to disperse. Such a mechanism is implicit in Wynne-Edwards's (1962, p.16) notion of epideictic displays, defined as those providing clues to population density. Whether such behaviors actually have evolved for this function is controversial. To the extent that they exist, however, they could result in pre-saturation dispersal.

There are several examples known from insects where gregarious behavior leads to subsequent dispersal. In the well-known case of migration in the desert locust, *Schistocerca gregaria,* crowding of young hoppers can induce development into the highly active gregarious phase. However, this crowding is not merely the result of high population densities but is actively reinforced by the developing gregarious behavior of the hoppers themselves. Hoppers that develop in noncrowded conditions avoid contact with other hoppers and mature into the solitary phase. However, as densities increase and more contacts are forced among the young nymphs, they become increasingly social, seeking contact with other individuals until they form highly cohesive bands of marching hoppers. It is in this context that crowding promotes the transformation into the gregarious phase (Kennedy, 1956). In a less well-known example, the sycamore aphid, *Drepanosiphum platanoides,* appears to exhibit gregarious behavior as it colonizes sycamore leaves. As numbers on a tree grow, the aphids are found to be concentrated on relatively few leaves and these aggregations result from the gregarious nature of the insects (Kennedy and Crawley, 1967). The high local densities resulting from this aggregation behavior inhibit reproduction and promote dispersal, the net result being that many aphids emigrate long before the entire tree is exploited (Kennedy 1969).

Other kinds of social interactions are less controversial. Some individuals are forced to disperse by others. This process is typified by territorial behavior which of course is extremely wide-spread (for vertebrates, see review by Watson and Moss, 1971). Where it can be established that territorial behavior has limited population numbers by inducing dispersal to something below carrying capacity, then pre-saturation dispersal has occurred. Such is the case even where territoriality is tied to some future carrying capacity that may be lower than that existing when forced dispersals occur. A classic example of this condition is provided by the beaver (*Castor*) where two-year olds are forced to disperse (Bradt, 1938) even though resources may be adequate to support the colony for many more decades. Perhaps a more typical example is provided by house mice *Mus musculus*, where juveniles tend to disperse at a certain age. A combination of an inherent ontogenetic propensity for dispersal is probably combined with social pressures from adults. However, if the juvenile does not find a suitable new home after making some exploratory excursions, it may return to its home territory and be accepted back by the rest of the group, and thereby increase group size (Lidicker, 1976). An interesting dynamic interaction therefore exists between the inherent properties of the juvenile, social pressures in its clan, resources, and availabilities of dispersal "sinks." In some cases, however, agonistic social encounters may actually reduce exploratory activity. Summerlin and Wolfe (1971) report on experiments with male Cotton Rats, *Sigmodon hispidus*, in which neophobia and reduced exploration characterized subordinate individuals following agonistic encounters with dominants. Fairbairn (1978b) also reports that dispersing male *Peromyscus maniculatus* are less aggressive than resident males in neutral arenas.

There is another kind of pre-saturation dispersal that can be considered as basically socially motivated, but that does not conform at all to the density-dispersal relationship described in Paper 11 (p. 105). We refer here to dispersal that is actually inversely related to population size and therefore will most likely occur at low densities. If the "normal" type of pre-saturation dispersal tends to be positively density-responsive or completely density-independent, this type is negatively density-responsive. It is found in colonial species where centripetal social forces increase with colony size. As a colony declines in size, social cohesion is dissipated and dispersal is increasingly likely. This situation has been described in the European Rabbit, *Oryctolagus cuniculus*, by Mykytowycz and Gambale (1965); for the Checker-spot Butterfly, *Euphydryas editha*, by Ehrlich (1961) and Gilbert and Singer (1973); and in the lesser milkweed bug, *Lygaeus kalmii*, by Caldwell (1974). Where social cohesion is important in this way, a possible corollary is that in extremely high densities when the social structure tends to disintegrate, dispersal will again be at a high rate. Fairbairn (1978a) provides an example of this in *Peromyscus maniculatus*. Somewhat surprising is the discovery of low-density dispersal in the Roman snail, *Helix pomatia,* a species not generally considered to be highly social (Woyciechowski,1980).

Finally, there exists another subcategory of pre-saturation dispersal that involves a fourth class of environmental motivating influences. Here dispersal depends on the composition of the gene pool; that is, individuals perceive and differentially respond to various genotypes in the population. Recall the research of Narise referred to in Part I in which he described how the relative dispersal propensity of various strains of *Drosophila* was dependent on which strains were kept together (Paper 15), and moreover how the intensity of dispersal stimulation was positively related to the degree of genetic difference between strains (Narise, 1969). The adaptive significance of these behavioral phenomena is unknown. To these fascinating examples we can add that the sex ratio of a population may influence the tendency to leave as well as the facility of immigration into it. Shapiro (1970) provides a discussion of how sex ratio influences dispersal in pierid butterflies. Any such differential responses to the presence and/or frequency of other genotypes would generally be superimposed on other dispersal motivating forces, and hence represent a second dimension of forces that could be related to resources, tactics, and/or social behavior.

Whatever factor (or factors) is providing the proximate motivation, the perception of better areas outside the current home range may be necessary, or at least desirable, precursor to successful dispersal. Such perceptions are probably based on exploratory excursions that seem to be more common than we generally think. Fisler (1962) found that *Microtus californicus* could readily home up to 200 m, males performing better than females, whereas the measured home range in this species is only about 13 m in diameter (Lidicker, 1979, p. 464). Moreover, Ford and Krumme (1979) have shown that when composite home ranges, which encompass 95% of movements, are calculated, an area of about 35 m across is enclosed. Males also have a larger peripheral, low-frequency area of utilization than do females. Collectively, this evidence seems to suggest that excursions are frequent, especially among males and that they probably account for the unexpectedly good homing ability of this species. Another particularly well-documented case is that of White-tailed Deer, *Odocoileus virginianus,* on the Texas coastal plain. Inglis et al. (1979) report that 30 out of 38 individuals radio-tracked for an average period of 2.2 months made excursions out of their home ranges. Excursions, therefore, may not only be preliminary to dispersal, but may provide the necessary information as well as valuable experience for a successful movement.

MIGRATION

At one extreme of our dispersal-migration continuum are what we call "true migrants" where either the individual initiating the movement or its immediate descendants return to the same area. The same factors that affect other types of dispersal—especially presaturation events—may also begin migrations. One of the major sets of adaptations characteristic of migrations, that they respond to predictable sequences of events using a variety of environmental and biological cues to leave from and return to their home areas at appropriate times, is particularly relevant to our discussion of the more proximate factors causing animals to initiate migrations. The other primary set of adaptations that characterize migrants, that movements are at least to some extent directional insuring the migrants' or their progeny's return to the home area, also may be affected by these same factors since local conditions and the timing of departure can determine the direction and distance traveled. We will concentrate here on proximate factors responsible for the initiation of migrations.

Many migrations begin prior to the deterioration of local conditions that would adversely affect the reproduction and/or survival of the migrants or their offspring. Should these animals wait for conditions to worsen before beginning their migration to more suitable habitats, it may be too late. Conditions may deteriorate so rapidly that they are unable to leave. Selection will favor those animals that can utilize cues allowing them to anticipate these conditions or that develop internal timing mechanisms initiating migration at the appropriate time. In many cases, however, stimuli reflecting worsening conditions may be highly unpredictable or temporally so closely associated with those conditions that they would not provide sufficient time for the animals to leave. Also, for many species, migrations are extremely demanding and require considerable changes in the organism's physiology. For example, fuel reserves must be laid down and this activity may require considerable shifts in a variety of parameters ranging from feeding behavior to metabolic pathways. These processes may require weeks or even months to properly prepare the animal for the migration (for reviews of the physiological aspects of migration in birds see Berthold, 1975 and Gwinner, 1977).

Editors' Comments on Papers 11 Through 16

The migrant must be able to predict well in advance when it must leave to allow time to properly prepare itself. Thus the monarch butterfly begins its southerly movement in the late summer before the food supply provided by the host milkweeds begins to decline and lethal freezing temperatures become likely. This schedule provides sufficient time for the long flight to more temperate overwintering sites in the extreme southern United States, California, and Mexico (Urquhart, 1960).

For most migrants perhaps more important than leaving to avoid deleterious conditions is timing their arrival to take advantage of conditions at their destinations. When short travel distances and times are involved, stimuli related to both may be the same. But for many animals where migrations may take weeks or even months and/or cover vast distances, cues to the suitability of local conditions at the arrival site are not available. This situation is particularly true in species such as arctic breeding birds for which weather conditions permitting nesting and food availability for feeding young may persist for only a few weeks. There is no room for error. Arrive too early and the lack of food, snow cover, and cold temperatures may so weaken the bird that when conditions improve, it will be unable to breed. Arrive too late and the young may be unable to fledge before food becomes unavailable and the weather deteriorates. This problem can be further compounded in species such as some birds and marine mammals where males establish breeding territories and the quality of the territory influences the acquisition of mates and subsequent reproductive success. A male must arrive sufficiently early to stake out a suitable territory, but again, if he arrives too early, he may become so weakened that he will be unable to defend it against males arriving later.

Synchronized arrival can be important for other reasons as well. For example, some species such as marine turtles, salmon, and some amphibia gather at specific times in specific locations to mate and reproduce. Advantages may include such factors as increased opportunity for sexual selection or the swamping of predators (see also Part V). Again, for these advantages to operate, timing is critical. In some species, such as caribou, synchronous arrival is assured because the animals gather together and travel in large groups. Most migrants, however, travel alone or in relatively small groups. Also, individuals may be arriving from over vast areas. For these animals, if they are to arrive synchronously, it may even be necessary that individuals intiate migrations at different times.

What we see, then, is the importance of initiating migratory movements at times that allow animals to avoid deteriorating conditions and/or to arrive when conditions are favorable elsewhere, and that these movements must sometimes be started temporally and spatially separated from the conditions necessitating them. Selection should favor those animals that can make accurate predictions. If reliable environmental cues are available, we would expect them to be used. For example, if changes at both the departure and arrival sites are highly correlated and predictable, animals may be able to use local conditions to time their departure. Temperature, food availability and quality, rainfall, and so forth may be used to prepare for and/or initiate migration. If conditions at the departure and arrival sites are variable, but seasonally predictable, photoperiod may be used, as often occurs in migrant birds and insects. Photoperiod is also often used where precise timing is required for synchronous arrivals since it is the most precise cue generally available (although other geophysical cues such as lunar rhythm may also be used). When cues for habitat variation are unreliable or unavailable, the time of migration may be genetically determined in relation to the development of the organism so that migrations are initiated at specific ages or states of ontogenetic development.

We have not considered here the immediate cues that actually trigger movement. These may be as varied as the specific changes in weather conditions that are known to cause some passerine birds and insects to take flight; the social stimulus of observing other members of the population beginning to move (as in caribou and spiny lobsters); or spontaneous initiation triggered by internal factors. In most migrants, whether due to changes in food, temperature, photoperiod, or a genetically determined internal program, there appears to be a gradual lowering of a response threshold to the specific stimuli that triggers migration. Often this lowering threshold is expressed by increased activity such as in the well-known flight restlessness or "zugenruhe" of migratory passerine birds or by assembly behavior as groups become more cohesive as witnessed in spiny lobsters (Herrnkind, Paper 16, 1980) and caribou. Paper 16 has been reprinted because the socially facilitated migrations of the spiny lobster, *Panulirus argus,* are sufficiently bizarre and important to warrant its inclusion.

While these timing adaptations allowing animals to prepare for and initiate movements at appropriate times are not unique to migrants, they are more likely to be present and highly developed in migrants since most migrations take place in response to fairly predictable sequences of events requiring the evolution of precise timing mechanisms.

REFERENCES

Berthold, P., 1975, Migration: Control and Metabolic Physiology, in *Avian Biology,* vol. 5, D. S. Farner and J. R. King, eds., Academic, New York, pp. 77-128.

Birdsell, J. B., 1957, Some Population Problems Involving Pleistocene Man, *Cold Spring Harbor Symp. Quant. Biol.* **22:**47-69.

Bradt, G. W., 1938, A Study of Beaver Colonies in Michigan, *J. Mammal.* **19:**139-162.

Caldwell, R. L., 1974, A Comparison of the Migratory Strategies of Two Milkweed Bugs, *Oncopeltus fasciatus* and *Lygaeus kalmii,* in *Experimental Analysis of Insect Behaviour,* L. B. Browne, ed., Springer-Verlag, Heidelberg, pp. 304-316.

Dingle, H., 1966, Some Factors Affecting Flight Activity in Individual Milkweed Bugs (*Oncopeltus*), *J. Exp. Biol.* **44:**335-343.

Dingle, H., 1968, The Influence of Environment and Heredity on Flight Activity in the Milkweed Bug *Oncopeltus, J. Exp. Biol.* **48:**175-184.

Ehrlich, P. R., 1961, Intrinsic Barriers to Dispersal in Checkerspot Butterfly, *Science* **134:**108-109.

Fairbairn, D. J., 1978a, Dispersal of Deermice, *Peromyscus maniculatus:* Proximal Causes and Effects on Fitness, *Oecologia* **32:**171-193.

Fairbairn, D. J., 1978b, Behaviour of Dispersing Deermice (*Peromyscus maniculatus*), *Behav. Ecol. Sociobiol.* **3:**265-282.

Fisler, G. F., 1962, Homing in the California Vole, *Microtus californicus, Am. Midl. Nat.* **68:**357-368.

Ford, R. G., and D. W. Krumme, 1979, The Analysis of Space Use Patterns, *J. Theor. Biol.* **76:**125-155.

Foster, W. A., 1978, Dispersal Behaviour of an Intertidal Aphid, *J. Anim. Ecol.* **47:**653-659.

Gilbert, L. E., and M. C. Singer, 1973, Dispersal and Gene Flow in a Butterfly Species, *Am. Nat.* **107:**58-72.

Green, G. W., 1962, Flight and Dispersal of the European Pine Shoot Moth, *Rhyacionia buoliana* (Schiff.). I. Factors Affecting Flight, and the Flight Potential of Females, *Can. Entomol.* **94:**282-299.

Grinnell, J., 1904, The Origin and Distribution of the Chestnut-backed Chickadee, *Auk* **21:**364-382.

Gwinner, E., 1977, Circannual Rhythms in Bird Migration, *Ann. Rev. Ecol. Syst.* **8:**381-405.

Herrnkind, W. F., 1980, Spiny Lobsters: Patterns of Movement, in *The Biology and Management of Lobsters,* vol. 1, *Physiology and Behavior,* J. S. Cobb and B. F. Phillips, eds., Academic, New York, pp. 349-407.

Inglis, J. M., R. E. Hood, B. A. Brown, and C. A. DeYoung, 1979, Home Range of White-tailed Deer in Texas Coastal Prairie Brushland, *J. Mammal.* **60:**377-389.

Kennedy, J. S., 1956, Phase Transformation in Locust Biology, *Biol. Rev.* **31:**349-370.

Kennedy, J. S., 1969, The Relevance of Animal Behaviour, in *Inaugural Lectures,* Imperial College of Science and Technology, London, pp. 91-106.

Kennedy, J. S., and L. Crawley, 1967, Spaced-out Gregariousness in Sycamore Aphids *Drepanosiphum platanoides* (Schrank) (Hemiptera, Callaphididae), *J. Anim. Ecol.* **36:**147-170.

Kluyver, H. N., and L. Tinbergen, 1953, Territory and the Regulation of Density in Titmice, *Arch. Neerl. Zool.* **10:**265-289.

Lavie, B., and U. Ritte, 1980, Correlated Effects of the Response to Conditioned Medium in the Flour Beetle, *Tribolium castaneum, Res. Pop. Ecol.* **21:**228-232.

Lidicker, W. Z., Jr., 1976, Social Behaviour and Density Regulation in House Mice Living in Large Enclosures, *J. Anim. Ecol.* **45:**677-697.

Lidicker, W. Z., Jr., 1979, Analysis of Two Freely-growing Enclosed Populations of the California Vole, *J. Mammal.* **60:**447-466.

Lidicker, W. Z., Jr., 1981, Organization and Chaos in Population Structure: Some Thoughts on Future Directions for Mammalian Population Genetics, in *Mammalian Populations Genetics,* M. H. Smith and J. Joule, eds., University of Georgia, Athens, pp. 309-320.

Łomnicki, A., and L. B. Slobodkin, 1966, Floating in *Hydra littoralis, Int. Assoc. Theor. Appl. Limnol. Proc.* **16:**1615-1619.

Mykytowycz, R., and S. Gambale, 1965, A Study of the Inter-warren Activities and Dispersal of Wild Rabbits, *Oryctolagus cuniculus* (L.), Living in a 45-ac Paddock, *CSIRO Wild. Res.* **10:**111-123.

Narise, T., 1969, Migration and Competition in *Drosophila*. II. Effect of Genetic Backgound on Migratory Behavior of *Drosophila melanogaster, Jpn. J. Genet.* **44:**297-302.

Pearson, O. P., 1966, The Prey of Carnivores during One Cycle of Mouse Abundance, *J. Anim. Ecol.* **35:**217-233.

Shapiro, A. M., 1970, The Role of Sexual Behavior in Density-related Dispersal of Pierid Butterflies, *Am. Nat.* **104:**367-372.

Stańczykowska, A., and A. Wasilewski, 1963, Przyczny i przebieg migracji [Causes and Process of Migration], *Ekol. Pol.* ser. B, **9:**151-160.

Summerlin, C. T., and J. L. Wolfe, 1971, Social Influences on Exploratory Behavior in the Cotton Rat, *Sigmodon hispidus, Comm. Behav. Biol.,* No. 2, Part A, pp. 105-109.

Topoff, H., and J. Mirenda, 1980, Army Ants on the Move: Relation between Food Supply and Emigration Frequency, *Science* **207:**1099-1100.

Tschinkel, W. R., and G. van Belle, 1976, Dispersal of Larvae of the Tenebrionid Beetle, *Zophobas rugipes,* in Relation to Weight and Crowding, *Ecology* **57:**161-168.

Urquhart, F. A., 1960, *The Monarch Butterfly,* University Press, Toronto, 361p.

Watson, A., and R. Moss, 1971, Dominance, Spacing Behaviour and Aggression in Relation to Population Limitation in Vertebrates, in *Animal Populations in Relation to their Food Resources,* A. Watson ed., Blackwell Sci. Pub., Oxford, pp. 167-220.

Woyciechowski, M., 1980, Experimental Studies of the Exploitation and Overcrowding of a Natural Population of the Roman Snail, *Helix pomatia* L., *Ekol. Pol.,* ser. A, **28:**401-421.

Wynne-Edwards, V. C., 1962, *Animal Dispersion in Relation to Social Behaviour,* Hafner, New York, 653p.

11

Copyright ©1975 by Cambridge University Press
Reprinted from pages 103-128 of *Small Mammals: Their Productivity and Population Dynamics*, F. B. Golley, K. Petrusewicz, and L. Ryszkowski, eds., Cambridge University Press, London, 1975, 451p.

The role of dispersal in the demography of small mammals

W. Z. LIDICKER, Jr

One feature of small mammal populations which is currently being given increasing attention is that of dispersal. The primary purpose of this chapter is to summarize the information currently available on dispersal in small mammals and to suggest some directions for future research. Commonly, dispersers have been either ignored or considered to be of little significance by population ecologists. In part this has been due to considerable difficulty in measuring dispersal, but it is also widely assumed that such movements are demographically unimportant. For example, dispersers are typically considered to be individuals that leave home only when conditions become intolerable, and then have a near-zero probability of surviving long enough to reproduce elsewhere. Widespread acceptance of the term 'gross mortality' to represent any losses from a population, whether from real mortality or from emigration, serves to reinforce the notion that emigrants are only briefly postponing their inevitably premature death. Moreover, population growth rates are all too frequently defined in terms of birth and death rates only. In fact, growth rates are a function of the rate of additions to a population, which are composed of births *and* immigration, minus the rate of losses, which are divided between deaths *and* emigration.

In contrast to these perfectly logical biases held by many ecologists, evolutionary theorists, including population geneticists, have generally considered 'migration' as a potentially important factor in their investigations. Differential movements of genes, for example, are acknowledged as one of the factors capable of disrupting Hardy-Weinberg equilibria. Similarly, gene flow can oppose selection acting to improve local adaptations; it can be one component in an adaptive strategy for a variable environment, oppose inbreeding, increase the probability of novel gene combinations, etc.

In my view it is now time for a re-appraisal of the role of dispersal in short-term population processes as well. Consequently, in this paper I will call attention to the accumulating evidence that strongly implicates dispersal as a generally important demographic parameter. In doing this, I wish to discourage the impression that I do not consider other factors important. Clearly such a view would be nonsense, and in fact I have advocated (Lidicker, 1973) a multi-factorial or community view of small mammal population dynamics. Of course, some ecologists have recognized the potential importance of dispersal, and I would like in particular to mention significant discussions by Andrewartha & Birch (1954; p. 124), Andrzejewski, Kajak & Pieczyńska (1963), Petrusewicz (1966a), and Krebs et al. (1973). In the present paper, dispersal will be shown to influence various population properties as well as, in some cases, the regulation of numbers. Clearly, if dispersal can affect the composition, size, and spacing of populations, it can influence the productivity of those populations as well. The focus will be on small mammal populations, but the conclusions drawn are, I believe, more generally applicable.

Additional objectives of this paper are: (1) to characterize two kinds of dispersal, each having different causes, demographic effects, and evolutionary bases; (2) to recommend the use of islands and enclosures to study the nature and influence of dispersal; and (3) to propose models for the possible role of dispersal in the regulation of numbers. Emphasis will thus be on the demographic aspects of dispersal. The extremely important evolutionary questions will not be dealt with adequately, although I have discussed some of the relevant issues in an earlier paper (Lidicker, 1962), and Gadgil (1971) has provided an extremely interesting discussion of the importance of environmental variations in space and time on the evolution of dispersal. Perhaps it will be sufficient to point out that models proposed here require only individual selection for their operation. However, they are also consistent with the possibility that group selection, in a manner such as proposed by Van Valen (1971), may have abetted individual selection in the evolution of dispersal behavior.

I should like to point out at the outset that I am using the term dispersal to refer to any movements of individual organisms or their propagules in which they leave their home area, sometimes establishing a new home area. This does not include short-term exploratory movements, or changes in the boundaries of a home range such that the new range includes at least part of the former. Dispersal thus produces homeless travelers (vagrants) who are in search of a new home.

When such movements result in individuals leaving or entering the population under study, this then becomes emigration and immigration respectively. These latter terms are thus defined in a demographic context, and are not the same as the geneticist's 'migration'. Migration usually implies that genetic material has moved (e.g., Parsons, 1963), whereas dispersal refers only to individuals. While dispersal is clearly necessary for gene flow, it is not sufficient.

Saturation and pre-saturation dispersal

Basic to the development of my propositions on the demographic role of dispersal is the notion that there are two qualitatively distinguishable kinds of dispersal, and these result in two corresponding kinds of emigration. One type I propose to call *saturation dispersal* and the other *pre-saturation dispersal* (both types come under the general definition of *emigration*). Previously (Lidicker, 1962) I had called attention to these two types, but used the term 'density responsive emigration' for the

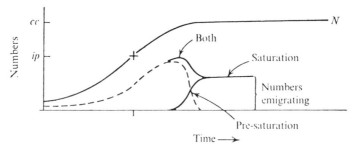

Fig. 5.1. Relation between changes in numbers of a population and both saturation and pre-saturation types of emigration; *cc* is carrying capacity and *ip* stands for inflection point.

latter. Fig. 5.1 illustrates the time course of these two kinds of emigration relative to changes in population numbers, and also suggests that both types may be exhibited by the same population.

Saturation emigration is the outward movement of surplus individuals from a population living at or near its carrying capacity. Such individuals (exhibiting saturation dispersal behavior) are faced with the immediate choice of staying in the population and almost certainly dying, or moving out and probably dying. They would represent, more often than not, social outcasts, juveniles, and very old individuals, those in poor condition, and in general those least able to cope with local conditions. For these reasons, they would be very susceptible to the various physical and biological hazards that they would surely encounter on their journeys, and would have only a very small chance of reaching a suitable location and reproducing successfully. Similarly, vagrants who do not become emigrants would not long survive under these conditions.

This is the traditional view of emigration, and contributes to the inclusion of such emigrants under the rubric 'gross mortality'. Wynne-Edwards (1962, p. 483) uses the term 'safety-valve' emigration for dispersal which relieves overpopulation, and Howard (1960) uses 'environmental dispersal'. These are similar but not identical notions to saturation emigration, as they are not tied so specifically to resource limitations. Many cases of mass movements of individuals (see Dymond, 1947, for examples) may be expressions of saturation dispersal.

Pre-saturation emigration is an exodus from a population before its habitat becomes saturated with that species. That is, it occurs during population growth, and may even begin very soon after growth starts. Moreover, such emigrants (exhibiting pre-saturation dispersal behavior) are not necessarily destitute either socially or economically, as there remains a surplus of resources in their home environment. Therefore, they will in general be in relatively good condition, and may include any sex and age group including pregnant females. Pre-saturation emigrants may be characterized by possessing a particular sensitivity to increasing densities, or they may have discovered some better home location during an exploratory excursion. Compared to saturation emigrants, such individuals can be expected to have a much greater chance of surviving and re-establishing themselves elsewhere.

Thus, these kinds of emigration differ not only in their timing with respect to population growth, but also produce emigrants of very different quality. In general they can be operationally identified by whether emigration occurs before carrying capacity is reached or after. Realistically, however, there should be some overlap as carrying capacity is approached, because at this stage individual variation and microgeographic variation in conditions will cause some individuals to leave because resources have run out for them, while others will still be leaving 'voluntarily'.

Saturation emigration does not seem to raise any difficulties with respect to its origin by natural selection. Such dispersal behavior is an act of desperation which prolongs life and very slightly increases the chances of longer-term survival. On the other hand, the evolution of pre-saturation dispersal behavior is not so easily explained. It involves, after all, the assumption of all the risks of leaving home when it is not really necessary. In an earlier discussion (Lidicker, 1962), I offered quantitative (increased opportunity for mating), qualitative (heterosis and increased chance for producing favorable genetic recombinants), and diplomatic (avoidance of population crashes and predator buildups) reasons why dispersal behavior may be favored by natural selection acting on individuals. To this we can add an economic reason. As populations become denser, intra-specific interference may result in inefficient utilization of remaining resources. For example, resources may be scattered where previously they were clumped, the most desired food or shelter may be scarce, more time and energy may be spent in social interactions, or activity periods may be shifted to less optimal times. Consequently, individual fitness may be improved by leaving rather than continuing to utilize available resources inefficiently. Of course, if circumstances permitted the operation of group selection, this too could lead to selection for dispersal behavior (cf. Van Valen, 1971).

In general, it can be claimed that where conditions favor the evolution of dispersal behavior, it will be the pre-saturation type that will appear. This is simply because it is this type that leads to a class of wanderers that has a relatively high probability of passing on its genetic make-up to future generations. If these arguments are valid, one would expect to find evidence that in many species dispersal behavior has a genetic component. This is not the place to document in detail the rapidly expanding evidence in support of this prediction. However, one can point to the existence among organisms of numerous adaptations for long-range movement, for example life history stages particularly suited for dispersal, as one line of evidence. Another kind of evidence is the nearly ubiquitous presence of sex, age and/or developmental stage differences in a tendency to move. This sort of evidence has been emphasized by Howard (1965) in distinguishing 'innate' from 'environmental' dispersals. Particularly elegant examples of this are provided by cases among migratory birds in which adults depart early from the breeding areas, and leave the remaining resources to juveniles (e.g., Pitelka, 1959). Strain differences in dispersal tendencies have also been described for a number of insects as well as house mice, *Mus musculus* (Incerti & Pasquali, 1967). Finally, there is the evidence for dispersal polymorphisms within a single population. This last includes a variety of invertebrates as well as vertebrates. Earlier reviews by Johnston (1961) on birds, and Howard (1960) and Anderson (1970) on mammals have been supplemented by recent evidence for several species of small mammals, for example, *Microtus pennsylvanicus* and *M. ochrogaster* (Myers & Krebs, 1971), and possibly *Peromyscus polionotus* (M. H. Smith *et al.*, 1972).

A further consequence of this view of dispersal is that pre-saturation emigration will in fact be common. Until recently, few investigators have thought to search for this kind of movement so that good evidence remains sparse. Nonetheless, Andrewartha & Birch (1954, p. 124) considered emigration from sparse populations to be ubiquitous, and Andrzejewski, Kajak & Pieczyńska (1963) seem to have recognized it as widespread.

Again, there are few examples from the invertebrates, and even humans (Birdsell, 1957), but I will focus my attention on evidence from species of small mammals. Perhaps the best known species in this regard is *Mus musculus*. It is now well-established that social pressures motivate dispersal long before economic saturation of the habitat is reached (Naumov, 1940; Strecker, 1954; Andrzejewski & Wrocławek, 1962; DeLong, 1967; Newsome, 1969a; Anderson, 1970; Lidicker, unpublished results). In addition, at least four species of microtines seem to show this behavior. These are: *Clethrionomys glareolus* (Evans, 1942; Smythe, 1968; Crawley, 1969), *Microtus californicus* (Lidicker & Anderson, 1962; Krebs, 1966; Fig. 5.3), *M. pennsylvanicus* (Van Vleck, 1968; Grant, 1971; Myers & Krebs, 1971; Ambrose, 1973), and *M. ochrogaster* (Myers & Krebs, 1971). Four other species of rodents with similar behavior are *Rattus villosissimus* (Newsome, 1975), *Spermophilus beecheyi* (Evans & Holdenreid, 1943), *Sigmodon hispidus* (Joule & Cameron, unpublished results), and *Myocaster coypus* (Ryszkowski, 1966). Of these last, *R. villosissimus* shows particularly spectacular dispersal often covering hundreds of miles across the deserts of central Australia. One insectivore, *Sorex cinereus* (Buckner, 1966), shows evidence of possibly exhibiting this kind of mobility.

Although this list is brief, consisting of only ten species, it should be greatly expanded when investigators are interested in searching for pre-saturation dispersal. As it is, it suggests a preliminary generalization about the circumstances that may be conducive to the evolution of this kind of behavior. In general, the rodents on this list fluctuate strongly in numbers and live in small demes when their numbers are low. It may be that this life history style leads to selection for pre-saturation dispersal. I suspect that it will also be found to be common amongst rare species.

Effects on population properties

Gross mortality

As already mentioned, the classic view of emigration is to lump it with real mortality under the heading of gross mortality, which then accounts for all losses to the population. Clearly emigration can contribute importantly to the total losses suffered by a population, and, to the extent that emigration is of the saturation type, it may not matter much how these losses are constituted. Emigration would then be merely a substitute for deaths as it only briefly delays mortality.

It is well known that vagrants carry a high risk of death from lack of shelter, starvation, predation (e.g., Errington, 1946; Pielowski, 1962; Metzgar, 1967; Varshavski, 1937, cited by Stoddart, 1970; Ambrose, 1972), etc. If they are further compromised by being in poor condition when they leave home, their chances are even less. Andrzejewski & Wrocławek (1961) have shown that dispersing *Clethrionomys glareolus* and *Apodemus flavicollis* die in live traps more often than do residents, and Janion (1961) has found that dispersing female house mice carry more fleas than do settled individuals.

Pre-saturation emigration, of course, also contributes to population losses, but is not so appropriately grouped with mortality. Not only may it occur when death rates are very low, but it may involve a component of the population which is very different from that most subject to mortality.

Age structure and sex ratio

Whenever emigration is differential with respect to age or sex, changes in these properties of the resident population are to be expected. The literature on small mammals is rich with references to dispersal dominated by juveniles. Certainly this is a common mammalian pattern, and may produce breeding populations with fewer young than would be expected on the basis of recruitment rates. Correspondingly, new colonies or low density groups may be more youthful than expected. This is not always the case, however. For example, Puček & Olszewski (1971) report that movements of *Clethrionomys glareolus* and *Apodemus flavicollis* into trapped-out areas were random with respect to age, at least in autumn, and Pearson et al. (1968) found a wide range of ages among *Ctenomys talarum* moving onto the frontiers of colonies.

A second common claim is that males disperse more often or to a greater distance than do females. In fact this seems to be characteristic of the entire family Sciuridae, and may be true of many other groups as well. Such sexual bias in dispersal may affect age specific mortality schedules for the two sexes, since males would be more often at greater risk. It would also mean that established colonies, or demes, would have more females relative to new ones. Such sex ratio biases could secondarily have profound effects on social structure or reproduction.

Of particular interest is the discovery that in at least five species of microtine rodents, adult or even pregnant females can play a major dispersal role along with young. These five are *Arvicola terrestris* (Stoddart, 1970), *Clethrionomys glareolus* (Kikkawa, 1964), *Microtus pennsylvanicus* (Grant, 1971; Myers & Krebs, 1971), *M. ochrogaster* (Myers & Krebs, 1971), and *M. oeconomus* (Tast, 1966). In the last species, it is even claimed that adult females normally disperse just prior to parturition. In the black-tailed prairie dog (*Cynomys ludovicianus*) adult females also frequently emigrate, and in this case they leave their recently weaned young in possession of the old territory (King, 1955). Adult females of *Perognathus formosus* are also known to disperse (French *et al.*, 1968; N. R. French, personal communication).

It is surprising to find significant dispersal among adult females, and hence future confirmation of the above reports will be important. Where such a pattern is established, we can predict that it will be among species with pre-saturation dispersal, simply because such dispersers have a higher probability of successful re-establishment. There is of course also the interesting possibility that adult females may disperse to provide ecological space for their offspring.

Growth rates

Perhaps it is not surprising that emigration has been given little credit as a suppressant on population growth rates since it requires the prior recognition of pre-saturation emigration. Only this type can be effective in this regard, because it is the only type which occurs during rapid growth. Saturation emigration occurs only at or near equilibrium densities so can exert no effect on the rate at which a population approaches that level. Moreover, even pre-saturation dispersal, if occurring only within the boundaries of the population being studied, will exert only local effects on growth rates.

However, if pre-saturation emigration is occuring, then dispersal may exert a profound effect on population growth rates. It follows logically that, if losses due to emigration are occurring during periods of growth, the resulting growth rate will be correspondingly less. Such an effect could be critical in a strongly seasonal environment where the period of time available for reproduction may be severely limited. This should be especially true for *r*-strategists (being short-lived and emphasizing high fecundities) who are attempting to make maximal use of temporarily very favorable conditions.

The best documented case for the importance of emigration in influencing growth rates is that of *Microtus californicus* (California vole). Relatively accurate density estimates are available for a number of rapidly growing populations of the species, and some of these are compared in Fig. 5.2. Four of these populations represent cases where emigration was absent or at least restricted. The fastest growth rates are shown by the population living on Brooks Island (curves E and F) in San Francisco Bay (Lidicker, 1973), and by a population (curve H) in an experimental enclosure (Houlihan, 1963). The short plateau shown by this last population occurred when the population temporarily ran out of food and shelter. Rapid growth continued when this limitation was artificially relieved. Only slightly less rapid growth is shown by the Brooks Island east transect (Lidicker & Anderson, 1962). These data (curve A) represent the growth pattern during the spring of 1959 on the side of Brooks Island distal to the point of colonization. Although the island was still being colonized during this period, the growth pattern shown was relatively little influenced by dispersal, as colonization was nearly completed by the time the east side was well populated. The transect, moreover, was in an area of very favourable habitat.

The four curves with lower slopes in Fig. 5.2, represent unenclosed populations. One (curve G) is from the mainland adjacent to Brooks Island (Batzli & Pitelka, 1971), and another (curve D) is from the Berkeley Hills (Krebs, 1966). The last two (curves B and C) are additional transects from Brooks Island. In this case the transects represent the pattern of growth while colonization of the island was in progress, the west transect being very close to the point of initial colonization. Both show the presumed effect of extensive emigration slowing population growth rate. When enclosed and unenclosed populations are compared with respect to monthly growth rates (increment of growth relative to density at start of month), differences are also apparent. The unenclosed populations had generally lower monthly rates which were less variable and showed a gradual downward trend as peak numbers were approached. Populations with limited emigration, on the other hand, had higher monthly rates which were more variable and declined more abruptly to zero as the peak was achieved.

The potential effect of emigration on growth rates is further emphasized by a comparison of growth rates on Brooks Island, not only between different parts of the island, but also between spring periods, when colonization was taking place, and afterwards (Fig. 5.3). Fig. 5.2 has already drawn attention to the dramatic difference in growth rates

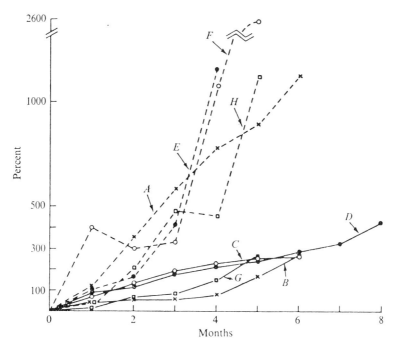

Fig. 5.2. Percentage increases (monthly intervals) from starting densities during periods of rapid increase to peaks in *Microtus californicus*. Solid lines refer to unenclosed populations and dashed lines to enclosed ones. *A*, Brooks Island, east transect (Lidicker & Anderson, 1962; Lidicker, 1973); *B*, Brooks Island, west transect (Lidicker & Anderson, 1962); *C*, Brooks Island, central transect (Lidicker & Anderson, 1962); *D*, Berkeley Hills, Tilden Park control (Krebs, 1966); *E*, Brooks Island, spring 1961 (Lidicker, 1973); *F*, Brooks Island, spring 1963 (Lidicker, 1973); *G*, Richmond Field Station (Batzli & Pitelka, 1971); *H*, enclosure (Houlihan, 1963).

on the east and west transects in 1959. This period of growth followed the arrival of the voles on the island in the summer of 1958. By the spring of 1960, all parts of the island showed similar rapid growth rates, the island then being completely colonized and net emigration presumably being nearly absent. The documentation presented here that populations of *Microtus californicus* in which emigration can occur grow at a much slower rate than similar populations in which emigration is inhibited, strongly supports the existence of a substantial amount of pre-saturation dispersal in this species.

Emigration has also been hypothesized to be one of six key factors explaining the three to four year population cycles exhibited by mainland populations of *M. californicus* (Lidicker, 1973). Its influence is manifest during the two years of the cycle when numbers are low. During this period, the species persists mainly in survival pockets or refuges of particularly favorable micro-habitat. Because of pre-saturation emigration, most of the reproduction which occurs during this time is channelled into colonization of empty habitat surrounding the refuges. Consequently, densities increase very little in the survival areas. Usually after two breeding seasons, the available habitat becomes colonized and the stage is set for an increase to peak densities. Clearly, there are other important factors involved, but without pre-saturation emigration the hypothesis proposes that the pattern of density changes in this species would be very different.

It seems possible that emigration may also be important in the demography of other microtines which show a three to four year cycle in numbers. By using enclosures Krebs *et al.* (1969) have shown that dispersal may be an important element in the demography of *Microtus pennsylvanicus* and *M. ochrogaster*, and in fact their enclosed populations generally had higher growth rates. Gentry (1968) has reported on a similar experiment with *M. pinetorum,* but does not compare growth

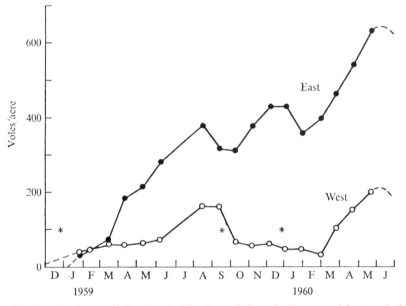

Fig. 5.3. Density changes in the Brooks Island population of *Microtus californicus* during the first two growing seasons following their arrival on the island. The west side of the island is where the original invaders established, and the east side is distal to that point. The east side also has a larger carrying capacity for voles. Asterisks represent times when heavy rains caused sprouting of new vegetation and thereby ended dry periods. Data from Lidicker & Anderson (1962) and Lidicker (1973).

rates between enclosed and unenclosed populations. The negative effect of emigration on population growth rates has also been demonstrated for *Clethrionomys glareolus* by Mazurkiewicz (1972). She has found that a population of this species living on Wild Apple Island (Wyspa Dzikiej Jabłoni) in Lake Bełdany grew to peak densities more rapidly, and reached higher numbers, than did an unrestricted population in the Kampinos National Park (Poland). Although other explanations for this difference are possible, the notion that it is due to pre-saturation emigration being frustrated on the island is appealing.

This last study also illustrates still another kind of effect that differential population growth rates can have. Mazurkiewicz (1972) has given evidence that in addition to the differences in population growth rates noted, individual growth rates were also different between her island and mainland populations. Individuals living in the unrestricted population grew more rapidly than did those from the island. A similar relationship may be present in other species as well. In such cases, emigration affects individual growth rates indirectly through its influence on population growth rates. Additionally, it would not be surprising if resident individuals were subject to modifications in fecundity and viability as a result of differing population growth rates.

Finally, we should consider that dispersal may produce immigrants as well as emigrants, and that an influx of immigrants into our subject population will produce enhancing effects on population growth rates. Clearly, if a population accepts immigrants they add directly to the growth rate by increasing the additions per unit of time. This is true whether or not the immigrants reproduce successfully in their new home.

Seasonal shifts of individuals between different habitats will clearly produce situations where immigration contributes massively to population growth. Many examples are known (see Brown, 1966, p. 131, for a good sample), but only a few will be mentioned here in order to indicate the nature and variety of the phenomenon. Regular seasonal movements are well known in *Mus musculus*, both in North America where winters are severe, and wheat fields in South Australia (Newsome, 1969a). A summer influx of *Apodemus sylvaticus* into grain fields has been described by several authors (see Crawley, 1970). Periodic re-invasions of seasonally flooded areas are characteristic of *Microtus oeconomus* (Tast, 1966), as well as of species associated with river plains (e.g., Sheppe, 1972; I. Linn, personal communication). Massive re-invasions can also be expected following burns (M. Delaney, personal communication). Finally, *Lemmus lemmus* seems to engage in regular spring and autumn dispersal between summer and winter habitats (Kalela, 1961; Kalela et al., 1971).

Various indirect effects of emigration on growth rate may also occur through influences on the social structure, reproductive activities, and/or genetic composition of the host population. Some of these will be considered below (see also Andrzejewski, Kajak & Pieczyńska, 1963).

Social structure

Dispersal movements can influence the social structure of both the residents remaining in the group from which dispersers have left and the host group into which the dispersers attempt, successfully or unsuccessfully, to immigrate. Little is known regarding the effects of emigration on residents. However, we can surmise that if emigration were combined with a resistance to immigration, social stability would be generally enhanced. This is because socially dominant individuals would be less likely to leave, those with a reduced inclination to move (physiologically or genetically) would tend to accumulate, and the addition of outsiders would be resisted. DeLong (1967) has suggested that this model applies to feral house mouse populations. He further points out that, if social stagnation becomes chronic, it can lead to reduced fecundity in the social group. Of course, the complete absence of emigration, as occurs in population cages, also leads in house mice to social stagnation and reduced reproduction (Southwick, 1955a; Crowcroft & Rowe, 1957; Petrusewicz, 1957, 1963; Lidicker, 1965). Additional effects on resident social structure might be expected to develop indirectly from emigration through influences on growth rate, sex ratio, age structure, individual growth and condition, etc.

More information is available on the influence of immigrants on host population social structure. We know, for example, that newcomers are generally added to the bottom of an existing social hierarchy so that there is a minimum of disruption (e.g., Andrzejewski, Petrusewicz & Walkova, 1963). An ingenious demonstration of the social avoidance of strangers by residents has been provided by Kołodziej et al. (1972) for *Clethrionomys glareolus*. Such avoidance or delegation to low rank may not always be the case, however, as immigrants may displace social dominants as well. One variable which can influence the success of immigrants in social integration is the number of immigrants relative to residents. If the number of immigrants is large, success is more likely as has been shown for *Mus musculus* by Andrzejewski, Petrusewicz &

Walkova (1963) and for *Peromyscus maniculatus* by Packer & Lidicker (unpublished results). In at least two studies, large numbers of colonizers have even produced a partial exodus of the residents (*Peromyscus maniculatus*, Healey, 1967; *Sigmodon hispidus*, Ramsey & Briese, 1971). Negative population growth rates were also reported to sometimes follow large introductions of aliens into established populations of *Rattus norvegicus* (Davis & Christian, 1956).

In addition to these direct effects on social structure, immigrants may become involved in social interactions which affect reproduction in important ways. One possibility is the disruption of pregnancies by strange males through olfactory contact with females in particularly sensitive stages of pregnancy (Bruce effect). This has been described for laboratory *Mus* (Bruce, 1959; Bruce & Parrott, 1960; Chipman & Fox, 1966*b*; Chipman *et al.*, 1966), feral *Mus* (Chipman & Fox, 1966*a*), and *Peromyscus maniculatus* (Eleftheriou *et al.*, 1962). A more common role appears to be that of stimulating reproduction. Such a role has already been mentioned for *Mus musculus*, and may be widespread among mammals. One such mechanism is the stimulation and/or synchronization of estrus by males (Whitten effect; see Whitten, 1966, for review), although it is not clear regarding the extent to which such stimulating males must be strangers. Among small mammals this effect is best known in house mice (e.g., Lamond, 1959; Marsden & Bronson, 1964; Whitten *et al.*, 1968). It has also been described for *Peromyscus maniculatus* (Bronson & Marsden, 1964), and the males of some microtines also stimulate behavioral estrous (e.g., Chitty & Austin, 1957, for *Microtus agrestis*; Richmond & Conaway, 1969, for *M. ochrogaster*).

Still another way in which immigrants may influence reproduction is through the release of reproductive inhibition in females which seems sometimes to accompany the act of dispersal. That is, when reproductively inhibited females (often subadults) leave home, they often rapidly become reproductively mature. This clearly is the case for *Mus musculus* (Crowcroft & Rowe, 1958; Lidicker, 1965), and may also be true for some species of *Microtus* (Richmond & Conaway, 1969; Myers & Krebs, 1971). It has also been found in *Peromyscus leucopus* (Sheppe, 1965), as well as various ungulates. It has been suggested by Richmond & Conaway (1969) that in these species any minor stress brings about an increase in FSH (follicle-stimulating hormone) release followed by estrous. Since dispersal is undoubtedly stressful, such a mechanism could explain the fact that dispersing females tend to rapidly become reproductively competent. In this way they not only increase their chances of social acceptance in a new home, but they are also immediately ready to contribute reproductively. A particularly extreme example of this strategy of reproductive readiness is the dispersal of pregnant females such as in *Microtus oeconomus* (Tast, 1966) and *M. pennsylvanicus* (Myers & Krebs, 1971). This kind of dispersal is probably an adaptation for rapid colonization and exploitation of temporarily favorable habitats.

The importance of frustrated dispersal

In order for dispersal to occur, individuals must not only be motivated to leave home but they must physically be able to do so. Thus physical barriers surrounding a population may prevent or severely reduce dispersal, even if strong motivation exists. Moreover, even in the absence of physical barriers, there generally needs to be some place that the potential disperser is willing to go to. These refuge areas I would like to refer to as dispersal sinks, modelling the term after the 'behavioral sink' proposed by Calhoun (1962). Such a sink will generally be some empty or unfilled suitable habitat, or perhaps marginal or even unsuitable habitat in which at least short-term survival is possible. In the absence of such a sink, a potential disperser will probably return home after making an exploratory excursion, although possibly a few individuals could persist for a limited time as vagrants while in search of a refuge.

Whenever motivation to disperse exists, but consummation is prevented by barriers or inhibited by the absence of an unfilled sink, a condition of frustrated dispersal exists. The level of this frustration will depend on the extent to which the stimulating forces providing the motivation exceed realized net dispersal. Dispersal does not, therefore, have to be absent to produce frustration, but only be at a rate less than would be required to relieve the pressure for dispersal that develops. Populations surrounded by severe physical barriers or filled dispersal sinks will be subject to a low emigration rate. Under these conditions, the appearance of almost any degree of motivation for dispersal will produce frustrated dispersal. Those few individuals who manage to survive as vagrants without even a temporary refuge or home would move randomly and not produce any reduction in the level of frustration. In species exhibiting pre-saturation dispersal, motivation to disperse will appear even at early stages in population growth, and at relatively low densities. With saturation dispersal, however, motivation to leave will only begin when the current carrying capacity is nearly reached. The inhibition of dispersal may, therefore, have quite different demographic effects in these two kinds of species.

It has become apparent to me that circumstances which serve to frustrate dispersal can be powerful, analytical tools for studying the normal role of dispersal, even when a population is ordinarily not subjected to such frustrations. In a manner analogous to the experimental removal of some organ in order to study its function, the population biologist can contrive to frustrate dispersal in an attempt to understand its normal role. An additional purpose for such manipulations is to permit the discovery of alternative density regulatory mechanisms which may be available to various species, but which are utilized only infrequently. We may thus be led to a better understanding of the ecology, behavior, and even physiology of our subject species. Experimental enclosures and natural islands readily serve these functions, and are to be recommended for further exploitation by ecologists studying small mammals.

Of course, the use of enclosures and islands by ecologists is not a new idea, and it will be of interest to consider some examples of such studies in order to illustrate what is known of the ways in which species of small mammals cope with frustrated dispersal. One tactic that is widely exploited in these circumstances is the increased usage of marginal habitats (especially where the absence of competitors makes such areas more suitable). This has been documented for *Microtus ochrogaster* on islands by McPherson & Krull (1972), for *M. pennsylvanicus* on islands by Webb (1965), for this species and *Peromyscus maniculatus* in enclosures by Grant (e.g., 1971), and for *P. leucopus* on islands by Sheppe (1965). If even this opportunity for dispersal is unavailable, however, there remain only two fundamental mechanisms for stopping growth; to decrease natality or to increase mortality rates. Both have been widely observed. Often, though, an isolated population is not able to achieve a long-term balance with its resources; it becomes too large, and then suffers a crash which may even result in its extinction. Direct evidence for this last point has been provided by Crowell (1973) involving *Microtus pennsylvanicus* living on islands in the Gulf of Maine, by Lidicker & Anderson (1962, p. 504) for *M. californicus* on a small islet in San Francisco Bay, and by Sheppe (1965) for *Peromyscus leucopus* on an island in Ontario.

The achievement of unusually high densities along with some form of reduction in birth rates has been found in several microtines and in *Mus musculus*. *Mus* has been most widely studied in this regard, and its ability to stop breeding under high density, socially stabilized conditions has already been mentioned. This is primarily accomplished by progressively increasing inhibition of reproductive maturation in females (see previous references and Lidicker, unpublished results). An increased frequency of pseudo-pregnancies may also be a factor to some extent (van der Lee & Boot, 1955). Lund (1970) placed *Arvicola terrestris* in enclosures and produced high densities and a drastic fall in reproductive rates. A particularly interesting case is that of *Clethrionomys glareolus* living on Wild Apple Island in the Mazurian Lake District, studied extensively by a group at the Polish Institute of Ecology. Bujalska (1970) has proposed that reproduction is controlled in this population by the territorial behavior of adult females, and there is also a strongly marked inhibition of growth in young born late in the breeding season (Bujalska & Gliwicz, 1968; Petrusewicz *et al.*, 1971). A long-term adaptation to island life is illustrated by the Skomer vole (*Clethrionomys glareolus skomerensis*) long isolated on a small Welsh island. This population lives at chronically high densities, and has an exceptionally short breeding season and slow maturation rate (Fullagar *et al.*, 1963; Jewell, 1966).

There are other examples in which increased mortality rates were emphasized. Louch (1956) found high mortality rates accompanying high densities of *Microtus pennsylvanicus* in small pens. Krebs *et al.* (1969) describe the fate of populations of both *M. pennsylvanicus* and *M. ochrogaster* in large enclosures (8372 m^2) in southern Indiana. These populations reached very high densities, ate out their food supply, and dropped back to control levels or below. No adjustments in reproduction were made by these populations, leading at first to excessive numbers and then to high mortality rates. In small population cages, increase in numbers of *Oryzomys palustris* and (to a major extent) *Peromyscus truei* was stopped by increasing mortality through aggression (Lidicker, 1965). Most of these deaths were of neonates within three days of birth. Occasionally a young animal survived and compensated for deaths among adults who became badly wounded. Also, house mice living in large complex enclosures characteristically cannibalize or desert their young under high densities (Brown, 1953; Southwick, 1955*b*; Anderson, 1961; Lidicker, unpublished results). Complete reproductive inhibition may not be possible under these conditions because of the continual stimulation stemming from interactions among the multiple social groups present. *Peromyscus maniculatus* is apparently another species which uses both reproductive inhibition and mortality of nestlings to cope with frustrated dispersal (Lidicker, 1965).

In spite of the variety of these examples, it still seems premature to draw generalizations from them regarding possible relationships between the kind of dispersal normally exhibited by each species and its mode of coping with dispersal inhibitions.

Emigration as a regulating factor

Although in theory emigration can act as a major factor regulating population density, there are few data directly bearing on this question. The potential for emigration to act in this way stems from its role in the population growth equation, as has already been pointed out. In considering this question, it is important to remember that the question of the manner of density regulation should be applied to the same defined population as is the measure of emigration. The term carrying capacity is used in the following discussion to specify the maximum numbers of individuals that an area can support based on the availability of essential resources such as space, food, water, and shelter.

In considering the role of emigration in the regulation of numbers we must, for convenience and clarity, subdivide our treatment into three levels of importance. The simplest role that emigration can play in this regard is as one of several factors which account for the total losses suffered by a population when its growth rate becomes zero at carrying capacity (i.e., a contributing factor). Under these conditions, if emigration should become frustrated, some other form of loss would soon increase to compensate for it. Thus, emigration is contributing to regulation of numbers, but is not an essential element in the process. This may be a common pattern among small mammal species, and a good example is provided by the Arctic ground squirrel, *Spermophilus undulatus* (Carl, 1971).

The second level of involvement is where emigration acts as a key factor in stopping population growth at carrying capacity. This is similar to the situation above, but now emigration makes up a very large fraction of the losses. It thus requires a large dispersal sink in order to accommodate long-term, large-scale emigration. If emigration becomes inhibited, under these conditions, population density would probably shoot up above carrying capacity, and then crash. Some long-term damage to the habitat may also be a consequence. This pattern would be enhanced under conditions in which demes were asynchronous in their fluctuations in number, because the presence of low populations in the vicinity of high ones would ensure the existence of unfilled habitat to accommodate the dispersers. Both these levels of involvement could in principle occur in species showing only saturation emigration or some combination of saturation and pre-saturation types.

An example of emigration playing a key role in regulation may have been provided by Pearson (1963) in his study of two outbreaks of feral house mice. He concluded that dissipation of the peak populations was primarily the result of emigration, although mortality from predation and disease played secondary roles. In these cases, the outbreaks were local leaving large areas of suitable habitat available as dispersal sinks.

The third potential role that emigration may play in density regulation is actually to prevent numbers from reaching carrying capacity (so that the equilibrium density, K, is then below carrying capacity). This possibility was suggested in an earlier paper (Lidicker, 1962), and is clearly the most controversial of the three proposed roles. Its existence, however, would seem to be predictable solely from the presumed reality of pre-saturation dispersal. Some of the difficulties inherent in understanding how this kind of dispersal behavior might evolve have already been discussed. However, if pre-saturation dispersal is real, and there seems to be good evidence that it is, then the potential exists for emigration to play this kind of role, at least occasionally. Several models for how this role can be fulfilled by emigration will be proposed; all require the involvement of pre-saturation dispersal.

The first model (I) combines the action of frustrated dispersal and a temporarily very favorable carrying capacity. Given these conditions, it is hypothesized that there will be a hyper-development of various behavioral and/or physiological traits which have evolved to stimulate pre-saturation dispersal. But, since emigration is inhibited and potential emigrants do not immediately become mortality statistics (the population not having yet reached its carrying capacity), the stage is set for overstimulation of all members of the group. These hyper-developed traits, such as aggressive behavior, are hypothesized to increase mortality or decrease natality sufficiently to stop population growth. The higher the carrying capacity, the greater would be the opportunity for this syndrome to develop. It can be thought of as a case where the existence of proximate regulating factors forces the population to be temporarily out of phase with its ultimate regulating forces.

Model I is illustrated in Fig. 5.4, which also contrasts the situation prevailing if saturation emigration only were to be present. In the model,

122 W. Lidicker, Jr.

emigration is shown at first to perform only a key factor role, density increasing to the carrying capacity of the habitat. The carrying capacity is then imagined to increase abruptly to a new higher level. When only saturation emigration is present, numbers increase promptly to the new

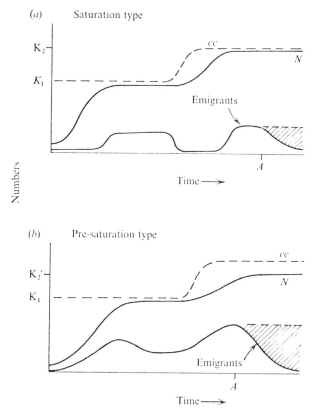

Fig. 5.4. Density changes (*N*) and numbers emigrating with increased carrying capacity (*cc*) when (*a*) only saturation dispersal is present and (*b*) when pre-saturation dispersal occurs along with the hyper-development of traits which stimulate such dispersal (Model I). Hatched areas indicate the extent of frustrated dispersal which is initiated at time *A*; K_i refer to equilibrium densities.

level. Even when emigration becomes frustrated (time *A* in the figure), no demographic effect is evident since individuals which cannot emigrate simply become mortality victims. If emigration were playing a key factor role, numbers might increase abruptly after *A*, and then crash. With pre-saturation dispersal, however, the population is shown to level off at a new *K*-level, this time below the carrying capacity, at a time shortly after emigration becomes frustrated.

The temporary nature of the proposed increase in carrying capacity is emphasized, because the situation in which numbers were being regulated below resource levels would not likely be stable in evolutionary time. This is simply because selection would favor individuals who were more tolerant of the conditions generated by frustrated dispersal and hence could more fully utilize available resources. Equilibrium densities would in time then approach the new carrying capacity. This notion is illustrated in Fig. 5.5. In this example it is imagined that a species displaying pre-saturation dispersal colonizes an island which has a very favorable carrying capacity for that species. Being an island, dispersal becomes frustrated as soon as colonization of all suitable parts of the

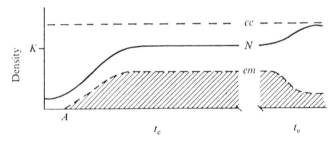

Fig. 5.5. Hypothetical regulation of numbers below carrying capacity in an island population of a species showing pre-saturation dispersal. Hatched area indicates extent of frustrated dispersal; t_c is ecological time, t_v evolutionary time, em number of potential emigrants, and other symbols as in Fig. 5.4.

island is complete. Because of this, and the extremely favourable level of resources, the population becomes regulated below carrying capacity. This condition would prevail through ecological time (t_c) as shown in the figure. After many years, however, selection would reduce the perceived level of frustrated dispersal, allowing the density to increase to carrying capacity.

A second model (II) contains the same basic elements as the first, (pre-saturation dispersal and its occasional frustration), but in this case large dispersal sinks along with large-scale pre-saturation dispersal keep low density populations from growing toward capacity. If and when the sinks become filled, dispersal is inhibited, and densities build up rapidly.

This model is illustrated in Fig. 5.6., and may apply to several species of *Microtus*. The enclosure experiments by Krebs and his students (Krebs *et al.*, 1969) mentioned earlier are also relevant here. Enclosed populations of *M. pennsylvanicus* and *M. ochrogaster* reached much higher densities than did unenclosed controls. Moreover, in one

124 W. Lidicker, Jr.

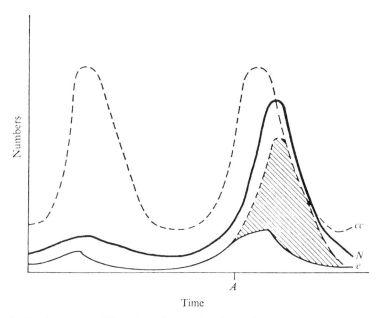

Fig. 5.6. Density changes (N) and numbers emigrating (e) illustrating Model II. A strongly fluctuating carrying capacity (cc) is coupled with strong pre-saturation dispersal. At time A dispersal becomes frustrated and a large increase in numbers follows. The extent of frustrated dispersal is indicated by hatching.

enclosure *ochrogaster* reached high numbers three years running. Hatt (1930) also reported that *M. pennsylvanicus* in enclosures reached very high densities and damaged vegetation. Similarly, Gentry (1968) enclosed two populations of *M. pinetorum* (in pens of two acres each) and found that they reached unusually high densities. These results suggest that when emigration is not prevented these species of *Microtus* are keeping their numbers significantly below the peaks in carrying capacity. This idea is supported by Myers & Krebs (1971) who report that 59 to 69 per cent of the *M. pennsylvanicus* in their study populations dispersed. In addition, I have suggested (Lidicker, 1973) that, in *M. californicus*, emigration is one important element in keeping densities from increasing to carrying capacities for two or more years at a time. Eventually, the dispersal sinks become filled, and numbers increase to peak densities, only to be stopped by seasonally declining carrying capacities. Perhaps in these species of *Microtus*, long-term evolution away from this pattern is prevented by the fact that the carrying capacity of their habitat varies greatly with season, making it only temporarily very favorable.

There is at least one example of a species which appears to show almost continuous regulation of population size below carrying capacity, that is, even temporary outbreaks do not occur. Moreover, the process is stable in evolutionary time. This is the North American beaver (*Castor canadensis*), for which we must provide still another model. Model III can be considered only a modification of II. In the beaver a population is equivalent to the family group. It is well-known that the numbers of beavers in any family group are very strictly regulated (e.g., Bradt, 1938; Taylor, 1970). There is typically one pair of adults, a litter less than one year old, and a litter of yearlings. When the young reach two years of age, they are firmly driven from the colony and disperse. Immigration into a population is impossible unless one of the adults should die. When a new colony is established, it is living far below its carrying capacity. As the area in the vicinity of the house and dam become exploited over the years, the carrying capacity of the area gradually diminishes. Eventually the site is abandoned, whereupon it gradually recovers. Thus, the beaver model requires only pre-saturation dispersal and a permanent dispersal sink, so that dispersal does not become frustrated permitting temporary high densities.

Since beavers exploit a resource which is only slowly renewable (they 'eat into the capital'), it would be a poor strategy to permit each population to build up in numbers until short-term carrying capacity was reached. If this were to happen, the carrying capacity of the site would soon drop to near zero, and would then recover only very slowly. *Microtus* also exploit the capital of their food supply, but since this is herbaceous rather than woody, they are able to recover more quickly following peak densities. Wiegert & Owen (1971) also argue that herbivores will evolve feeding styles that avoid long-term oscillations in their food populations.

Model III may also be applicable to prairie dogs (*Cynomys*). According to King (1955) these social rodents produce two kinds of emigrants (yearlings which are mostly males and older adults which are mostly females), and do not ordinarily destroy their food resources. A very large dispersal sink is assured by the widely spaced arrangement of their colonies.

Discussion and summary

Emphasis in this chapter has focused on the potential importance of dispersal movements as demographic agents. This represents a significant shift in perspective from the widespread view that such movements can be largely ignored, either because they merely substitute for other forms of mortality or because they cause internal perturbations more likely to have social than demographic effects. In part this change has been a consequence of the accumulating evidence that many species of small mammals are organized in a demic dispersion pattern, even in the centers of distribution (cf. Anderson, 1970). With this has come the realization that knowledge of dispersal may be critical to an understanding of demographic phenomena. It must be cautioned, however, that the relative importance and nature of dispersal will undoubtedly be found to vary greatly among different species (see also Chapter 4). I expect further that not infrequently variation within species will also be found. Particularly is this to be sought in comparisons of populations on the edges of a species' range with those from the center. A demic structure, for example, is likely to be found on the periphery of ranges even where it is not characteristic elsewhere.

A second reason for giving increased attention to dispersal is the recognition that there are two kinds of dispersal (pre-saturation and saturation), each having a different timing with respect to phase of population growth and producing qualitatively quite different kinds of emigrants. Moreover, pre-saturation dispersal can have greatly different demographic effects compared to the saturation type. The available evidence for pre-saturation dispersal in small mammal species has been outlined, but evidence is also available for numerous other organisms as well. An intriguing aspect of the existence of pre-saturation dispersal is the probability that it has a genetic basis, and that a population may even be polymorphic for dispersal tendencies. The evidence for this has been summarized. However, it is of further interest to consider whether or not the phenomenon of cyclomorphosis exhibited by some species which have multiple generations per year may incorporate such a polymorphism. Reviews of the phenomenon have been provided by Shvarts *et al.* (1964) and Anderson (1970), and the latter suggests that a colonizing and survival cohort can often be recognized. Of course we do not know if the differences observed among such seasonal cohorts are phenotypic, genotypic, or both. The possibility that such shifts in the average properties of individuals in a population are at least in part genetic must be seriously considered in view of the recently available evidence that seasonal shifts in gene frequencies can occur in small mammals (Semeonoff & Robertson, 1968, for *Microtus agrestis*; Tamarin & Krebs, 1969, for *M. pennsylvanicus* and *M. ochrogaster*;

Berry & Murphy, 1971, for *Mus musculus*; and McCollum, 1975, for *Microtus californicus*).

One generalization about pre-saturation dispersal may tentatively be offered even at this very early stage in our knowledge, namely, that this type of dispersal may often be associated with (1) colonizing species such as *Mus musculus*, and (2) those species with feeding styles such that they can affect the future supply of their food. In the latter case, I am referring to species which can, and do, consume more than the surplus production of their food supply. These would include herbivores which can damage their food plants to the extent that the plant's reproductive abilities are reduced, or carnivores who kill prey individuals with high reproductive values. In such species, high densities imperil future carrying capacities and natural selection may then favor individuals who disperse before crashes occur (this is in part the 'diplomatic advantage' of Lidicker, 1962, and is also suggested by Evans, 1942, and Andrzejewski, Kajak & Pieczyńska, 1963). The fact that the meager evidence for pre-saturation dispersal so far available turns up in genera such as *Microtus* and *Castor*, and not in mast-feeders such as *Apodemus* and *Peromyscus* lends support to this generalization (see also discussion of herbivory by Wiegert & Owen, 1971).

Among the population properties which can be influenced by dispersal are 'gross mortality', age structure, sex ratio, growth rates, and social structure. In considering these effects, it is important to distinguish the role of emigrants from that of immigrants. This is especially true when social effects and secondary influences on reproduction are being studied.

Two new concepts are considered essential for discussion of the role of dispersal in the regulation of numbers. One is the notion of 'frustrated dispersal' which is thought to exist whenever dispersal is inadequate relative to its motivation. Space which provides an outlet for wandering impulses is termed a 'dispersal sink'. If inadequate sinks exist or if physical barriers prevent access to them, dispersal becomes frustrated. Such circumstances can have interesting demographic consequences, and where pre-saturation emigration occurs, can even lead to equilibrium numbers which are below current carrying capacity. The study of populations exhibiting frustrated dispersal is recommended as one means of understanding the normal demographic role of dispersal.

Dispersal is thought to act at three levels of involvement in the regulation of densities: as a contributing factor, as a key factor, and in the establishment of equilibrium densities below carrying capacity. Only the last is considered controversial, and three models are proposed suggesting how emigration may act in this manner. All require the presence of pre-saturation emigration, and only one involves a situation in which the separation between numbers and resources is both chronic and evolutionarily stable.

The initial impetus for the preparation of the theoretical portion of this paper came from Drs G. N. Cameron and M. H. Smith when they invited me to participate in a symposium on population regulation for the 1973 annual meeting of the American Society of Mammalogists. I am grateful for their confidence and contagious enthusiasm.

The further development of this paper, and my participation in this IV Symposium of the Working Group on Small Mammals, was only possible by the encouragement and assistance of Drs K. Petrusewicz and K. Andrzejewska. I am extremely appreciative of their efforts on my behalf. Dr Petrusewicz further stimulated my thinking on dispersal and provided many helpful suggestions in the course of several intensive discussions.

I am particularly grateful to my colleagues Drs O. P. Pearson and F. A. Pitelka for carefully reading an early draft of this paper and making many helpful suggestions. Mr R. Glenn Ford gave me the idea for Fig. 5.2, and Mr S. F. Smith not only critically read the manuscript but also provided some help with the literature. Warm thanks are due to many of the participants in the Symposium at Dziekanów, and especially to Dr G. O. Batzli, for stimulating discussion of this chapter.

ERRATA

Page 105, lines 9 and 10 should read: *"pre-saturation dispersal.* Previously (Lidicker, 1962) I had called attention . . . ."

Page 113, line 1 should read: "this period, the species persists mainly in survival pockets or refugia . . . ."

Page 113, line 4 should read: ". . . into colonization of empty habitat surrounding the refugia."

Page 115, line 8 from the bottom should read: "Walkowa, 1963). An . . . ."

Page 116, line 1 should read: "Walkowa (1963) . . . ."

Page 125, line 26 should read: "herbaceous rather than woody, it recovers more. . . ."

LITERATURE CITED FOR CHAPTER 5

(These are not part of Chapter 5,
but are listed at the end of the volume.)

Ambrose, H. W., III, 1972. Effect of habitat familiarity and toe-clipping on rate of owl predation in Microtus pennsylvanicus. Jour. Mamm. 53(4): 909-912.

_____, 1973. An experimental study of some factors affecting the spatial and temporal activity of Microtus pennsylvanicus. Jour. Mamm. 54(1): 79-110.

Anderson, P. K., 1961. Density, social structure, and nonsocial environment in house-mouse populations and the implications for regulation of numbers. Trans. N. Y. Acad. Sci., Ser. II, 23(5): 447-451.

_____, 1970. Ecological structure and gene flow in small mammals. Symp. Zool. Soc. London, no. 26: 299-325.

Andrewartha, H. G. and L. C. Birch, 1954. The distribution and abundance of animals. Univ. Chicago Press, Chicago; 782 p.

Andrzejewski, R. and H. Wrocławek, 1961. Mortality of small rodents in traps as an indication of the diminished resistance of the migrating part of a population. Bull. Acad. Polonaise Sci., Cl. 2, 9(12): 491-492.

_____, 1962. Settling by small rodents a terrain in which catching out had been performed. Acta Theriol. 6: 257-274.

Andrzejewski, R., A. Kajak, and E. Pieczyńska, 1963. Efecty migracji. Ekol. Polska, Ser. B, 9(2): 161-172.

Andrzejewski, R., K. Petrusewicz, and W. Walkova, 1963. Absorption of newcomers by a population of white mice. Ekol. Polska, Ser. A, 11(7): 223-240.

Batzli, G. O. and F. A. Pitelka, 1971. Condition and diet of cycling populations of the California vole, Microtus californicus. Jour. Mamm. 52(1): 141-163.

Berry, R. J. and H. M. Murphy, 1970. The biochemical genetics of an island population of the house mouse. Proc. Roy. Soc. Lond., Ser. B, 176: 87-103.

Birdsell, J. B., 1957. Some population problems involving Pleistocene man. Population studies: animal ecology and demography, Cold Spg. Harbor Symp. Quant. Biol. 22: 47-69.

Bradt, G. W., 1938. A study of beaver colonies in Michigan. Jour. Mamm. 19(2): 139-162.

Bronson, F. H. and H. M. Marsden, 1964. Male-induced synchrony of estrus in deermice. Gen. and Comp. Endocr. 4: 634-637.

Brown, L. E., 1966. Home range and movements of small mammals. Symp. Zool. Soc. Lond., no. 18: 111-142.

Brown, R. Z., 1953. Social behavior, reproduction, and population change in the house mouse (Mus musculus). Ecol. Monog. 23: 217-240.

Bruce, H. M., 1959. An exteroceptive block to pregnancy in the mouse. Nature 184(4680): 105.

Bruce, H. M. and D. M. V. Parrott, 1960. Role of olfactory sense in pregnancy block by strange males. Science 131: 1526.

[*Editors' Note:* In Andrzejewski, Petrusewicz and Walkova reference, "Walkova" should be "Walkowa."]

Buckner, C. H., 1966. Populations and ecological relationships of shrews in tamarack bogs of southeastern Manitoba. Jour. Mamm. 47: 181-194.

Bujalska, G., 1970. Reproduction stabilizing elements in an island population of Clethrionomys glareolus (Schreber, 1780). Acta Theriol. 15(25): 381-412.

Bujalska, G. and J. Gliwicz, 1968. Productivity investigations of an island population of Clethrionomys glareolus (Schreber, 1780). III. Individual growth curve. Acta Theriol. 13(25): 427-433.

Calhoun, J. B., 1962. A "behavioral sink". In Roots of behavior, E. L. Bliss (ed.), p. 295-315; Harper and Bros., N. Y.

Carl, E. A., 1971. Population control in Arctic ground squirrels. Ecology, 52(3): 395-413.

Chipman, R. K. and K. A. Fox, 1966a. Oestrous synchronization and pregnancy blocking in wild house mice (Mus musculus). Jour. Reprod. Fertil. 12: 233-236.

_____, 1966b. Factors in pregnancy blocking: age and reproductive background of females: numbers of strange males. Jour. Reprod. Fertil. 12: 399-403.

Chipman, R. K., J. A. Holt, and K. A. Fox, 1966. Pregnancy failure in laboratory mice after multiple short-term exposure to strange males. Nature 210(5036): 653.

Chitty, H. and C. R. Austin, 1957. Environmental modification of oestrus in the vole. Nature 179(4559): 592-593.

Crawley, M. C., 1969. Movements and home-ranges of Clethrionomys glareolus Schreber and Apodemus sylvaticus L. in north-east England. Oikos 20: 310-319.

_____, 1970. Some population dynamics of the bank vole, Clethrionomys glareolus and the wood mouse, Apodemus sylvaticus in mixed woodland. Jour. Zool., Lond. 160: 71-89.

Crowcroft, P. and F. P. Rowe, 1957. The growth of confined colonies of the wild house-mouse (Mus musculus L.). Proc. Zool. Soc. Lond. 129: 359-370.

_____, 1958. The growth of confined colonies of the wild house-mouse (Mus musculus L.): the effect of dispersal on female fecundity. Proc. Zool. Soc. Lond. 131: 357-365.

Crowell, K. L., 1973. Experimental zoogeography: introductions of mice to small islands. Amer. Nat. 107(956): 535-558.

Davis, D. E. and J. J. Christian, 1956. Changes in Norway rat populations induced by introduction of rats. Jour. Wildlife Mgmt. 20: 378-383.

DeLong, K. T., 1967. Population ecology of feral house mice. Ecology 48(4): 611-634.

Dymond, J. R., 1947. Fluctuations in animal populations with special reference to those of Canada. Trans. Roy. Soc. Canada 41 (Ser. 3, sect. 5): 1-34.

Eleftheriou, B. E., F. H. Bronson, and M. X. Zarrow, 1962. Interaction of olfactory and other environmental stimuli on implantation in the deermouse. Science 137: 764.

Errington, P. L., 1946. Predation and vertebrate populations. Quart. Rev. Biol. 21: 144-177, 221-245.

Evans, F. C., 1942. Studies of a small mammal population in Bagley Wood, Berkshire. Jour. Anim. Ecol. 11: 182-197.

Evans, F. C. and R. Holdenried, 1943. A population study of the beechey ground squirrel in central California. Jour. Mamm. 24(2): 231-260.

French, N. R., T. Y. Tagami, and P. Hayden, 1968. Dispersal in a population of desert rodents. Jour. Mamm., 49(2): 272-280.

Fullagar, P. J., P. A. Jewell, R. M. Lockley, and I. W. Rowlands, 1963. The Skomer vole (Clethrionomys glareolus skomerensis) and long-tailed field mouse (Apodemus sylvaticus) on Skomer Island, Pembrokeshire in 1960. Proc. Zool. Soc. Lond. 140: 295-314.

Gadgil, M., 1971. Dispersal: population consequences and evolution. Ecology 52(2): 253-261.

Gentry, J. B., 1968. Dynamics of an enclosed population of pine mice, Microtus pinetorum. Res. Pop. Ecol. 10(1): 21-30.

Grant, P. R., 1971. The habitat preference of Microtus pennsylvanicus, and its relevance to the distribution of this species on islands. Jour. Mamm. 52(2): 351-361.

Hatt, R. T., 1930. The biology of the voles of New York. Roosevelt Wildlife Bull., 5(4): 505-623.

Healey, M. C., 1967. Aggression and self-regulation of population size in deermice. Ecology 48: 377-392.

Houlihan, R. T., 1963. The relationship of population density to endocrine and metabolic changes in the California vole Microtus californicus. Univ. Calif. Pub. Zool. 65(5): 327-362.

Howard, W. E., 1960. Innate and environmental dispersal of individual vertebrates. Amer. Midl. Nat. 63: 152-161.

_____, 1965. Interaction of behavior, ecology, and genetics of introduced mammals. In The genetics of colonizing species, H. G. Baker and G. L. Stebbins (eds.), p. 461-484; Academic Press, N. Y.

Incerti, G. and A. Pasquali, 1967. Esperimenti sulle attivita' esploratorie in topi domestici e selvatici (Mus musculus). Istituo Lombardo (Rend. Sc.) B, 101: 19-46.

Janion, S. M., 1961. Studies on the differentiation of a house mice population according to the occurrence of fleas (Aphaniptera). Bull. Pol. Acad. Sci., Cl. II, 9: 501-506.

Jewell, P. A., 1966. Breeding season and recruitment in some British mammals confined on small islands. In Comparative Biology of Reproduction in Mammals, Zool. Soc. Symp. no. 15 (I. W. Rowlands, ed.): 89-116.

Johnston, R. F., 1961. Population movements of birds. Condor 63: 386-389.

Kalela, O., 1961. Seasonal change of habitat in the Norwegian lemming, Lemmus lemmus (L.). Ann. Acad. Sci. Fenn. 4A(55): 1-72.

Kalela, O., L. Kilpeläinen, T. Koponen, and J. Tast, 1971. Seasonal differences in habitats of the Norwegian lemming, Lemmus lemmus (L.), in 1959 and 1960 at Kilpisjärvi, Finnish Lapland. Suomalainen Tied. Toim. (Ann. Acad. Sci. Fenn.), Ser. A, IV Biol., no. 178: 1-22.

Kikkawa, J., 1964. Movement, activity and distribution of the small rodents Clethrionomys glareolus and Apodemus sylvaticus in woodland. Jour. Anim. Ecol. 33: 259-299.

King, J. A., 1955. Social behavior, social organization, and population dynamics in a black-tailed prairie-dog town in the Black Hills of South Dakota. Contr. Lab. Vert. Biol., Univ. Michigan, no. 67: 1-123.

Kolodziej, A., I. Pomianowska, and E. Rajska, 1972. Differentiation of contacts between specimens in a Clethrionomys glareolus population. Bull. Acad. Polonaise Sci., Cl. 2, 20(2): 97-102.

Krebs, C. J., 1966. Demographic changes in fluctuating populations of Microtus californicus. Ecol. Monog. 36: 239-273.

Krebs, C. J., B. L. Keller, and R. H. Tamarin, 1969. Microtus population biology: demographic changes in fluctuating populations of M. ochrogaster and M. pennsylvanicus in southern Indiana. Ecology 50(4): 587-607.

Krebs, C. J., M. S. Gaines, B. L. Keller, J. H. Myers, and R. H. Tamarin, 1973. Population cycles in small rodents. Science 179: 35-41.

Lamond, D. R., 1959. Effect of stimulation derived from other animals of the same species on oestrous cycles in mice. Jour. Endocr. 18: 343-349.

Lidicker, W. Z., Jr., 1962. Emigration as a possible mechanism permitting the regulation of population density below carrying capacity. Amer. Nat. 96 (886): 29-33.

_____, 1965. Comparative study of density regulation in confined populations of four species of rodents. Res. Pop. Ecol. 7(2): 57-72.

_____, 1973. Regulation of numbers in an island population of the California vole, a problem in community dynamics. Ecol. Monog. 43(3): 271-302.

Lidicker, W. Z., Jr. and P. K. Anderson, 1962. Colonization of an island by Microtus californicus, analysed on the basis of runway transects. Jour. Anim. Ecol. 31: 503-517.

Louch, C. D., 1956. Adrenalcortical activity in relation to the density and dynamics of three confined populations of Microtus pennsylvanicus. Ecology 37: 701-713.

Lund, M., 1970. Diurnal activity and distribution of Arvicola terrestris terrestris L. in an outdoor enclosure. EPPO Public Ser. A, no. 58: 147-158.

Marsden, H. M. and F. H. Bronson, 1964. Estrous synchrony in mice: alteration by exposure to male urine. Science 144: 1469.

Mazurkiewicz, M., 1972. Density and weight structure of populations of the bank vole in open and enclosed areas. Acta Theriol. 17 (34): 455-465.

McCollum, F. C., 1975. Biochemical variation in populations of Microtus californicus. PhD. thesis Univ. Calif., Berkeley.

McPherson, A. B. and J. N. Krull, 1972. Island populations of small mammals and their affinities with vegetation type, island size and distance from mainland. Amer. Midl. Nat. 88(2): 384-392.

Metzgar, L. H., 1967. An experimental comparison of screech owl predation on resident and transient white-footed mice (Peromyscus leucopus). Jour. Mamm. 48(3): 387-391.

Myers, J. H. and C. J. Krebs, 1971. Genetic, behavioral, and reproductive attributes of dispersing field voles Microtus pennsylvanicus and Microtus ochrogaster. Ecol. Monog. 41: 53-78.

Naumov, N. P., 1940. The ecology of the hillock mouse, Mus musculus hortulanus. Jour. Inst. Evol. Morph. 3: 33-77 (in Russian).

Newsome, A. E., 1969. A population study of house-mice temporarily inhabiting a South Australian wheatfield. Jour. Anim. Ecol. 38: 341-359.

Newsome, A. E. and L. K. Corbett, 1975. Outbreaks of rodents in semi-arid to arid Australia: causes, preventions, and evolutionary considerations. In Rodents in desert environments, I. Prakash and P. K. Ghosh, eds., p. 117-153; Monographicae Biologicae, Junk Pub.

Parsons, P. A., 1963. Migration as a factor in natural selection. Genetica 33(3): 184-206.

Pearson, O. P., 1963. History of two local outbreaks of feral house mice. Ecology 44(3): 540-549.

Pearson, O., N. Binsztein, L. Boiry, C. Busch, M. Di Pace, G. Gallopin, P. Penchaszadeh, and M. Piantanida, 1968. Estructura social, distribucion espacial y composicion por edades de una poblacion de tuco-tucos (Ctenomys talarum). Inv. Zool. Chilenas 13: 47-80.

Petrusewicz, K., 1957. Investigation of experimentally induced population growth. Ekol. Polska, Ser. A, 5: 281-301.

_____, 1963. Population growth induced by disturbance in the ecological structure of the population. Ekol. Polska, Ser. A, 11: 87-125.

_____, 1966. Dynamics, organization and ecological structure of population. Ekol. Polska, Ser. A, 14(25): 413-436.

Petrusewicz, K., G. Bujalska, R. Andrzejewski, and J. Gliwicz, 1971. Productivity processes in an island population of Clethrionomys glareolus. Ann. Zool. Fenn. 8: 127-132.

Pielowski, Z., 1962. Untersuchungen über die ökologie der Kreuzotter (Vipera berus L.). Zool. Jb. Syst. 89: 479-500.

Pitelka, F. A., 1959. Numbers, breeding schedule, and territoriality in pectoral sandpipers of northern Alaska. Condor 61: 233-264.

Pucek, Z. and J. Olszewski, 1971. Results of extended removal catches of rodents. Ann. Zool. Fenn. 8: 37-44.

Ramsey, P. R. and L. A. Briese, 1971. Effects of immigrants on the spatial structure of a small mammal community. Acta Theriol. 16(13): 191-202.

Richmond, M. and C. H. Conaway, 1969. Induced ovulation and oestrus in Microtus ochrogaster. Jour. Reprod. Fert., Suppl. 6: 357-376.

Ryszkowski, L., 1966. The space organization of nutria (Myocaster coypus) populations. Symp. Zool. Soc. London, no. 18: 259-265.

Shvarts, S. S., A. V. Pokrovski, V. G. Istchenko, V. G. Olenjev, N. A. Ovtschinnikova, and O. A. Pjastolova, 1964. Biological peculiarities of seasonal generations of rodents, with special reference to the problem of senescence in mammals. Acta Theriol. 8: 11-43.

[*Editors' Note:* In Newsome and Corbett reference, the article title should read: "Outbreaks of rodents in semi-arid and arid . . . ."]

Semeonoff, R. and F. W. Robertson, 1968. A biochemical and ecological study of plasma esterase polymorphism in natural populations of the field vole, Microtus agrestis L. Biochem. Gen. 1: 205-227.

Sheppe, W., 1965. Island populations and gene flow in the deer mouse, Peromyscus leucopus. Evolution 19(4): 480-495.

_____, 1972. The annual cycle of small mammal populations on a Zambian floodplain. Jour. Mamm. 53(3): 445-460.

Smith, M. H., J. L. Carmon, and J. B. Gentry, 1972. Pelage color polymorphism in Peromyscus polionotus. Jour. Mamm. 53(4): 824-833.

Smythe, M., 1968. The effects of the removal of individuals from a population of bank voles Clethrionomys glareolus. Jour. Anim. Ecol. 37(1): 167-183.

Southwick, C. H., 1955a. The population dynamics of confined house mice supplied with unlimited food. Ecology 36(2): 212-225.

_____, 1955b. Regulatory mechanisms of house mouse populations: social behavior affecting litter survival. Ecology 36(4): 627-634.

Stoddart, D. M., 1970. Individual range, dispersion and dispersal in a population of water voles (Arvicola terrestris (L.)). Jour. Anim. Ecol. 39: 403-425.

Strecker, R. L., 1954. Regulatory mechanisms in house-mouse populations: the effect of limited food supply on an unconfined population. Ecology 35(2): 249-253.

Tamarin, R. H. and C. J. Krebs, 1969. Microtus population biology. II. Genetic changes at the transferrin locus in fluctuating populations of two vole species. Evolution 23(2): 183-211.

Tast, J., 1966. The root vole, Microtus oeconomus (Pallas), as an inhabitant of seasonally flooded land. Ann. Zool. Fenn. 3: 127-171.

Taylor, D., 1970. Growth, decline, and equilibrium in a beaver population at Sagehen Creek, California. Ph.D. thesis, Univ. California, Berkeley; 162 p.

van der Lee, S. and L. M. Boot, 1955. Spontaneous pseudo-pregnancy in mice. Acta Physiol. Pharmac. Neerl. 4(3): 442-444.

Van Valen, L., 1971. Group selection and the evolution of dispersal. Evolution 25(4): 591-598.

Van Vleck, D. B., 1968. Movements of Microtus pennsylvanicus in relation to depopulated areas. Jour. Mamm. 49(1): 92-103.

Webb, W. L., 1965. Small mammal populations on islands. Ecology 46: 479-488.

Whitten, W. K., 1966. Pheromones and mammalian reproduction. Adv. Reprod. Physiol. 1: 155-177.

Whitten, W. K., F. H. Bronson, and J. A. Greenstein, 1968. Estrus-inducing pheromone of male mice: transport by movement of air. Science 161: 584-585.

Wiegert, R. G. and D. F. Owen, 1971. Trophic structure, available resources and population density in terrestrial vs. aquatic ecosystems. Jour. Theoret. Biol., 30: 69-81.

Wynne-Edwards, V. C., 1962. Animal dispersion in relation to social behaviour. Hafner Publ. Co., N. Y.; 653 p.

Dispersal in relation to carrying capacity

P. R. Grant*

ABSTRACT Dispersal of the herbivorous vole *Microtus pennsylvanicus* from grassland to woodland was studied in an experimental field system during spring to autumn 1969. Dispersal first occurred when there was at least 100 times more energy available than was required by the population. Sodium and phosphorus were in short supply in the food. By feeding selectively or copiously, voles could make up nutrient deficits and still consume only 10% of what was available. However, calculations show that depletion of the food was potentially severe in the forthcoming winter; consumption of energy- and nutrient-sufficient food had the potential of approaching 100%. These results suggest the following explanation of presaturation dispersal. Nutrients may be more limiting to herbivores than is total energy. Selective or copious harvesting becomes increasingly necessary as density increases. Natural selection, acting upon known genetic variation in dispersal propensity, has favored a dispersal response to environmental conditions that presage food shortage. Aggressive behavior and other forms of interactive behavior are the means by which land-tenured breeding individuals control their access to food resources and by which nontenured individuals are excluded and induced to disperse. Herbivore populations in general are limited well below carrying capacity. Carrying capacity may have been overestimated by ignoring chemical quality of the food, but the relationship is still probably true. The above explanation helps us to understand it. Because animals need to feed selectively, exploitation is based on cost-benefit balances and is less than total.

As ecologists grapple with increasing degrees of complexity in real and model systems, the importance of dispersal is becoming apparent. Dispersal is instrumental in animal population regulation as a means of lowering density (1, 2). It is crucial to the persistence of a species in heterogeneous and varying environments (3). Finally, it has an evolutionary significance in being the means by which new environments are entered (4), and therefore it is the initial condition for allopatric speciation.

For animals seeking specific, spatially restricted, resources (e.g. seeds, nest sites), dispersal first occurs when all resources are fully exploited or nearly so (e.g., see ref. 5). In contrast, for animals seeking more evenly dispersed resources, dispersal may occur well before the carrying capacity is reached; Lidicker (6) refers to this as presaturation emigration. Presaturation dispersal has been a puzzle that has not been adequately explained. Lidicker (7) suggested an explanation in terms of natural selection. There are postulated advantages to dispersing as opposed to remaining. These are genetic improvement (outbreeding) and enhanced survival. However, it is not clear why dispersal first occurs so far ahead of the point at which carrying capacity is reached. For example, dispersal is a frequent and important process in populations of voles in the genus *Microtus* (2, 8), yet these herbivorous rodents usually consume less than 3% of available food (9, 10, 11).

Hitherto the relationship between the initiation of dispersal from a breeding population and available food in relation to consumption has not been estimated in a single study. In this paper I present such an estimate with data from a field study of the grassland-dwelling microtine rodent *Microtus pennsylvanicus pennsylvanicus*. I then offer an interpretation of dispersal in terms of an avoidance of deteriorating foraging conditions. It is similar to Lidicker's hypothesis, but gives nutrient limitation an explicit role. Inasmuch as the explanation is not dependent upon genus-specific characteristics of *Microtus* biology, it should help in the understanding of dispersal in other taxa. The interpretation is then extended to questions of herbivore population limitation.

METHODS AND RESULTS

Terms. *Dispersal* will be used in the restricted sense of movement from one habitat to another, regardless of whether the animal returns or not. I shall use the term *carrying capacity* to mean the maximum biomass of an exploiter population that a specified environment can support at a particular time without suffering long-term damage as connoted by the phrase *over-exploitation* (e.g., see ref. 12). I use the term with regard to biomass because maximum supportable biomass is closer to a fixed quantity than is maximum supportable numbers.

Dispersal. A 0.4-hectare (ha; 1 ha = 10^4 m^2) enclosure containing equal and adjacent blocks of grassland and deciduous woodland was used in southern Quebec, Canada, in 1969 (13). Resident animals were trapped and removed. Sixteen *Microtus pennsylvanicus* were introduced to the grassland in April. The population increased through reproduction. Dispersal into the woodland first occurred in late August or early September, when the density in the grassland reached 170–200/ha. Thereafter, and until the end of the experiment in mid-October, the frequency of occurrence of animals in the woodland paralleled variations in the population size (63 animals at maximum). Subadults were proportionately more frequent among the dispersers than were adults, and males predominated.

Grass. It is difficult to simultaneously monitor properties of the population and properties of the environment without affecting the former. During the course of the experiments all that was known quantitatively was the percent cover of the various plant species in the enclosures (14). Cover was uniform and abundant; grass stems were as much as 2 m tall (15) and there were no signs of overgrazing, in contrast to longer experiments with confined populations of *Microtus* (16). There were plenty of cut grass stems and leaves in the runways. Alfalfa (*Medicago sativa*) was present, and elsewhere may be a preferred food (17), but there were scarcely any signs of its being eaten (see also ref. 18). Therefore grass is assumed to have been the sole food.

To find out how much food was present at different points

---

* Address after Sept. 1, 1977: Department of Ecology and Evolutionary Biology, University of Michigan, Ann Arbor, MI 48109.

Table 1. Energy available in enclosure and consumed during the 1969 experiment

Date	Dry wt of grass in 1973, g/m²*	Estimated total dry wt of grass in 1969, kg†	% new growth	kcal/g‡	Total kcal	Digestible kcal available (i.e., supply)§	Estimated consumption, kcal digestible grass/wk, in 1969 (i.e., demand)¶
April	25.98 ± 2.03	45.8	100	4.40	202,000	102,000	1060
May	264.25 ± 24.39	469	100	4.40	2,060,000	1,030,000	730
June	530.87 ± 25.61	943	80	4.35	4,010,000	1,860,000	770
July	752.53 ± 61.73	1320	45	4.26	5,660,000	2,260,000	1180
August	650.13 ± 65.81	1140	30	4.23	4,850,000	1,940,000	2360
September	821.35 ± 41.64	1450	25	4.21	6,100,000	2,440,000	4350
October	590.63 ± 112.89	1040	20	4.20	4.380,000	1,750,000	5190
April 20 (dead grass)	704.00 ± 41.44	1240	0	4.15	5,150,000	1,540,000	—

* Five random samples were taken every two weeks, Apr. 17 to Oct. 26, 1973 (dead grass in 1974). Results are expressed as monthly averages ± SEM. $n = 5$ for April, October; $n = 10$ for June–August; $n = 15$ for May.
† 1973 dry weights are multiplied by 0.84 to correct for percent coverage difference between years, then multiplied by 2.10×10^3 (the number of m² in the enclosure).
‡ Determined by bomb calorimetry. New (green) growth has an average of 4.40 kcal/g, old stems have an average of 4.15 kcal/g. Total sample size is 80; no seasonal changes in energy content within these two categories were detected.
§ Digestibility for *Microtus* is less than 100% and varies seasonally with the fiber and lignin content (19–21). Calculations here are based on a digestibility of 50% for new growth up to the end of May, 40% for the remainder of the experiment, and 30% for dead material.
¶ Weight-specific, daily, consumption rates of animals of known or estimated weights and reproductive condition are summed and converted to weekly values. Calculations are based upon estimates of energy requirements of *Microtus* species in ref. 22 and allow for the higher requirements of pregnant and lactating females.

within the experimental period, a program of destructive sampling of the vegetation and chemical analysis was carried out in 1973 and 1974 over the experimental period April–October. Results of the analyses can then be compared with estimates of animal requirements to see how the pattern of dispersal was related to the proportional exploitation of the food (Tables 1 and 2). This procedure assumes that the properties of the grassland studied in 1973 and 1974 faithfully reflect the unstudied properties in 1969 when the experiment was performed. Some confidence in this assumption is given by the results of an analysis of variance, which showed that chemical composition was much more pronounced among months than between years for each nutrient studied (N, Ca, Mg, K, P, Na, carbohydrates, and lipids), as well as for dry weight and water content. Furthermore, the vegetation in 1969 and 1974 was at the same stage of development, i.e., four years after plowing and seeding, and species composition was the same, although proportions differed. It is not known if the voles had a differential effect upon the vegetation in 1969 (high density) and 1973–74 (low density).

Food Available and Consumed during the Experiment. Total supply of energy exceeded demand throughout the 25-week experiment (Table 1). At no time did the weekly requirement of the population exceed 1% of the standing crop of grass, and it was usually 0.1%. More specifically, at the time of dispersal (August–October), supply exceeded demand by at least two orders of magnitude. Therefore the energy demand upon the environment appears to have been minimal. Total consumption of digestible dry matter throughout the entire experiment is estimated to be equivalent to 5.0×10^4 kcal. This is about 2% of that available in the maximum standing crop of grass. The figure of 2% agrees with estimates from other studies of *Microtus* cited earlier.

Estimates of average daily nutrient requirements and intake are given in Table 2. These show the amount of nutrients consumed by the animals in meeting their energy requirements.

Table 2. Nutrients and water required and consumed by the population of voles while meeting minimal energy needs

Date	Daily consumption of grass, g dry wt	H₂O, g/day		Nutrients, mg/day										
				Mg		Ca		K		Na		P		N	
		Req.	Cons.	Req.	Cons.	Req.	Cons.	Req.	Cons.	Req.	Cons.	Req.	Cons.	Req.	Cons.
April	76.8	113	190	22	99.8	324	414.7	216	3738.5	180	384.0	324	299.5	1280	1550.2
May	48.2	63	144	12	51.8	181	190.4	120	1415.5	74	43.2	181	133.3	714	706.8
June	62.2	56	141	11	63.8	162	247.3	109	1377.1	90	46.5	162	113.2	640	498.4
July	92.4	79	126	16	92.4	233	412.5	156	1084.6	129	36.6	233	123.5	900	619.6
August	195	177	253	35	169.1	532	714.2	354	1668.8	295	66.9	532	226.7	2013	1148.7
September	362	311	446	64	217.0	933	1228.4	622	3235.0	518	114.7	933	480.4	3536	2121.0
October	446	400	455	77	312.4	1160	1472.9	773	3976.1	644	105.3	1160	493.0	4544	2508.3
Winter	590	395	560	76	295.0	1134	1711.0	756	2006.0	630	118.0	1134	354.0	4480	3082.7

Req., required; Cons., consumed. Daily nutrient requirements of all but nitrogen have been taken from ref. 23 as listed in ref. 24. Because these values are for young males, and adults have lower requirements per gram of body weight, an adjustment factor of 0.9 has been applied to animals of 20 g or more. Nitrogen requirements have been taken from ref. 25. Water requirements have been taken from ref. 26. Consumptions are based on data in Table 1. Digestibility is known for nitrogen only (55%; ref. 19). The figures for this energy element listed under Cons. are digestible amounts consumed.

The assumption is now made that all nutrients except N can be completely digested; N has a digestibility marginally greater than total energy (19). This assumption will be reconsidered later. On this basis Mg and K requirements were met by the daily intake of a random selection of energy-sufficient food at all times. Ca was in short supply during mid-May only. However, from May to the end of the experiment Na was insufficient by a factor of between 2 and 7, and N was insufficient by a factor of up to 2. P was insufficient at all times by a factor of $1/2$–3. Initiation of dispersal did not coincide with a sharp increase in nutrient insufficiency in the grass. Water supply was sufficient at all times.

Na insufficiency could possibly be made up by eating soil (27), but soil is a poor source of P. Alternative ways of making up the deficits are eating more of the same (20), eating selectively either within or between grass species or on nutrient-rich parts of individual grasses (28), or eating nongrass species rich in these nutrients. Because, apparently, the animals did not resort to the third alternative, they must have adopted one or both of the first two. The probable reason why they did not consume alfalfa to any appreciable extent is that it contained only 30% on average ($n = 5 \times 1$ m^2) of the amount of P in the grasses. House mice can select for Na (27), as can rabbits (29) and moose (30), and rabbits and hares are known to select for both N and P (31). So selective feeding by *Microtus* to make up the deficiencies of Na, N, and P is a plausible hypothesis.

Whatever means was adopted to overcome the nutrient deficiencies, the ratio of supply to demand of food (100:2) must be decreased to allow for the fact that an energy-sufficient sample of food taken randomly from the grasses did not have enough Na, N, and P to meet animal requirements from May onwards. The adjustment is made as follows. From the time of initial dispersal to the end of the experiment a maximum of 2.44×10^6 kcal was available, and approximately 4.0×10^4 kcal or 1.6% was consumed. Multiplying 1.6 by 5, the average factor by which Na was in short supply in the grasses during the period of dispersal, we get an effective consumption of 8%. If we further assume that P deficiencies were not met in meeting Na requirements, but had to be met separately, the effective consumption should be doubled to 16%. It should then be raised to around 20% to compensate for N insufficiency. This is an extreme maximum. It says that dispersal occurred when there was approximately 5 times more nutrient-sufficient food available than was required.

In fact, a realistic percent consumption is distinctly lower. Grass regrows following cropping (see below). Thus the amount maximally available to the animals should be elevated above the maximum standing crop. It is therefore probably safe to conclude that during the whole dispersal period there was at least 10 times more food available than was required.

Food Available and Consumed during the Remainder of the Year. Given that there was an abundance of food when dispersal took place, we now ask how demand would have related to supply if the experiment had been prolonged through the winter months.

To estimate energy and nutrient budgets for the remainder of a 12-month period, we have to make further assumptions. We assume no mortality, reproduction, or growth over winter, a minimum body weight of 20 g, and a winter population size constant at 60, i.e., close to the October maximum. Summing the weights of survivors in mid-October, and scaling young animals up to 20 g, gives a vole biomass of 1400 g. This is reasonable. In the spring of 1967, when reproduction had only just started, a neighboring enclosure grassland of the same size and grass composition had a vole biomass of 1390 g (14). A vole biomass of 1400 g with the mid-October age-size structure has a weekly energy requirement of approximately 5.0×10^3 kcal. We assume that temperature effects on winter energy demand (22) are cancelled by the customary decrease (32) in body sizes then. Finally, we assume constant nutrient concentration over the winter, because concentrations in standing grass in October and in dead grass the following April were almost identical.

It is important to know the period of dependence upon old growth. The period of complete dependence is the time from cessation of grass growth in late autumn to resumption in early spring. For a few weeks more, voles are partially dependent on old growth while consuming some of the new growth. Consumption of new growth by voles stimulates regrowth (33). The period of combined partial and complete dependence upon old growth spans the time between the autumn and spring dates at which regrowth balances the energy needs of the voles. Weekly, replicated ($n = 5$), clipping experiments in April and May show that regrowth is rapid at this time, and suggest that the energy needs of the population would be met by regrowth alone before April, i.e., within two weeks of the onset of spring growth, even if weekly needs rose to 7.0×10^3 kcal per week as a result of increasing reproductive energy costs. Apr. 1 may be taken as the spring balance point. Nutrient concentrations in regrowth are not known, but water content is the same as in original growth. More limited experiments in September to November suggest that the autumn balance point is about Nov. 7, i.e., one week before growth ceases altogether. Thus 17 weeks is the period of complete dependence upon old growth, and partial dependence extends this by about three weeks.

The population energy requirements for these periods are 0.85×10^5 kcal and 1.00×10^5 kcal, respectively. We take an arithmetic mean of 0.92×10^5 kcal and relate that to the 1.54–1.88×10^6 kcal available (Table 1) to obtain a potential depletion of 5–6%. It could be even less than this if roots were added to available food. These represent approximately 1.00×10^6 kcal. But being the "capital investment" that determines the "interest" in the next growing season, it is a bad strategy for the voles to deplete the capital in their home ranges to the detriment of the interest. Hence roots should be, and usually are, ignored.

As in summer and autumn, the grass does not have sufficient average concentrations of Na, N, and P (Table 2). At this time of low soil temperatures beneath the snow it is unlikely that much of the Na deficit can be made up by eating soil. In some years the soil freezes before snow accumulates.

Deficit factors are 5 for Na, 3 for P, and $1\frac{1}{2}$ for protein. Multiplying the energy consumption of 5–6% by 5 to correct for the largest (Na) imbalance gives an adjusted consumption of 25–30%. The consumption could approach 100% if the other nutrient deficits had to be made up separately. This may not be realistic. Nevertheless, the calculations are sufficient to show that effective consumption overwinter during the period of dependence upon old growth is at least 25% and could be much higher.

Augmentation of Potential Food Depletion. The above calculations show that depletion of food resources was potentially more severe overwinter than during the experiment. This is consistent with the views of Grodziński (34) and Hansson (28) that the end of the winter can be the period of greatest environment-induced stress on the voles. In arriving at this conclusion we have had to make several assumptions in the calculations; see Gentry *et al.* (35) for cautionary remarks about such calculations. Future experimental work may show that the assumptions are unrealistic and hence the calculations are in error. A major goal of future work should be the replacement of estimates with measurements. But there are reasons for believing that the calculations presented here are conservative. They did not take into account the following four factors whose effects are to elevate the effective demand upon the resources.

First, the maximum density of about 300/ha is not as high as that recorded for some unenclosed populations of this species elsewhere (36). Therefore the population may have had the potential in autumn for further increase. Under favorable climatic and food conditions voles increase in numbers during winter by breeding beneath the snow (14). Second, if aggressive conflicts increased in frequency and intensity as population density increased, individual metabolic requirements probably increased (e.g., see ref. 37), and time available for feeding and for digestion correspondingly decreased. Under such conditions efficiency in foraging would become increasingly important. Third, digestibility of the five macronutrients was unrealistically assumed to be 100%. In fact, digestibility is likely to be lower (24, 35), but how much lower is not known. Increase in nutrient requirements for breeding was not considered, nor was the possibility that micronutrients or trace elements were limiting. Finally, voles do not consume all of the plant material that they cut. This has never been satisfactorily explained. One possible explanation is that they sample stems and leaves, find some to be unsuitable, and abandon them. Whatever the reason, the amount of potential food damaged by vole activity of one sort or another and not consumed is surprisingly high. Ryszkowski *et al.* (11) estimated the damaged portion to be 1.8 times the amount consumed by a population of *Microtus arvalis* at high density.

None of these factors were taken into account in the calculations. It is difficult to attach numbers to them. In combination they strengthen the contention that depletion of food resources overwinter was potentially severe.

PRESATURATION DISPERSAL: AN EVOLUTIONARY INTERPRETATION

The necessity to feed selectively or copiously provides the basis for an evolutionary interpretation of presaturation dispersal. Instead of being surrounded by food of equal and high suitability as is generally believed, a vole has an array of foods of graded suitability to choose from. In seeking and harvesting the most suitable grasses it incurs a metabolic cost in proportion to distance moved. In addition, it also incurs a predation risk in proportion to time spent above ground. Its harvesting will be governed by the law of diminishing returns. In the presence of abundant suitable food its harvesting costs are low and its benefits, including reproduction, are high. Harvesting costs increase as the abundance of suitable food decreases, and there comes a point when high costs preclude further gain in benefits; thereafter benefits decline. This point is reached well before exploitation is total. When the population as a whole is at the point of balance between costs and benefits, and is just replacing itself, it is at carrying capacity.

Strategic considerations suggest why dispersal starts before this point is reached. The evolutionary goal of the individual is to maximize reproductive success. Choice of a place to breed is a crucially important contributor to this success. As a vole reaches maturity it has to make a decision; to stay in the vicinity of its birthplace to breed or to move out and seek breeding space elsewhere. It is genetically programmed under the influence of past selection regimes to make that decision on the basis of present conditions. But it lives in a changing and unpredictable environment in which present conditions are not a precise indicator of future conditions. There is uncertainty about how much food will be available and how many competitors will be seeking it. Some allowance has to be made for this uncertainty.

For a vole in an expanding population in autumn, conditions are likely to deteriorate until grass starts growing in the following spring, because grass is removed and not replaced in winter, and because food harvesting and processing costs (energy and time) increase as digestibility, energy content, and nutrient content all decrease. Spring conditions can be gauged imprecisely in the autumn by means of environment and population cues. Difficulties of finding suitable food provide one cue. Behavioral interaction with conspecifics, and the daily rate of increase in such interactions, provide another. Overt aggression, scent-marking, and vocalization are possible modes of interaction. These food and behavior cues interact, because aggressive behavior is likely to vary with the degree of crowding in relation to food and cover resources (38). Both sets of cues will vary spatially within the area occupied by the population, and in all likelihood so will the genetically determined (2, 6) dispersal responses to them, which is why dispersal starts with a trickle and not with a rush.

A future low probability of survival is only part of the reason why voles disperse. The fate of their genotypes is dependent upon their ability to breed, and upon the ability of their offspring to breed. Under deteriorating environmental conditions the probabilities of these two reproductive episodes occurring successfully in the home environment may be vanishingly small, and the selective advantage will shift from the residents to the dispersers.

Herbivore Limitation. The idea of nutrient limitation is not new (10, 39, 40). Several authors have suggested that population regulation is achieved by behavioral adjustment to food supply (39, 41). But others have noted inconsistencies in field data which have led them to doubt the importance of food supply. For example, Krebs (42) found dispersal to be frequent at peak densities of dense and sparse populations of the same species, even in the presence of experimentally provided excess food. It is therefore logical to take the view that food plays a minor role in determining population size and how it is regulated (e.g. 42, 43). Thus food has a debated role in determining herbivore population limits and regulation.

Watson and Moss (44) concluded a comprehensive review of vertebrate demography with the suggestion that populations may be limited in the presence of excess food by dominance and spacing behavior, but cautioned that "the measurements of food in any one study in the wild are usually inadequate to establish the point fully." This caveat contains the essence of the problem of understanding the importance of food.

I suggest that quantity and quality of the food, together with its pattern of distribution in time and space, set an upper limit to the size of herbivore populations whose members incur a metabolic cost in harvesting. Aggressive behavior is the mechanism of adjustment to food resources, with dispersal, mortality, and inhibited reproduction being the consequences at high density. Genetic feedback mechanisms (2, 45) are possible but inessential adornments to this scheme.

Supportive evidence for the suggestion that numbers adjust to resources is that different dispersal thresholds are observed in environments of different carrying capacities. In an enclosure in Saskatchewan, dispersal of *M. pennsylvanicus* from grassland to woodland first occurred at a density in the grassland of 26 animals per ha (46). The dispersal threshold in the Quebec experiment was 170–200/ha, about 7 times higher. Species composition of the *Stipa-Bouteloua*, prairie short grass, community in the Saskatchewan enclosure is known (15), but energy, water, and nutrient content is not. For an estimate of energy available (but not nutrients) we have to turn to another study. Redman's (47) figures for the annual, above-ground, net primary production of *Stipa-Bouteloua* grassland at seven sites in North Dakota lie between 77 and 200 g/m². The maximum standing crop in the Quebec experiment was 820 g/m^2, about 7 times higher. Therefore both energy and dispersal thresholds differed by a factor of 7. This correspondence may just be a coincidence, and it might break down if nutrients were considered, but it suggests a similar adjustment of two populations to different resource levels. In the Quebec study area, maximum population in woodland with a sparse cover of grasses and sedges was much lower than in neighboring grassland (48).

The adjustment is far from perfect, owing to time delays in population response to a carrying capacity that varies seasonally and perhaps annually. Time delays make it difficult for the investigator to detect the adjustment by simple correlation techniques (49), particularly where environmental conditions vary irregularly (28). In addition, population increase may be checked before the food limit is reached by other factors, such as predation or shortage of burrows. Adjustment is likely to be better in optimal habitats where the social hierarchy is stable than in suboptimal habitats that are populated from the optimal habitat by emigrants that disrupt the developing social hierarchy there (50).

The above hypothesis suggests an explanation of why diets shift with increasing density (51). As preferred foods are consumed and become rarer, foraging attention must turn to the chemically less preferred (20, 51). Lemmings are exceptions to the general foliovore pattern. Like granivores, they may consume 90% of their food (10, 52). In this way they differ fundamentally from voles even though, like them, they are subject to nutrient limitation in time and space (10, 41). They consume a lot and assimilate relatively little, while voles consume less and assimilate relatively more (10). This suggests an inability of the lemmings to feed selectively. But perhaps in their more exposed situation on arctic tundra, where they are subject to heavy predation at times (53), they cannot afford the time above ground to forage selectively, so compensate for nutrient insufficiency by feeding copiously. Thus lemmings, voles, and mice have different feeding strategies, and these should have different demographic correlates.

I am grateful to G. Davidson, E. Donefer, D. Luk, D. Macleod, and P. Mineau for technical assistance, and to G. A. C. Bell, F. B. Golley, S. L. Iverson, C. J. Krebs, R. D. Montgomerie, K. Myers, A. R. E. Sinclair, and B. A. Wunder for comments on the manuscript. This research was supported by the National Research Council (Canada) Grant A2920.

1. Kalela, O. (1949) *Ann. Zool. Soc. Zool. Bot. Fenn. Vanamo* **13**, 1–90.
2. Krebs, C. J. & Myers, J. H. (1974) *Adv. Ecol. Res.* **8**, 267–399.
3. Gadgil, M. (1971) *Ecology* **52**, 253–261.
4. Christian, J. J. (1970) *Science* **168**, 84–90.
5. Tinkle, D. W., McGregor, D. & Dana, S. (1962) *Ecology* **43**, 223–229.
6. Lidicker, W. Z., Jr. (1975) in *Small Mammals: Their Productivity and Population Dynamics*, eds. Petrusewicz, K., Golley, F. B. & Ryszkowski, L. (Cambridge Univ. Press, New York), pp. 103–128.
7. Lidicker, W. Z., Jr. (1962) *Am. Nat.* **96**, 29–33.
8. Myers, J. H. & Krebs, C. J. (1971) *Ecol. Monogr.* **41**, 53–78.
9. French, N. R., Grant, W. E., Grodziński, W. & Swift, D. M. (1976) *Ecol. Monogr.* **46**, 201–220.
10. Batzli, G. O. (1975) in *Small Mammals: Their Productivity and Population Dynamics*, eds. Petrusewicz, K., Golley, F. B. & Ryszkowski, L. (Cambridge Univ. Press, New York), pp. 243–268.
11. Ryszkowski, L., Goszczyński, J. & Truszkowski, J. (1973) *Acta Theriol.* **18**, 125–165.
12. Pimentel, D. & Soans, A. B. (1971) in *Proceedings of the Advanced Study Institute on Dynamics of Numbers in Populations*, eds. den Boer, P. J. & Gradwell, G. R. (Wageningen, Holland), pp. 313–326.
13. Grant, P. R. (1971) *J. Mammal.* **52**, 351–361.
14. Grant, P. R. (1969) *Can. J. Zool.* **47**, 1059–1082.
15. Grant, P. R. & Morris, R. D. (1971) *Can. J. Zool.* **49**, 1043–1052.
16. Krebs, C. J., Keller, B. L. & Tamarin, R. H. (1969) *Ecology* **50**, 587–607.
17. Thompson, D. Q. (1965) *Am. Midl. Nat.* **74**, 76–86.
18. Zimmerman, E. G. (1965) *J. Mammal.* **46**, 605–612.
19. Johannigsmeier, A. G. (1966) Dissertation (Purdue University, Lafayette, IN).
20. Evans, D. M. (1973) *J. Anim. Ecol.* **42**, 1–18.
21. Keys, J. E., Jr. & van Soest, P. V. (1970) *J. Dairy Sci.* **53**, 1502–1508.
22. Grodziński, W. & Wunder, B. A. (1975) in *Small Mammals: Their Productivity and Population Dynamics*, eds. Petrusewicz, K., Golley, F. B. & Ryszkowski, L. (Cambridge Univ. Press, New York).
23. Spector, W. S. (1956) *Handbook of Biological Data* (Saunders, Philadelphia, PA), p. 196.
24. Davis, D. E. & Golley, F. B. (1963) *Principles in Mammalogy* (Rheinhold, New York).
25. Sadleir, R. M. F. S., Casperson, K. B. & Harling, J. (1973) *J. Reprod. Fert. Suppl.* **19**, 237–252.
26. Getz, L. L. (1963) *Ecology* **44**, 202–207.
27. Aumann, G. D. & Emlen, J. T. (1965) *Nature* **208**, 198–199.
28. Hansson, L. (1971) *Viltrevy* **8**, 267–378.
29. Myers, K. (1975) *Aust. Wildl. Res.* **2**, 135–146.
30. Botkin, D. B., Jordan, P. A., Dominski, A. S., Lowendorff, H. S. & Hutchinson, G. E. (1973) *Proc. Natl. Acad. Sci. USA* **70**, 2745–2748.
31. Miller, G. R. (1968) *J. Wildl. Manage.* **32**, 849–853.
32. Iverson, S. & Turner, B. N. (1974) *Ecology* **55**, 1030–1041.
33. Smirnov, V. S. & Tokmakova, S. G. (1972) in *Proceedings International Meeting on the Biological Productivity of Tundra, 6th* eds. Wielgolaski, F. E. & Rosswall, T. (Leningrad, U.S.S.R.) pp. 122–127.
34. Grodziński, W. (1963) in *Proceedings International Congress of Zoology, 16th* (National Academy of Sciences, Washington, DC), Vol. 1, p. 257.
35. Gentry, J. B., Briese, L. A., Kaufman, D. W., Smith, M. H. & Wiener, J. G. (1975) in *Small Mammals: Their Productivity and Population Dynamics*, Petrusewicz, K., Golley, F. B. & Ryszkowski, L. (Cambridge Univ. Press, New York), pp. 205–222.
36. Aumann, G. D. (1965) *J. Mammal.* **46**, 594–604.
37. Brown, R. Z. (1963) *Bull. Ecol. Soc. Am.* **44**, 129.
38. Warnock, J. E. (1965) *Ecology* **46**, 649–664.
39. Pitelka, F. A. (1964) in *Grazing in Terrestrial and Marine Communities*, ed. Crisp, D. J. (Blackwells, Oxford), pp. 55–56.
40. Sinclair, A. R. E. (1975) *J. Anim. Ecol.* **44**, 497–520.
41. Batzli, G. O. & Pitelka, F. A. (1971) *J. Mammal.* **52**, 141–163.
42. Krebs, C. J. (1966) *Ecol. Monogr.* **36**, 239–273.
43. Chitty, D. H., Pimentel, D. & Krebs, C. J. (1968) *J. Anim. Ecol.* **37**, 113–120.
44. Watson, A. & Moss, R. (1970) in *Animal Populations in Relation to Their Food Resources*, ed. Watson, A. (Blackwell, London), pp. 167–220.
45. Chitty, D. H. (1967) *Proc. Ecol. Soc. Aust.* **2**, 51–78.
46. Morris, R. D. & Grant, P. R. (1972) *J. Anim. Ecol.* **41**, 275–290.
47. Redman, R. E. (1975) *Ecol. Monogr.* **45**, 83–106.
48. Grant, P. R. (1975) *Can. J. Zool.* **53**, 1447–1465.
49. Armitage, K. B. (1975) *Oikos* **26**, 341–354.
50. Kock, L. L., Stoddart, D. M., & Kacher, H. (1969) *Z. Tierpsychol.* **26**, 609–622.
51. Ashby, K. R. (1967) *J. Zool.* **152**, 389–513.
52. Kalela, O. & Koponen, T. (1971) *Ann. Zool. Fenn.* **8**, 80–84.
53. Maher, W. J. (1970) *Wilson Bull.* **82**, 130–157.

PHYSIOLOGICAL FACTORS IN INSECT MIGRATION BY FLIGHT

By Dr. C. G. JOHNSON

MOST migratory flight in insects is prereproductive, occurring at a particular time in adult life. This suggests a physiological and experimental approach hitherto lacking in migration investigations[1].

The modern view of migration sees the locomotory drive as of first and of general importance and the mechanisms of orientation secondary and specialized; many migrants are displaced by wind, and well-documented instances where the major direction of prolonged displacement is controlled by the insect are few. It is the drive, therefore, which is the main concern here, and particularly in females for reasons discussed later on.

Migration in Females in Relation to Ovary Development

Most, if not all, flights generally recognized as migratory by their simultaneity, undistractedness and duration start at the breeding sites and usually begin either immediately after the teneral period, as with aphids, or relatively early in adult life[1]; the females nearly always have undeveloped or only partially developed ovaries—a fact known for many species, but not fully recognized as a generality. Examples are butterflies[2-5], moths[6-8], mosquitoes[9], syrphids[10], aphids[11], leafhoppers[12], aleyrodids[13], Heteroptera[14,15], coccinellids[16,17], scolytids[18-21], chrysomelids[22], thrips[23], locusts[24], dragonflies[25], ants and termites[26]. Exceptions are insects probably at the very end of migration or species, discussed later, that make prolonged flights in between bouts of oviposition.

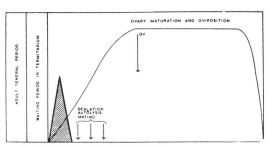

Fig. 1. Ants and termites. Migration flight shaded: *ov.*, oviposition.

Primary migration from the breeding site. Five examples are given: others can be obtained from the insects already listed.

The flight of termites has the behavioural and ecological characters of migration. Though short, it is not distracted by mating, feeding or oviposition, is made simultaneously by sexually immature females on the maiden flight and is followed by settling, mating, wing-muscle autolysis and ovary development. It is evidently adapted to transfer part of the population to another breeding site (Fig. 1*).

The almost irrepressible exodus flight of alate, parthenogenetic, female alienicolæ of *Aphis fabae* L. occurs, as with termites, on the maiden flight, but aphids contain partially developed embryos. After flight, which can

* All six figures in this article are diagrammatic and do not imply scales or quantities accurately.

range from a few seconds to many hours, aphids settle on a suitable host, feed, complete the development of the embryos and deposit larvæ. They may fly several times during the next three or four days before the flight muscles autolyse and flying ends. They spend the rest of their lives reproducing (Fig. 2)[11].

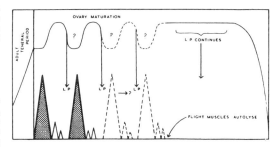

Fig. 2. Migratory aphids. Migratory flights may be repeated a variable number of times (dotted peaks). Small peaks represent more local flights. The course of ovary maturation between flights is uncertain. *LP*, larviposition

The butterfly *Ascia monuste* L. (Pieridae) in Florida (Fig. 3) migrates when females are sexually immature; migratory flights end on the first or second day of adult life. Thereafter, maturation and relatively short flights between feeding and oviposition sites are evident[2]. Many other butterflies begin to migrate before ovaries are fully developed; none is known to start otherwise.

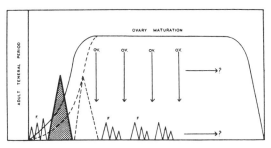

Fig. 3. *Ascia monuste* (Lepitopt: Pieridae). Normally with one migration flight, but with delayed ovary development an additional one occurs (dotted lines). *F*, flights at feeding sites; *ov.*, oviposition

Gregarious locusts begin to migrate while sexually immature and ovaries ripen during migration. Considerable migrations occur after initial oviposition, though how long migration and oviposition continue to alternate is controversial (Fig. 4). Three or four batches of eggs seem to be laid at intervals of a few days; but it is uncertain whether migration ends before all the eggs are laid[27-30].

Adults of some dragonflies[25] and chafers[31,32] leave their birth-place while sexually immature and fly straight to a feeding site a mile or two away, where the ovaries develop. They return later in the season, often to the original

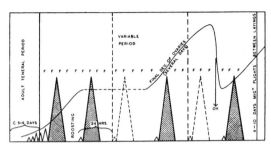

Fig. 4. *Schistocerca gregaria*. Migration flights occur often, during several weeks. Each is preceded by local 'milling and surging' flights (small peaks). F, feeding; ov., oviposition. The number of successive ovipositions is variable and migration may occur in between

habitat, to oviposit (Fig. 5). This type of migration resembles ecologically those from breeding sites to hibernation or æstivation sites and the return to oviposition sites next season, as with classical migrants like the Monarch butterfly[33]. Perhaps migrations interrupted by diapause evolved from the non-diapausing type by winter intercepting ovary development.

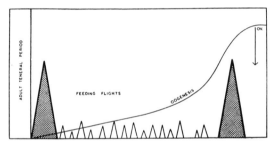

Fig. 5. Chafers, Dragonflies: initial migration from site of imaginal ecdysis (shaded), to a feeding site. Small peaks are flights within feeding site. Remigration to oviposition site (shaded); ov., oviposition

Migration in hibernating or æstivating insects. Females of all migrants that hibernate (or æstivate) make the primary migration to hibernation quarters with the ovaries undeveloped, and they hibernate in this state. When diapause ends, the re-migration flight appears to end with oviposition (Fig. 6). Thus, in hibernating and æstivating insects migration is interrupted and temporarily suppressed[6,14,17,22,23].

Physiological Factors in Migratory Flight

Neurophysiological processes cause the characteristic prolonged and undistracted migratory flight, but ecological factors determine when it occurs.

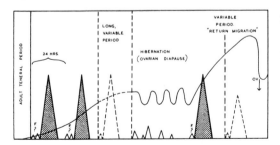

Fig. 6. Migrants with adult diapause: for example, *Danaus plexippus*. Migratory displacement flights shaded: F, feeding; ov., oviposition

Because migratory species, or genotypical migrants, migrate with immature ovaries, factors that control ovary development probably control migration; a plausible hypothesis can be made on these lines explaining why some populations migrate at some times and not at others. It is more difficult to explain how migratory and non-migratory individuals differ physiologically and neurophysiologically. However, some of the factors involved are considered in the next section, before the evocation of migratory behaviour in nature is discussed.

Factors affecting flight efficiency. Sexually immature migrant females often have more fuel in the fat body, lighter bodies and relatively longer wings (making a lower wing-loading) than non-migratory individuals of the same species. These attributes may increase duration and perhaps efficiency of flight; but they do not cause the characteristic persistence and undistractedness of migratory flight, which is not only a matter of increased efficiency or lack of need to oviposit. The thresholds of response to stimuli that promote take-off and continuous flight appear to be lowered, or alternatively thresholds for settling are raised.

Interrelation of flight and settling. The antithesis of migration is a stable and prolonged rest: the physiological reasons for one illuminate those of the other. Flight and settling inevitably alternate in time, but they also have a functional relationship; for flight or any intense muscular activity is a physiological necessity for settling, feeding, mating or oviposition, at least with some, perhaps with all species.

Thus new, unflown alatæ of migratory aphids will not stay on the plant on which they were born and will not settle or feed until they have flown[11,35]. The Scolytid wood-boring beetle *Dendroctonus* behaves similarly[21]. Some mosquitoes[9,36] and *Hypoderma* (Diptera)[37] mate only after a short flight, and oviposition by noctuids[38] and muscids[39] alternates with bouts of flying; oviposition by the wax moth *Achroia* is preceded by intense walking[40].

The duration of settling and flight are related quantitatively. The longer an aphid flies the longer it later remains settled[11]; stability of the settling response in *Automeris* (Saturniidae) (as measured by an accompanying rocking motion) and the waggle dance of honey bees after settling are directly related to the length of the previous flight[41,42]. Incomplete settling induced by a succession of unsuitable hosts is reported to prolong flights of aphids[35].

This functional interplay is thought to be caused by reciprocating reflex systems which alternately release and suppress each other[35], and prolonged function of settling reflexes gradually lowers the flight thresholds so that flight replaces settling and vice versa; this is apparently a matter of reflex relationships, not of fatigue. With an insect in a migratory state flight exceeds settling; thus Lepidoptera feed hurriedly during migration flight[43], and settled immature locusts are said to join in a passing swarm more readily than those about to oviposit[28,29]; Weis-Fogh noticed that migratory locusts near to oviposition were reluctant to fly in the laboratory[44].

Thresholds of response to various general stimuli (such as sounds, movements, touch, wind, light) which initiate take-off and to others such as feed-back from the wind-sensitive organ on the head, needed to maintain continuous flight, appear to be abnormally low in migrants and evidently vary with the sexual state of females. This is perhaps why exceptionally vigorous individual migrants of *Ascia*[2] and *Aleyrodes*[13] produced in very crowded populations continue to fly at night at light-intensities that would normally inhibit flight.

That neural response thresholds are affected by endocrine balance was shown by Haskell and Moorhouse[45]. Locomotor activity in locust hoppers reaches its peak some days after moulting; the subsequent decline as the next moult approaches coincides with increased titre of the moulting hormone, ecdyson, from the prothoracic glands (or with some associated chemical process). That thresholds are so changed is shown by the lack of response by isolated hoppers to stimuli that cause orientated locomotory reactions in hoppers bred in crowds; and because injection of blood from crowded into isolated hoppers lowers their response thresholds so that they respond to stimuli which before were ineffective, and vice versa. Haskell and Moorhouse showed a still more intimate effect. When the metathoracic ganglion of mature adult male *Schistocerca* is bathed with ecdyson or with blood from moulting hoppers electrical activity in the motor neurones supplying the extensor tibialis muscles of the hind legs decreases. Thus the drive of migratory flight may well be caused by a lowering of the neural response thresholds by changes in endocrine balance acting selectively on the particular reflex systems involved in wing-beat, as it does in hind-leg movements of hoppers.

Migratory flight occurs frequently, though not always, immediately after metamorphosis; presumably the ecdyson titre is then decreasing, but it is not minimal and hence other associated factors may be involved, as Haskell and Moorhouse suggest for other reasons. Clearly, the analysis of flight response thresholds in relation to the sexual cycle and the endocrine system controlling it will be of great interest.

Phototaxis and migratory flight. Many migrants such as aphids, scolytids and frit flies[46] fly upwards at exodus with a strong, positive phototaxis, apparently an adaptation ensuring effective wind-borne travel. Eventually, however, the sign of phototaxis is reversed or the insects respond to the green-yellow of the earth rather than ultra-violet light of the sky[47]. So long as insects are high in the air this reversal may not end flight, for aphids in such a state will fly to exhaustion when free from visual distractions[47]. Descent on downward currents gives opportunities for settling, and, the insects being correspondingly receptive, their flight ends.

The change of phototaxis and the development of settling responses can be interpreted as a successive induction of reflex systems. However, the endocrine system is also involved for when the corpus allatum and corpus cardiacum are removed from the adult Colorado beetle a negative phototaxis and burrowing (which in nature precedes oviposition) is induced[48]. When prothoracic glands from *Locusta* (the moulting hormone, ecdyson) are implanted in photo-positive larvæ of the Eyed-Hawkmoth these become photonegative; neotinine, the juvenile hormone from the abdomen of *Cecropia*, causes a reversal back to a positive sign[49]. However, in migrants a negative phototaxis often develops long before ovaries are fully mature, when corpus allatum activity is presumably still small, as in scolytids after flight or when many insects hibernate with sexually immature ovaries. Phototactic responses also change in *Trypodendron* (Scolytidae) from positive to negative as air is swallowed during flight and pressure in the proventriculus increases. Artificial release of this air provokes positive phototactic flight once more[21]. Pressure on the abdominal ganglia in *Choristoneura* larvæ has a similar effect[50].

The end of migration often coincides with oviposition, and oviposition responses can be induced by hormones. Neurosecretory cells, or blood from ovipositing females of the heteropteron *Iphita* introduced into non-ovipositing females, induces them to make oviposition responses; blood from fertilized *Bombyx* females induces oviposition in non-fertilized females; the neurosecretory cells in ovipositing phasmids secrete actively[51].

Thus, while the short-term oscillations of flight and settling within the migratory period may depend on successive inductions of reflexes, the balance in favour of one response compared with another appears to be associated with endocrine balance.

Differences between migrant and non-migrant species. The internal factors that control migratory activity are partly humoral, partly neurophysiological, partly use of fuel, perhaps partly a lessened wing-load. All flying species possess some of these elements yet not all migrate. Thus the saturniid, *Cerodirphia speciosa*, has such a high threshold for settling that it never completely settles after flight and even slight stimuli cause it to fly again[52]. The insect is extremely restless, but not a migrant. Locomotory activity at the expense of settling goes a stage further without becoming migration when insects like chironomids and *Simulium* swarm for considerable periods over a stationary object.

Migrants in contrast possess a factor leading to displacement, such as a positive phototaxis which leads to wind-borne displacement, or an orientation which leads to a 'straightening out'[70] of flight, and they pass it from generation to generation; they are genotypical migrants. Nevertheless, individuals with the equipment to migrate often fail to do so. The evocation of this latent ability now needs considering.

Evocation of Migratory Flight in Nature

Variation in the pre-oviposition period and its effect on migration. The length of the pre-oviposition period affects the duration of migratory activity, for it can be so brief as to obliterate it or prolonged to increase it. Information on the relation of migration to the speed of ovary development is scanty, but it suggests a quantitative relationship.

Observations with *Ascia monuste* indicate that lengthening the pre-oviposition period prolongs, and shortening it suppresses migratory flight. When the insects emerge too late in the day they are not flight-mature when it is time to migrate next day. The day afterwards, they are sexually mature and cannot migrate[2]. The migration flights of hundreds of miles to overwintering sites by *D. plexippus* is possible only because of prolonged ovarian immaturity, for individuals that mature quickly oviposit, die *en route* and never reach the hibernation sites[33].

Recent work by El Khidir[13] on *Aleyrodes brassicae* in England shows that during summer adults develop their ovaries quickly and fly only a few inches or feet to another and younger leaf before they settle and oviposit. In the autumn, ovaries develop slowly and sexually immature females accumulate as cold inhibits flight. On a warm day, however, they all fly off together and apparently unlike the summer adults fly for considerable distances *en masse* to overwintering sites.

The early generations of *D. plexippus* apparently do not migrate long distances presumably because of a short pre-oviposition period, but later generations do; they therefore resemble whiteflies except for distances travelled.

Internal and external factors affecting the pre-oviposition period. The speed of ovary development is controlled by the corpus allatum and the associated endocrines. The activity of the corpus allatum is diminished by environmental stimuli associated with a short day, insufficient or deficient food, crowding or high temperature, especially before metamorphosis.

Decrease in length of day during larval development slows ovary development, prolongs the preoviposition-period and causes ovarian diapause in many insects[53-57].

Removal of the corpus allatum from the adult has the effects of a short day during development, and inhibits or delays egg formation.

Food affects oogenesis, and the selection of specific feeding sites by syrphids, dragonflies, chafers and many others confirms this. Food supply is affected by crowding and the effects of one may be confused with the other, as in the literature on alate production in aphids.

Starvation or a diet of water or sugar solution delays oogenesis in *Oncopeltus* (Heteroptera) and *Calliphora* (Diptera) and removal of the corpus allatum has the same effect[58,59].

A most striking example of food affecting migration through ovary development, however, occurs with the spruce budworm *Choristoneura fumiferana*: first instars in spring feed on old leaves and young staminate flowers of balsam fir and move to new leaves, producing gravid adults too heavy to fly. Large populations, however, destroy new leaves and caterpillars then feed on old ones, so becoming small sexually immature adults, capable of upward phototactic flight. Here crowding appears to act by changing the food[8].

Either crowding or diminished food or both together also affect the bean beetle, *Callosobruchus maculatus* (F.), which has two alate forms. A bean with one larva produces an adult with rapidly ripening ovaries, a large wing-loading, unable to fly. With two larvæ in each bean the adults are sexually immature, have a small wing loading and can fly[60-62].

Crowding alone during larval stages lengthens the pre-oviposition period of some adults (*Locusta migratoriodes* (R. and F.)[63], *Leucania unipunctata* Haworth[64] and *Laphygma exigua*[7], *Pieris brassicae* L.,[65] *A. monuste*) and shortens it in others (*Schistocerca gregaria*[66], *Nomadacris septempunctata* Serville[67], *Plusia gamma*[65] and *Barethra brassica*[68]).

Low temperature during the adult stage slows metabolism and ovary development, but might stop flight. Such a retardation is not necessarily accompanied by a diminished corpus allatum activity which seems to produce migratory flight. But a relatively high temperature in pre-adult development decreases corpus allatum activity and produces an adult of abnormal sexual immaturity[56]. A relatively high temperature even after metamorphosis might then permit migration before the ovaries fully mature.

Displacement during the oviposition cycle. Though migration is commonly pre-reproductive, some species (aphids, locusts, muscids) make considerable flights between bouts of oviposition. Spruce budworm adults fly on convective storms after first ovipositing[34]. Some chafers make additional migratory flights after ovipositing back to feeding sites where oogenesis is resumed[32]. There seems to be some doubt whether *Aedes taeniorhyncus* resumes migration after ovipositing[9], but other mosquitoes fly scores of miles after the blood-meal that follows post-teneral flight but precedes oogenesis[69].

The volume of the corpus allatum, at least in *Calliphora*, waxes and wanes as eggs develop periodically during adult life; some chafers, locusts and many other insects also take several days to mature successive egg batches; evidently the physiological condition associated with endocrine activity necessary for long migration occurs during these inter-oviposition periods, as in the pre-reproductive stage after imaginal ecdysis. Migration can be interreproductive as Kennedy points out[70].

Environmental-endocrine system in relation to migration and habitat. Southwood[71] showed that migratory species live in temporary habitats and migration can be regarded as an adaptation to ensure that such places are relinquished before they disappear or become untenable. But there is no evidence that migration is a reaction to current adversity and when this kind of behaviour has been claimed ('alimental' or 'climatic' migration as when locusts leave a drying habitat)[72,73] the ontogenetic hypothesis developed here seems more plausible. The factors associated with a habitat doomed to decay are the same as those that delay ovary development, namely, shortening day, senescing food, crowding and high temperatures, particularly with the later summer and autumn generations in temperate climates. These then may be supposed to produce an individual with the necessary sexual immaturity to evoke and prolong migratory flight.

In tropical countries day-length is more constant and food may be more ubiquitous; it may then be dryness[74] or heat during larval growth that delays sexual development. Many token stimuli may affect corpus allatum activity.

The operation of an eco-physiological system controlling ovary development and the migration associated with it in genotypical migrants will not be identical for all species, and many adaptations to ensure that migration is controlled ontogenetically might be expected. Sexual immaturity delayed for days or weeks is geared to long distance travel with locusts and the Monarch, when it is disadvantageous for oviposition to overtake migrants before they have reached their destination. With aphids the hosts of which are scattered but ubiquitous, and immense populations are broadcast among them, the migration period can afford to be shorter, and indeed migration occurs while embryos are nearly mature.

Structural, physiological and behavioural 'polymorphism'. All the alate adults of migratory aphids are obligatory migrants and their non-migratory adults are wingless. It is easy to accept the ability or inability to migrate when it is rigidly associated with such gross structural differences as the presence or absence of wings[77]. But most migrant insects normally produce only winged adults some of which migrate and some do not. It is these that have made migration seem mysterious. Many of these species nevertheless have various degrees of alary polymorphism. Thus *gregaria* locusts have relatively longer wings than *solitaria*; with *Plusia gamma*, *Pieris brassicae*, *Callosobruchus quadrimaculatus*, crowding produces a smaller wing-loading. But although locust *gregaria* migrate in the classical swarms and so migration is connected with crowds and swarm development, *solitaria* also migrate. Indeed migration by many insects is popularly associated with dense populations and although this association may be real, and fits the hypothesis developed here, there is in fact little direct evidence for it[76]. There is an immense literature on structural polymorphism and its association with a short day, crowding and other factors affecting the endocrine system but differences in flight behaviour are rarely considered[75]. Migration studies suggest that polymorphism does not stop short at structural differences but extends also to behaviour and that this, like form, is also a matter of ontogeny whether linked with morphological differences or not.

We can therefore expect a gradation from one extreme, of structural polymorphism and obligatory migration as in aphids and termites[77], to the other extreme where little or no visible morphological differences are apparent and the only differences are physiological, neurophysiological and behavioural, nevertheless controlled by the endocrine associated with development and metamorphosis and evoked by environmental factors.

Migration in Males

Any mention in the literature of the state of the gonads in migrants almost always refers to the ovaries; therefore this article has dealt mostly with females. Females of all species have the same function, to ensure that offspring are deposited in suitable habitats; migration is a way of doing this in special circumstances. But the males are needed only to fertilize the females, and that can be done at the source without migrating, during migration, or at its end. Male migration is therefore much more variable from species to species than female migration.

At one extreme, as with the European pine shoot moth (*Rhyacionia buoliana* Schiff.), females fly readily only after mating and migrate without males accompanying them[78].

In *Aedes taeniorhyncus* mating often occurs at initial exodus and males accompany females probably only for a relatively short distance[9]. Some males accompany females to hibernation sites[14,17,33,43], but remain immature during hibernation[14]; others are mature soon after pupal emergence and migrate while sexually mature, with females, retaining live sperms in the seminal vesicles during hibernation, while the females are wholly immature[17,79]. With locusts, immature males and females migrate and mature together[24]. Thus the relation of endocrines and gonad development to thresholds of flight response does not show the obvious association in males as in females. Males probably have features peculiar to the sex and need separate investigation in each species.

The difference between the sexes, in general migration studies, has not been given the attention it deserves; usually both sexes have been considered as if belonging to the same causative system and this is not generally true.

Conclusion

(1). Females of most migratory species migrate while the ovaries are immature. The migration from the breeding site usually occurs soon after imaginal ecdysis; re-migrations are also made before ovaries mature, often after imaginal diapause.

(2) This suggests two experimental approaches to migration: (a) an examination of the physiological and neurophysiological factors responsible for the migratory drive; (b) to investigate the hypothesis that migration is evoked, prolonged or suppressed in genotypical migrants by environmental factors affecting ovary development.

(3) Flight and settling thresholds are functionally independent and vary reciprocally over short periods. But an endocrine mechanism evidently controls the neural thresholds so that flight is emphasized while ovaries are immature and settling is emphasized when ovaries ripen. That increase in the moulting hormone raises the neural response threshold for locomotor activity in locusts hoppers supports this.

(4) The following hypothesis explains why genotypical migrants migrate at some times and places and not at others.

Migratory flight can be prolonged by a lengthened pre-oviposition period and obliterated by a short one. Therefore the factors which prolong sexual immaturity also probably evoke and prolong migration in females. These factors are crowding, too little food or food of the wrong kind, a short day and high temperature acting through the corpus allatum and associated endocrines, especially during pre-adult development. Extended photoperiod, lack of crowding and enough of the right food would have an opposite effect, and tend to shorten or suppress migratory flight.

(5) It is suggested that these factors, singly or together, cause adults in some generations of genotypical migrants to develop into obligatory migrants and to suppress migration in other generations. This could explain seasonal and 'climatic' and 'alimental' migration and why populations usually migrate before a habitat obviously deteriorates. The same eco-physiological system that produces morphometric differences associated with migratory behaviour may be expected also to produce behavioural differences alone.

(6) Males of some species migrate while sexually immature, accompanying immature females all the way; females of other species are fertilized before migration and the males do not migrate. Some males migrate while sexually mature. There is a gradation between these extremes. Migration of females is physiologically similar between species, but that of males is more variable and requires a separate investigation.

I thank Sir Boris Uvarov, Prof. V. B. Wigglesworth, Mr. F. C. Bawden, Dr. P. T. Haskell, Dr. Z. Waloff, Dr. T. R. E. Southwood and Mr. L. R. Taylor for their criticism of the manuscript.

[1] Johnson, C. G., *Nature*, **186**, 348 (1960).
[2] Nielsen, E. T., *Biol. Medd. Dan. Vid. Selsk.*, **23**, 1 (1961).
[3] Tilden, J. W., *J. Res. Lepid.*, **1**, 43 (1962).
[4] Williams, C. B., *Insect Migration* (Collins, London, 1958).
[5] Roer, H., *Z. angew. Entomol.*, **44**, 272 (1959).
[6] Common, I. F. B., *Austral. J. Zool.*, **2**, 223 (1954).
[7] Faure, J. C., *Un. S. Afr. Dep. Agr. and For. Sci. Bull.*, 234 (1943).
[8] Blais, J. R., *Canad. Entomol.*, **85**, 446 (1953).
[9] Provost, M. W., *Mosq. News.*, **17**, 233 (1957).
[10] Schneider, F., *Mitt. Schweiz. Entomol. Ges.*, **31**, 1 (1958).
[11] Johnson, B., *Anim. Behav.*, **6**, 9 (1958).
[12] Lawson, F. R., Chamberlain, J. C., and York, G. T., *U.S. Dept. Agri. Tech. Bull.*, No. 1030 (1951).
[13] El Khidir, E., Ph.D. thesis, Univ. London (1963).
[14] Feodotov, D. M., *C. R. Acad. Sci. U.R.S.S.*, **42**, 408 (1944).
[15] Kelly, E. O. G. L., and Parks, T. H., *U.S. Dept. Agric. Bur. Entomol Bull.*, No. 95, 23 (1911).
[16] Hagen, K. S., *Ann. Rev. Entomol.*, **7**, 289 (1962).
[17] Hodek, J., and Cerkasov, J., *The Ontogeny of Insects* (Prague, 1959).
[18] Henson, W. R., *Ann. Entomol. Soc. Amer.*, **55**, 524 (1962).
[19] Atkins, M. S., and Farris, S. H., *Canad. Entomol.*, **94**, 25 (1962).
[20] Chapman, J. A., and Kinghorn, J. M., *Canad. Entomol.*, **90**, 362 (1958).
[21] Graham, K., *Nature*, **191**, 519 (1961).
[22] Tower, W. L., *Carnegie Inst. Publ.*, 48, Washington, D.C. (1906).
[23] Lewis, T., *Entomol. Exp. and App.*, **2**, 187 (1959).
[24] Rainey, R. C., and Waloff, Z., *Anti-Locust Bull.* (London), No. 9 (1951).
[25] Corbet, P. S., *A Biology of Dragonflies* (London, 1962).
[26] Imms, A. D., *A General Textbook of Entomology* (Methuen, London, 1960).
[27] Waloff, Z. (personal communication; and in the press).
[28] Popov, G. B., *Entomol. Mon. Mag.*, **94**, 179 (1958).
[29] Popov, G. B., *Trans. Entomol. Soc. London.*, **105**, 65 (1954).
[30] Bodenheimer, F. S., *Govt. Iraq: Min. Econ. Bull.*, No. 29 (1944).
[31] Schneider, F., *Mitt. Schweiz. Entomol. Ges.*, **25**, 111 (1952).
[32] Couturier, A., and Robert, P., *Ann. Epiph.* (6 Année) No. 1, 52 (1955).
[33] Urquhart, F. A., *The Monarch Butterfly* (Univ. Toronto Press, 1960).
[34] Greenbank, D. O., *Canad. J. Zool.*, **35**, 385 (1957).
[35] Kennedy, J. S., *Proc. Tenth Intern. Congr. Entomol.*, *1956*, **2**, 397 (1958).
[36] Nielsen, E. T., and Haeger, J. S., *Miss. Publ. Entomol. Soc. Amer.*, **1**, 71 (1960).
[37] Weintraub, J., *Canad. Entomol.*, **93**, 149 (1961).
[38] Makings, P., Thesis for Dip. Imp. Coll. Lond. (1956).
[39] Macleod, J. (personal communication: and in the press).
[40] Makings, P., *Proc. Roy. Entomol. Soc. Lond.*, A, **33**, 136 (1958).
[41] Bastock, M., and Blest, A. D., *Behaviour*, **12**, 243 (1958).
[42] Blest, A. D., *Behaviour*, **16**, 188 (1960).
[43] Williams, C. B., *The Migration of Butterflies* (Oliver and Boyd, London, 1930).
[44] Weis-Fogh, T., *Phil. Trans. Roy. Soc.*, B, **237**, 1 (1952).
[45] Haskell, P. T., and Moorhouse, J. E., *Nature*, **197**, 56 (1963).
[46] Johnson, C. G., Taylor, L. R., and Southwood, T. R. E., *J. Anim. Ecol.*, **31**, 373 (1962).
[47] Kennedy, J. S., and Booth, C. O., *Discovery*, 311 (August 1956).
[48] De Wilde, J., in *The Ontogeny of Insects* (Prague, 1959).
[49] Beetsma, J., de Ruiter, L., and de Wilde, J., *J. Insect Physiol.*, **8**, 251 (1962).
[50] Wellington, W. G., *Canad. Entomol.*, **80**, 56 (1948).
[51] Nayar, K. K., *Proc. Ind. Acad. Sci.*, B, **47**, 233 (1958).
[52] Blest, A. D., *Zoologica*, **45**, 81 (1960).
[53] Hodek, J., *Acta Soc. Entomol. Czechosl.*, **57**, 1 (1960).
[54] De Wilde, J., *Ann. App. Biol.*, **50**, 606 (1962).
[55] Bonnemaison, L., and Missonier, J., *Ann. Inst. Nat. Rech. agron.*, Ser. C: *Ann. Épiphyt.*, **6**, 457 (1955).
[56] Wigglesworth, V. B., *The Physiology of Insect Metamorphosis* (Cambridge, 1954).
[57] Lees, A. D., *The Physiology of Diapause on Arthropods* (Cambridge, 1955).
[58] Strangeways-Dixon, T., in *The Ontogeny of Insects* (Prague, 1959).
[59] Johansson, A. S., in *The Ontogeny of Insects* (Prague, 1959).
[60] Caswell. G. H., *Rev. App. Entomol.*, A, **46**, 212 (1956).
[61] Utida, S., *Rev. App. Entomol.*, **47**, 389 (1959 1954).
[62] Utida, S., *Contrib. Entomol. Lab. Kyoto Univ.*, No. 279, 93 (1956).
[63] Norris, M. J., *Anti-Locust Bull.* (London), No. 6 (1950).
[64] Iwao, S., *Contrib. Entomol. Lab. Kyoto Univ.*, No. 277, 60 (1956).
[65] Zaher, M. A., and Long, D. B., *Proc. Roy. Entomol. Soc. London.*, A, **134**, 7 (1959).
[66] Norris, M. J., *Anti-Locust Bull.* (London), No. 13 (1954).
[67] Norris, M. J., *Anti-Locust Bull.* (London), No. 36 (1959).
[68] Kirata, J., *Contrib. Entomol. Lab. Kyoto Univ.*, No. 278, 79 (1956).
[69] Garrett-Jones, C., *WHO/MAL/298* (1961).
[70] Kennedy, J. S., *Nature*, **189**, 785 (1961).
[71] Southwood, T. R. E., *Biol. Revs.*, **37**, 171 (1962).
[72] Kennedy, J. S., *Biol. Revs.*, **31**, 349 (1956).
[73] Volkonsky, M., *Arch. Inst. Pasteur. d'Alg.*, **20**, 236 (1942).
[74] Southwood, T. R. E., *Proc. Roy. Entomol. Soc. Lond.*, A, **36**, 63 (1961).
[75] Muller, H. J., *Z. angew. Ent.*, **47**, 7 (1960).
[76] Key, K. H. L., *Quart. Rev. Biol.*, **25**, 363 (1950).
[77] Lees, A. D., in *Insect Polymorphism*, Roy. Entomol. Soc. Lond., Symposium No. 1 (1961).
[78] Green, G. W., and Ponting, P. J., *Canad. Entomol.*, **94**, 299 (1962).
[79] Downes, J. A., *Trans. Roy. Entomol. Soc. Lond.*, **92**, 101 (1942).

14

Copyright ©1942 by the British Ecological Society
Reprinted from pages 194–197 of *J. Anim. Ecol.* **11**:182–197 (1942)

STUDIES OF A SMALL MAMMAL POPULATION IN BAGLEY WOOD, BERKSHIRE

F. C. Evans

[*Editors' Note:* In the original, material precedes this excerpt.]

5. THE RELATIONSHIP OF POPULATION FLUCTUATIONS TO HABITAT OCCUPATION

From the data presented in Table 4, differences in the population density of both *Apodemus* and *Clethrionomys* appeared to exist. In the case of *Clethrionomys*, four periods were selected to represent these different states, as follows: (1) from December 1936 through May 1937, a period of medium density, (2) from November 1937 through February 1938, a period of high density, (3) from March through June 1938, a period of low density, and (4) from September through December 1938, another period of medium density. Each period was comprised of the same number of trapping periods, and the different densities are relative to each other. The distribution of captures in these four population states is shown in Figs. 4 and 5.

It seemed likely that the differences in population density were variables which might affect the extent to which the habitat was occupied and might therefore be associated with the problem of distribution. Table 12 analyses the distribution of *Clethrionomys* captures as 'bracken' and 'non-bracken' trap locations in each of the four population states mentioned above. In period (1) there were considerably more captures in 'bracken' locations than in 'non-bracken' locations. In period (2) there were more captures at 'bracken' locations on area A, but fewer at similar locations on area B. In period (3)

Table 12. *Distribution of* Clethrionomys glareolus *captures associated with the distribution of bracken based on different states of population density.* (*Four trapping periods in each group*)

	Area A		Area B	
	Bracken	Non-bracken	Bracken	Non-bracken
No. of trap locations	18	18	18	18
No. of catches in different states of population density:				
(1) Medium: Dec. 1936–May 1937 incl.	26	6	64	23
(2) High: Nov. 1937–Feb. 1938 incl.	89	25	68	71
(3) Low: Mar.–June 1938 incl.	12	16	31	47
(4) Medium: Sept.–Dec. 1938 incl.	43	23	62	59
'Percentage of survival', i.e. (3)/(2)	13	64	46	66

'non-bracken' locations produced more captures than 'bracken' locations. In period (4) there was a return to greater occupancy of bracken habitats. This suggests that relative density and distribution are associated.

It will be observed that period (3) follows period (2) in immediate chronological order, and it will further be recalled from Table 4 that an unusually high number of *Clethrionomys* deaths was recorded during period (2). Evidence of any large-scale mortality is lacking, yet it is possible that some disease factor may have been responsible for the relatively rapid decline of the population from the apparent high density of period (2) to the low density of period (3). Such a factor would operate more thoroughly in a dense population than in a scattered one, and the proportion of those which survived would be correspondingly less. Table 12 suggests that such may have been the case; the 'percentage of survival' from period (2) to period (3) shows a correspondingly higher figure at locations where the density had been relatively low during the state of medium, and for area A at least, of high density.

This interpretation of the relationship of population fluctuations to habitat occupation differs from other theories. For example, Naumov (1936) has suggested that when a population is at its lowest density it occupies only the most favourable habitats; that as the density increases, the pressure of numbers forces the population into less favourable habitats until a maximum density is reached when all possible habitats are occupied;

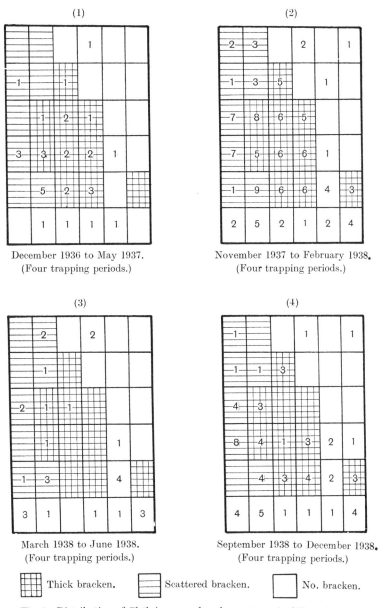

Fig. 4. Distribution of *Clethrionomys glareolus* captures in different states of population density. Area A.

and that periodic decreases of the population caused by migration or disease result in the occupation of only the most favourable habitats. The use of the term *favourable* seems somewhat misleading here. The new interpretation suggests that habitats which will permit high densities of animal populations will also permit high densities of predators and parasites whose decimating effect may be so rapid as virtually to destroy those populations; habitats which will maintain only low densities may in the long run be essential to the survival of the species.

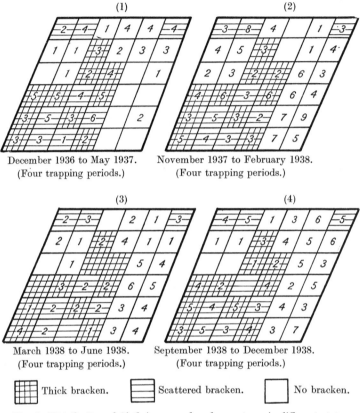

Fig. 5. Distribution of *Clethrionomys glareolus* captures in different states of population density. Area B.

6. Summary

1. A small mammal population in Bagley Wood, Berkshire, was studied from October 1936 through February 1939 by trapping and marking techniques.

2. The areas in which trapping was carried on provided a relatively large number of distinct and easily separated habitat types within a few acres. Two types of trapping grid were employed, and four nights of trapping completed the monthly census in each area. Individual mice were marked on the leg with a nickel ring and were also weighed.

3. The chief species studied were the wood-mouse (*Apodemus sylvaticus*) and bank-vole (*Clethrionomys glareolus*). Fluctuations in the numbers of both species were observed, and factors contributing to these changes are discussed. (*a*) About one-half of the population disappeared within one month after first capture. (*b*) The decrease in numbers caught to a very low relative density in the spring of 1938 (subsequent to a relatively high density and an increased number of deaths in traps) suggested some mortality factor. (*c*) *Apodemus*

was found to be a wider-ranging animal than *Clethrionomys*, but both species tended to restrict their activity to a fairly small home range. (*d*) Immigration and emigration apparently played a definite role in the population fluctuations. (*e*) It was suggested that weather changes may have affected the numbers of animals caught. (*f*) The sex ratio approximated to 1:1 for the total number of individuals caught, but males preponderated in the spring and autumn. (*g*) The populations of both species showed a general increase of weight with the onset of the breeding season.

4. Statistical analysis showed that the distribution of the population in the trapping areas was not random. Certain habitat types appeared to be avoided. The distribution of *Apodemus* could not be associated with any one habitat factor, but the greatest number of *Clethrionomys* were taken at trap locations in bracken (*Pteridium aquilinum*) areas.

5. The distribution of *Clethrionomys* appeared to be associated with its relative density. The population 'survival' from a period of high density to one of low density appeared to be greatest in those habitats which had normally maintained a low population density. It is suggested that habitats which will maintain only low densities may be essential to the ultimate survival of a species.

REFERENCES

Allen, A. A. (**1921**). 'Banding bats.' J. Mammal. 2: 53–7.
Baker, J. R. (**1930**). 'The breeding-season in British wild mice.' Proc. Zool. Soc. Lond. 1: 113–26.
Burt, W. H. (**1940**). 'Territorial behavior and populations of some small mammals in southern Michigan.' Misc. Publ. Mus. Zool. Univ. Mich. 45: 1–58.
Chitty, D. (**1937**). 'A ringing technique for small mammals.' J. Anim. Ecol. 6: 36–53.
Elton, C., Ford, E. B., Baker, J. R. & Gardner, A. D. (**1931**). 'The health and parasites of a wild mouse population.' Proc. Zool. Soc. Lond.: 657–721.
Fisher, R. A. (**1938**). 'Statistical methods for research workers.' 7th ed. rev. Edinburgh.
Green, R. G. & Larson, C. L. (**1938**). 'A description of shock disease in the snowshoe hare.' Amer. J. Hyg. 28: 190–212.
Johnson, M. S. (**1926**). 'Activity and distribution of certain wild mice in relation to biotic communities.' J. Mammal. 7: 245–77.
Kalabukhov, N. & Raevskii, V. (**1935**). ['A study of the migrations of ground squirrels (*Citellus pygmaeus* Pall.) in the steppe areas of Northern Caucasus by means of the banding method.'] Problems of Ecology and Biocenology [2]: 170–95. (In Russian; summary in English.)
Naumov, N. P. (**1936**). [On some peculiarities of ecological distribution of mouse-like rodents in Southern Ukraine.] Zool. Zh. 15: 675–96. (In Russian; summary in English.)
Sumner, F. B. (**1922**). 'Longevity in *Peromyscus*.' J. Mammal. 3: 79–81.
Zverev, M. D. (**1928**). ['Materials on the biology of the red-cheeked ground squirrel.'] [Cited by Kalabukhov & Raevskii (1935).]

The Mode of Migration of *Drosophila ananassae* Under Competitive Conditions.[1]

TAKASHI NARISE[2]

INTRODUCTION

According to Patterson and Stone (1952), *Drosophila ananassae* is distributed widely in tropical and sub-tropical areas. In such a widely distributed species, it is expected that there is an abundance of genetic types in competition in populations with some flow of migrants. Under competitive conditions, what is the pattern of this migration? Although there are many problems in relation to migration, one of the more important is the interaction among genotypes in the species under competitive conditions. With regard to migration of *Drosophila*, Sakai et al. (1958) and Narise (1962) found that the migratory activity of *Drosophila melanogaster* is under genetic control. However, the experiments were made using single isolated strains, not mixed strains. In a polymorphic population, the mode of migration may be quite different from the migration in single strains, even if each strain in the mixture has its own genetic migratory activity. Narise (in preparation) has conducted competition experiments with two genotypes of *D. melanogaster* in an open and a closed population. The experimental results showed that the migratory activity of a weaker competitor was stimulated by the strong competitor and the stimulation of migratory activity prevented the elimination of the weaker competitor in the open population. The problem arises, then, whether such stimulation of migratory activity occurs in other species under competitive conditions. To attack this problem, an experiment was conducted with four strains of *Drosophila ananassae*, collected on islands in the South Pacific.

MATERIALS AND METHODS

Four strains of *Drosophila ananassae* collected by Dr. Wilson S. Stone and Dr. Marshall R. Wheeler on some islands of the South Pacific were used in this experiment. Two of them were so-called "light *ananassae*" which has yellow body color, and were collected in Pago Pago, Tutuila, American Samoa, and Majuro, Marshall Islands. The other two strains were "dark *ananassae*" having black body color, and were collected in Pago Pago, Tutuila, American Samoa, and Rarotonga, Cook Islands.

1) Migratory activity of two kinds of *ananassae* in a mixed population.

One hundred flies, consisting of light and dark *ananassae* in different frequencies were introduced into a migration-tube and the tube (called "original tube") was kept 24 hours before three fresh tubes were connected with it. Flies which migrated to the three new tubes were counted after 6 hours. The relative fre-

---

[1] This investigation was supported by Public Health Service Research Grants No. GM-06492 and GM-11609 from the National Institutes of Health.

[2] Present address: National Institute of Genetics, Mishima, Shizuoka-ken, Japan.

quency of the dark form in different experiments was 0.0, 0.1, 0.2, 0.3, 0.4, 0.5, 0.6, 0.7, 0.8, 0.9, and 1.0, and each relative frequency was replicated 10 times. The migratory activity for a strain at each relative frequency presented is calculated as the per cent of that strain in the new tubes to that strain in the original mixture.

2) The mode of migration for dark and light *ananassae* in mixed populations.

Seventy-five pairs of light and seventy-five pairs of dark *ananassae* were introduced into the original tube and another fifteen fresh tubes were connected with the original tube as shown schematically in Figure 1, after 24 hours. Two tubes which were connected directly with the original tube were called the first-step-tubes, and the four tubes which were connected with the two first step-tubes were called the second-step-tubes, and so on. Only one tube was used for the fourth step-tube. The migrant flies were counted 8 hours after the connection, and the number of both strains which migrated to each step-tube was counted. The experiment was replicated five times for each combination.

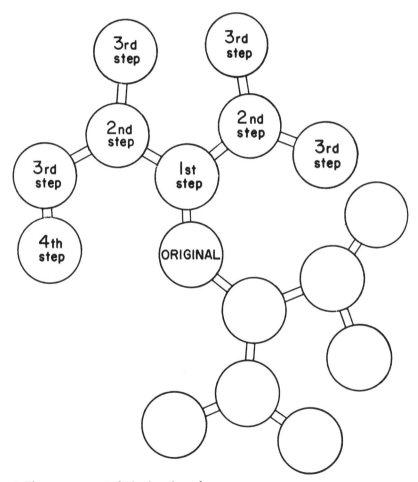

Fig. 1. The arrangement of 16 migration tubes.

All experiments were conducted at 25° ± 1°C in a dark room. The following abbreviations were used for the four strains: *

Dark *ananassae* from Pago Pago	: D-pp
Dark *ananassae* from Rarotonga	: D-rar
Light *ananassae* from Pago Pago	: L-pp
Light *ananassae* from Majuro	: L-maj

Experimental Results

1) The migratory activity of dark and light *ananassae* in mixed population.

Table 1 and Figure 2 show the migratory activity of dark and light *ananassae* at each relative frequency, and Table 2 shows the results of the analysis of variance on the migratory activities in the four different combinations.

a) Migratory activity of L-pp and D-pp strains in a mixed population.

The migratory activity of the original D-pp strain is 33.48% and that of L-pp is 34.45%. Much difference regarding the migratory activity between the two strains under the single condition is not detected. However, the activity of D-pp

Table 1

The migratory activity of dark and light *ananassae* under competitive condition

Combination	Relative frequency of dark *ananassae*										
	0.0	0.1	0.2	0.3	0.4	0.5	0.6	0.7	0.8	0.9	1.0
L-pp and	34.45	39.11	39.81	41.43	41.42	41.60	42.00	44.50	38.25	44.00	....
D-pp	....	24.50	24.75	25.33	22.38	30.05	26.67	25.21	28.69	31.44	33.48
L-pp and	34.45	32.83	36.75	37.29	36.50	36.15	32.84	39.50	33.00	41.50	....
D-rar	....	24.50	30.00	25.83	26.13	27.60	25.17	30.71	29.99	34.44	33.10
L-maj and	36.78	41.48	42.88	32.93	32.75	40.40	37.38	36.83	34.00	35.50	....
D-rar	....	35.50	32.00	28.83	32.50	34.40	39.42	40.29	31.38	32.78	33.10
L-maj and	36.78	37.11	42.94	30.80	37.67	37.40	35.38	33.42	41.25	39.50	....
D-pp	....	25.00	29.25	26.77	28.63	29.30	23.00	24.29	23.08	24.22	33.48

Table 2

Analysis of variance on migratory activity of dark and light *ananassae* under competitive condition

Source of deviation	d.f.	M.S.			
		D-pp and L-pp	D-pp and L-maj	D-rar and L-pp	D-rar and L-maj
Between strains	1	7243.67**	4575.03**	2599.71**	368.03**
Between frequency within strain	18	113.72	88.95	220.16**	126.54**
Between single and mixed	2	508.45**	162.15*	121.82	65.02
Between frequency within mixed	16	64.38	79.80	232.45**	134.42**
Error	380	87.88	83.98	65.96	59.02

* Significant at 5% level.
** Significant at 1% level.

* Editor's note: Futch (This Bulletin) shows that L-pp is probably a sibling species of *D. ananassae*; L-maj, D-pp, and D-rar are true *ananassae*.

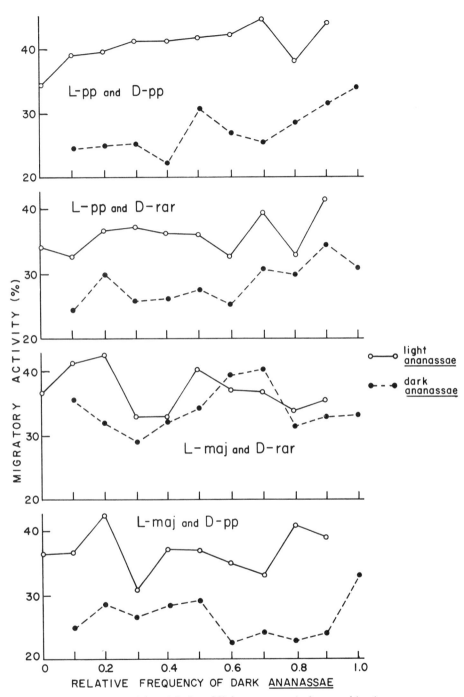

Fig. 2. The migratory activity of dark and light *ananassae* in four combinations.

strain decreases to 26.7% on average from 33.48% under the mixed condition. On the contrary, the migratory activity in L-pp strain increases from 33.48% to 41.35% on average. As seen in Table 2, there is no significant variation in migratory activity between frequencies within strains under mixed condition, although the activities between single and mixed conditions are different in both strains. Thus, the relative frequency of the strains in the original tube has no effect on migratory activity, but the presence of the other strain decreases activity of the D-pp strain and increases the activity of the L-pp strain.

b) Migratory activity of L-pp and D-rar strains in a mixed population.

The migratory activity in the D-rar strain decreases due to mixtures from 33.10% to 28.15% on the average, while a small increase of the activity takes place in L-pp strain in mixed populations. In Table 2, statistically significant deviations are found between strains, between frequencies within strains, and between frequencies within mixed conditions. Therefore, from Table 1 and Table 2 it can be said that the activity of D-rar strain decreases with decreasing of the relative frequency of the strain in the original tube, and the L-pp strain shows lower migratory activity when the relative frequency of the strain is higher.

c) Migratory activity of L-maj and D-pp strains in a mixed population.

In unmixed condition, the L-maj strain has an activity of 36.78%. When the strain is mixed with D-rar strain, a small increase of the activity is detected. However, the activity of the D-rar strain decreases due to mixing with L-maj strain. It is of interest that the activity of the strain decreases from 33.48% to 27.41% on the average, if the relative frequency of the strain is less than 0.5, while the activity decreases to 23.24%, when the frequency is more than 0.6. Consequently, there is a possibility that the migration of D-rar strain occurs as the threshold reaction to the relative frequency in the original tube in this combination, although there is no significant variance among relative frequencies within the strains.

d) Migratory activity of L-maj and D-rar strains in a mixed population.

In mixed populations of these two strains, the migratory activity of both strains is quite different from the other three combinations. The migratory activities of D-rar strain are 39.42% and 40.29% at 0.6 and 0.7, respectively. These activities are higher not only than that in the pure tests, but also higher than those at 0.4 and 0.3 in the L-maj strain, although the activities at other frequencies of D-rar strain are lower than those in the L-maj strain in mixed populations. As seen in Table 2, highly significant deviation is detected between strains as well as between frequencies. In the L-maj strain it is also found that the migratory activities of the strain are quite different for different relative frequencies. The facts suggest that the migratory activities of both strains are affected by the relative frequency of both strains in the original tube. However, in general, the migratory activity of D-rar strain seems to be lower than that of L-maj strain in combination.

Figure 3 shows the migratory activity of one strain when the strain is mixed with another strain, and Table 3 presents the increment of migratory activity in the four strains under mixed condition. As seen in Table 3, the average increment of the L-pp mixed with D-pp is 6.90% and mixed with D-rar is 1.98%. The difference between these two increments is highly significant. On the other hand, a decrease of activity is found in D-pp strain. The average decrement of activity in this strain when mixed with L-pp is 6.92%, and with L-maj is 7.75%. How-

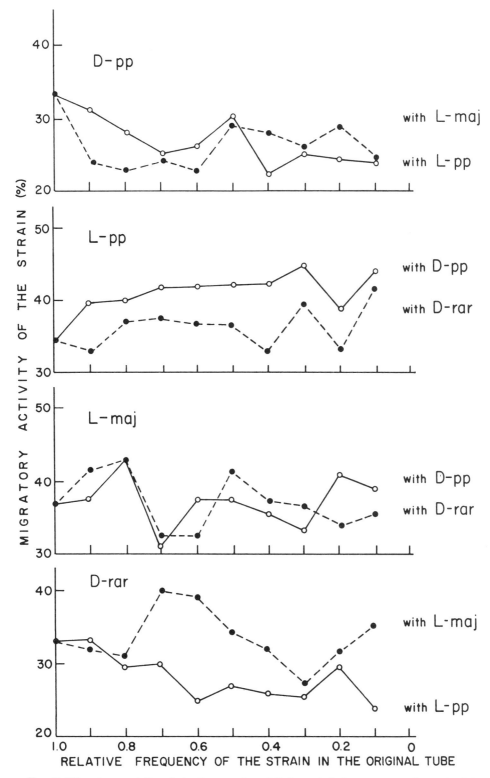

Fig. 3. Migratory activity of the four strains of light and dark *ananassae* when mixed in varying frequencies with strains of the opposite color.

TABLE 3

The increments of the migratory activity in four strains under competitive condition

Relative frequency of the strain	L-pp mixed with		L-maj mixed with		D-pp mixed with		D-rar mixed with	
	D-pp	D-rar	D-pp	D-rar	L-pp	L-maj	L-pp	L-maj
0.9	4.66	—1.62	0.33	4.70	—2.04	—9.26	0.34	—0.32
0.8	5.39	2.30	6.16	6.10	—4.79	—10.40	—3.11	—1.72
0.7	6.98	2.84	—5.98	—3.85	—8.27	—9.19	—2.39	7.19
0.6	6.97	2.05	0.89	—4.03	—6.81	—10.48	—7.93	6.32
0.5	7.15	1.70	0.62	3.62	—3.43	—4.18	—5.50	1.30
0.4	7.55	—1.61	—1.40	0.60	—11.10	—4.85	—6.97	—0.60
0.3	10.05	6.60	—3.36	0.05	—8.15	—7.71	—7.27	—4.27
0.2	3.80	—1.45	4.47	—2.78	—8.73	—4.23	—3.10	—1.10
0.1	9.55	7.05	2.78	—1.28	—8.98	—9.48	—8.60	2.40
Average	6.90	1.98	0.50	0.28	—6.92	—7.75	—4.95	1.02

ever, neither decrement is significant. In the L-maj strain, activity is slightly stimulated by D-pp and by D-rar (0.50% and 0.28%), but the average increments of the activity with D-pp and D-rar are not different. An interesting fact is found for the D-rar strain. If the strain is mixed with the L-pp strain, decreasing of activity occurs except at a frequency of 0.9. However, the activity increases when the strain is mixed with L-maj. The average increment of the strain with L-maj is 1.02%, and the average decrement of the strain with L-pp is 4.95%. From Figure 3, it is also clear that migratory activity in the L-pp strain is stimulated by dark *ananassae* strains, while the activity in the D-pp strain decreases due to mixing with two light *ananassae* strains. In the D-pp as well as the L-maj strains, the increase or decrease of the migratory activity is not much affected by the other strains, but the D-rar strain shows quite different activities with different strains.

Apparently, the migratory activity of light *ananassae* is stimulated by dark *ananassae*. However, the increment of the activity in light *ananassae* is quite different depending on which strain coexists in the population. In dark *ananassae*, the rate of decrease of the activity is also different not only due to relative frequency, but also due to different competitor strains.

2) The mode of migration under mixed conditions.

The number of migrant flies of dark and light *ananassae*, as well as the relative frequency of both strains in each step-tube, are given in Table 4. In general, it can be said that light *ananassae* shows more migratory activity than does dark *ananassae*. For example, in a D-pp/L-pp combination, 25.5 flies out of 150 flies of light *ananassae* migrate to the first step-tubes, 9.0 to the second, 4.8 to the third, and 1.0 fly to the fourth step-tube, while in "dark *ananassae*" the figures are 21.0 flies to the first step-tube, 7.2 to the second, 2.4 to the third, and 0.4 flies to the fourth. In this combination, light *ananassae* dominates in all tubes. The same tendency in the mode of migration is found in the D-pp and L-maj combination, except in the fourth step-tube as presented in Table 4. Only one fly of dark *ananassae* migrated to the fourth step-tube in five replications. On the other hand, the mode of migration in the D-rar/L-pp combination as well as the D-rar/L-maj combination is distinct from that in the previous combinations. As presented in

TABLE 4

The number of migrated flies from the original tube to the surrounding tubes in four kinds of combinations between light and dark *ananassae*

Combination strain	D-pp and L-pp Light	Dark	D-pp and L-maj Light	Dark	D-rar and L-pp Light	Dark	D-rar and L-maj Light	Dark
1st-step	25.4	21.0	40.6	26.2	26.4	22.2	33.2	20.4
	(0.5471	0.4526)	(0.6078	0.3922)	(0.5432	0.4568)	(0.6194	0.3806)
2nd-step	9.0	7.2	18.2	10.2	7.6	8.0	6.8	9.0
	(0.5556	0.4444)	(0.6408	0.3592)	(0.4972	0.5128)	(0.4304	0.5696)
3rd-step	4.8	2.4	8.2	4.6	4.6	7.4	5.0	8.2
	(0.6667	0.3333)	(0.6406	0.3594)	(0.3833	0.6167)	(0.3788	0.6212)
4th-step	1.0	0.4	0.0	0.2	0.0	0.2	0.0	0.0
	(0.7143	0.2859)	(0.0000	1.0000)	(0.0000	1.0000)	. . . . .	. . . . .
Total	40.2	31.0	67.0	41.2	38.6	37.8	45.0	37.6
	(0.5646	0.4354)	(0.6192	0.3808)	(0.5052	0.4948)	(0.5448	0.4552)

The numbers in the parentheses are the relative frequency of light and dark *ananassae*.

Table 4, there are 26.4 flies of light *ananassae* in the first step-tube, 7.6 in the second, 4.6 in the third, and no light *ananassae* in the fourth step-tube, while 22.2 flies out of 150 flies of dark *ananassae* migrated to the first step-tube, 8.0 to the second, 7.4 to the third, and 0.2 flies to the fourth step-tube in the D-rar/L-maj combination. In this combination, light *ananassae* dominates in central tubes and dark *ananassae* in the surrounding tubes, although the total number of migrant flies to the connected tubes is greater in light *ananassae* than in dark *ananassae*. The same tendency is found in the D-rar/L-pp combination as seen in Table 4. The interesting fact is that the total number of migrant flies in the combinations between different populations is greater than in the combination of strains from the same population.

Discussion

1) The stimulation or diminution of migratory activity under competitive condition.

As shown in these experiments with two kinds of *ananassae*, the migratory activity of light *ananassae* is stimulated when coexisting with dark *ananassae*, while the activity in dark *ananassae* is diminished by light *ananassae*. Narise (in preparation) also found the stimulation and diminution of the activity in the experiment with two genotypes of *D. melanogaster*. Therefore, it can be said that the stimulation or diminution of migratory activity takes place when many strains coexist in a population under migratory and competitive condition. In this experiment, two important facts are found. One of them is that the rate of stimulation or diminution of the activity of a strain depends on the strain which coexists with it in the population. As seen in Table 4, the average increment of activity in the L-pp strain is quite different when it is mixed with D-pp and with D-rar. In this connection, it is interesting to find that the migratory activity in the D-rar strain is stimulated by L-maj but the L-pp strain causes a diminution of the activity of D-rar. Furthermore, there are strains which are not much affected by another strain.

The other fact of interest is the relation between the migratory activity of a strain and the relative frequency of the strain in the original tube. As shown in the D-pp/L-pp combination as well as the D-pp/L-maj combination, the activity of a strain may be independent of the relative frequency of both strains in the original tube. On the other hand, in some cases the activity depends on the relative frequency of both strains as seen in the D-rar/L-pp combination as well as the D-rar/L-maj combination. In the former two combinations, the rate of migration does not change within the strain under competitive conditions. However, the rate of migration in the latter two combinations is quite variable not only due to the strain coexising in the population, but also the relative frequencies of both strains. Of special interest is the D-rar strain whose migratory activity is higher than that of the L-maj strain when its frequency is 0.6 and 0.7, but lower at other frequencies.

2) The mode of migration in two kinds of *ananassae* in mixed populations.

As far as this experiment is concerned, two modes of migrations are detected. In the first mode, the strain whose migratory activity is stimulated by another strain, dominates in all tubes as seen in the combination between D-pp and light *ananassae* strains. This result coincides with the previous experimental results that the migratory activity of L-pp and L-maj is stimulated by the D-pp strain, and the migratory activities of light *ananassae* strains are higher than those of the D-pp strain at any frequency. In such combinations, light *ananassae* may have more opportunities to invade other populations to which the strain migrates, or dominate either the original or invaded population under migratory and competitive conditions.

In the other mode, a strain whose migratory activity is diminished by another strain dominates in surrounding tubes, even if the *total* migration to the connected tubes is less than that in the coexisting strain. This is seen in the D-rar strain. From the previous experiment it was not expected that light *ananassae* would dominate only in central tubes, since it has higher migratory activity than D-rar strain when the relative frequency of light *ananassae* strain is less than 0.5 as shown in Figure 2 and Table 1. In order to explain the experimental result, let us consider two kinds of migratory activity: mass-migratory activity caused by population density and random-migratory activity due to random movement of individual flies as described by Sakai et al. (1958) and Narise (1962). Further, let us assume that the mass- and random-migratory activity of four strains under mixed conditions are as follows:

$$\text{Random-migratory activity} \quad \text{D-rar} > \text{L-pp} \geq \text{L-maj} > \text{D-pp}$$
$$\text{Mass-migratory activity} \quad \text{L-pp} = \text{L-maj} > \text{D-rar} > \text{D-pp}$$

It is probable that the migration to the first step-tubes from original tube occurs due to mass-migration, and random-migration takes place in the outer tubes with reduced population density. Then, the migration of light *ananassae* in the D-pp/light *ananassae* combination, takes place due to the stimulation in the mass- and random-migratory activity, with the result that light *ananassae* strain is superior in the number of migrant flies in all tubes. In D-rar and L-pp as well as the D-rar/L-maj combination, a smaller number of flies of the D-rar strain migrates to the first step-tubes because of lower mass-migratory activity under competitive condition. However, the strain has higher random-migratory activity than light *ananassae* strain, resulting in the excess of that strain in the surrounding tubes. From this point of view, the relation between mass- and random-migratory activity should be another important factor in migration under competitive condition.

Needless to say, there are many factors controlling migration aside from strain interaction. However, the interaction among strains which exist in the population seems to be one important factor. As pointed out by Lidicker (1962), the immigrant spreads his genetic materials more widely and there would be more opportunity for advantageous recombination of genetic materials. In such a situation, it is expected that such new recombinations of genes may bring about new stimulating effects on migratory activity. However, in a genetically heterogeneous and open population, the mode of migration is complicated so that the interaction among strains becomes important. From this standpoint, migration should be determined not only by genetic migratory activities of strains in isolation, but also by the genetic structure of the population.

Summary

1. Four strains of *D. ananassae* collected on islands of the South Pacific were used in this experiment. Two of them were light *ananassae* and two were dark *ananassae*.

2. The migratory activity of light *ananassae* is stimulated by dark *ananassae*, but dark *ananassae* is reduced in activity when coexisting with light forms. The rate of stimulation or diminution of activity depends on the particular strains that coexist in the population. The migratory activity of the strain is also affected by the relative frequencies of strains as shown in the D-rar/L-pp combination as well as the D-rar/L-maj combination. However, no relation between migratory activity and relative frequency was detected in the combination between D-pp/L-pp or D-pp/L-maj.

3. The light *ananassae* dominates in all tubes when mixed with D-pp. However, dark *ananassae* dominates in the surrounding tubes in the combination between D-rar and light *ananassae* strains. This is probably due to high random-migratory activity in the D-rar strain under competitive conditions.

4. Migration under competitive conditions is dependent on the strains which coexist in the population, and the interaction of the strains is a very important factor in migration.

Acknowledgment

The author is very grateful to Professor Wilson S. Stone for his helpful suggestions and comments, and critical reading of this manuscript. The author is also grateful to Dr. Richard C. Lewontin, Department of Zoology, University of Chicago, for his valuable comments. I wish to thank Mr. David E. Briles, Mr. Jay P. Mumma, and Mr. Stephen Goldfarb for their help throughout the experiments.

Literature Cited

Lidicker, W. Z., Jr. 1962. Emigration as a possible mechanism permitting the regulation of population density below carrying capacity. Amer. Nat. 96(886): 29–33.

Narise, T. 1962. Studies on competition in plants and animals. X. Genetic variability of migratory activity in natural population of *Drosophila melanogaster*. Jap. J. of Genet. 37(6): 451–461.

Narise, T. (in preparation). The competition between wild and vestigial flies of *Drosophila melanogaster* in an open and a closed population.

Patterson, J. T., and W. S. Stone. 1952. Evolution in Genus Drosophila. The Macmillan Company. p. 51.

Sakai, K. I., T. Narise, Y. Hiraizumi, and S. Iyama. 1958. Studies on competition in plants and animals. IX. Experimental studies on migration in *Drosophila melanogaster*. Evolution 12(1): 93–101.

[*Editors' Note:* The "light" and "dark" forms of *D. ananassae* referred to on page 148 are considered to be separate species by Bock, I. R. and M. R. Wheeler, 1972, The *Drosophila melanogaster* species group, *Stud. Genet. (Univ. Texas) No. 7, 102p.;* see pages 37-40.]

QUEUING BEHAVIOR OF SPINY LOBSTERS
William Herrnkind
Department of Biological Science, Florida State University, Tallahassee 32306

Abstract. *Autumnal mass migrations of spiny lobsters,* Panulirus argus, *involve diurnal movements of thousands of individuals in single-file queues. Initiation, posture, and alignment of a queue can be effected entirely by tactile cues received through antennular inner rami, pereiopods, and antennae. Since spiny lobsters queue when deprived of shelter, this behavior may serve a defensive function. Specimens captured while migrating maintain the queue indoors for up to several weeks, whereas at other times the queue lasts only a few hours. Hence, the migratory behavior probably depends in part upon environmentally induced neurohormonal changes.*

The occurrence of "columns" or "trains" of spiny lobsters, *Panulirus argus*, moving over open areas was reported as early as 1922 (*1–3*), but the extent and significance of the phenomenon has remained inadequately described. In autumn, thousands of lobsters migrate diurnally in parallel single-file queues across shallow areas near Bimini, Bahamas, and the Florida east coast (*4, 5*). This activity is remarkable since spiny lobsters are usually seclusive by day, remaining in crevices on the reef and emerging at night to feed.

The widespread occurrence, periodicity, and large numbers of individuals involved, as well as the stereotyped behavioral character of the queues, clearly defines this phenomenon and suggests previously unrecognized significance of these migrations in the life history of the species. Furthermore, the mode of mass movement, which outwardly resembles migrations of army ants and certain bird flocks, and its abrupt nature make it unique among benthic marine crustaceans (*3*). I now present some major behavioral characteristics of the single-file formation, evidence for internal influences, and offer hypotheses to explain causal and functional aspects of the migrations.

The following analysis of postural and behavioral elements was derived from observations of 325 queues involving 20 captive lobsters (7.0 to 12.0 cm carapace length) in a 2 m diameter vinyl-lined pool. Particular attention was directed at determining the mode of queue formation and mechanism of queue maintenance.

Normal appearing lobsters with all appendages intact most often queued by approaching the posterior of another moving individual until antennal contact was made (Figs. 1 and 2a). Such behavior suggested initial visual orientation during queue formation. The approaching lobster then turned until both antennular inner rami touched the other's abdomen, thus completing alignment. Constant physical contact was maintained with the preceding individual by either the antennular inner rami, anterior pereiopods, or antennae (Figs. 1 and 2b). The antennular inner rami were most often involved (88 percent) and made intermittent contact every few seconds for the duration of the queue (6). The pereiopods either touched or hooked around the telson of the lobster ahead in 51 percent of queuing individuals and always during the times the antennules were not in contact. In 19 percent of the cases one or both antennae were brought into contact for several seconds but were otherwise held slightly forward of a line perpendicular to the direction of movement. I did not observe normal lobsters maintaining alignment without physical contact.

The relative importance of visual versus tactile stimuli was further investigated on specimens blinded by opaque tape. After an acclimation period of several hours, the blinded lobsters readily queued but only after their antennae or antennules were touched by another lobster (Fig. 2c). After initial contact, a blinded individual turned neatly into alignment and maintained appendicular contact, posture, and position equivalent to that of unblinded specimens (Fig. 2d). These results tended to rule out the influence of waterborne chemicals in queue formation and emphasized the importance of tactile stimuli.

The effect of loss of the extensively used antennules was investigated by taping them to the posterior surface of the antennae. Individuals so treated queued in the same way as normal lobsters (Fig. 2e), but they aligned themselves by hooking the anteriormost pereiopods around the telson of the preceding lobster and usually maintained contact in this way for the duration of the queue (Fig. 2f, 92 percent). Position was maintained as long as the path was reasonably straight but misalignment occurred during sharp turns when the pereiopods often lost contact. Hence, either pereiopod or antennular contact was necessary for proper queue maintenance although the antennules alone were more effective than the walking legs.

Evidence presented thus far suggests that queuing in light was apparently initiated most frequently by visual cues associated with movement of other lobsters within the visual field of an individual. Supplementary experiments indicated that acoustic or waterborne olfactory cues were not responsible for queue formation or maintenance (7). Tactile stimuli, taste or touch or both, seemed necessary for proper maintenance of the queue at all times and also served to initiate column formation in the absence of visual cues. This suggests that spiny lobsters are able to effectively form and maintain the single file at night and under conditions of poor visibility such as during the autumnal mass migrations when the water is often cloudy.

Appendicular-receptor usage in the queue was hierarchially arranged with antennular inner rami playing the major role, followed by the anterior pereiopods and the antennae (Fig. 2). Accordingly, deprivation of antennular sensitivity resulted in substitution by the pereiopods while removal of each pair of anterior pereiopods was followed by substitution of the next posterior pair. Loss of antennae or vision, seldom used by normal specimens, had little effect.

Lobsters captured during nonmigratory periods, as well as those captured from migrating columns, queued under controlled conditions. However, the overall activity patterns and responses to certain stimuli differed markedly between the two groups.

One dozen nonmigratory lobsters (7 to 12 cm carapace length) in rectangular tanks tended to gather in the corners, or other sheltered areas, and became inactive. However, when released in a large circular vinyl pool (2 m diameter, 20 cm depth), they immediately formed queues of two to six individuals. The queues moved over the open sand substrate in tortuous paths or around the perimeter for periods averaging about 40 seconds. Individuals then broke away and wandered to join other queues or became inactive for

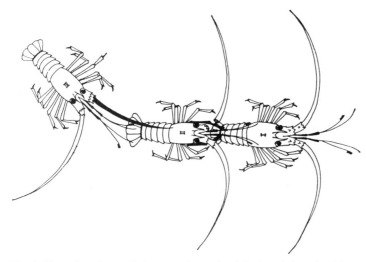

Fig. 1. Illustration of appendicular usage in queuing behavior of the spiny lobster *Panulirus argus*; appendages involved are blacked in. Lobster III joins the queue after touching lobster II with an extended antenna and turning until both antennular inner rami touch the sides of II. Lobster II maintains queue alignment either by frequent intermittent flicks of the antennules against lobster I, grasping or touching I with extended anterior pereiopods, or both. Antennae, antennular inner rami, and pereiopod tips are sensitive to tactile stimulation.

several minutes. The frequency of queue formation decreased after 1 hour until most of the lobsters gathered in stationary clusters along the pool perimeter. However, queuing could be reinitiated by transferring the animals for several hours to tanks providing shelter, then reintroducing them to the pool.

Lobsters captured during a mass migration (transported within 3 hours to a pool) also queued but, in contrast to the nonmigratory specimens, continued this activity for an extensive period. Four lobsters (7.0 to 10.0 cm carapace length) continuously marched around the perimeter at a rate of approximately 6 m/min for the first 6 hours after introduction. Subsequent monitoring at various intervals each day revealed that these lobsters filed nearly continuously for 33 days, both during the day and at night (in dim light).

Responses to food and shelter by the migrating lobsters were markedly different from those of nonmigratory individuals. The latter readily took up residence in concrete block shelters and fed heavily after several days of starvation. By contrast, the migratory specimens ate part of the cut shrimp provided or briefly investigated the concrete blocks, then resumed the queue. They finally took up residence in the shelters on the 33rd day after a gradual increase over the final 2 weeks in periods of erratic individual wandering and quiescence.

The function and causation of queuing behavior and of the mass single-file migrations cannot, as yet, be explained fully although the available evidence yields several conclusions and suggests reasonable hypotheses. Field observations and the above information from captive specimens indicate that spiny lobsters are induced to queue when moving over relatively open areas. At such times the formation may serve a defensive function. The vulnerable abdomen, ordinarily protected while a lobster faces outward from a crevice, is "protected" in the queue by the spinous cephalothorax of each trailing individual (with the obvious exception of the last in line). The defensive function of the file may be particularly important during the mass migrations when the lobsters are exposed for extensive periods.

Internal physiological changes are probably responsible, at least in part, for the mass movements as indicated by the contrast in the behavior between captive migratory and nonmigratory individuals under similar external conditions. The continuous file activity is interpreted as a form of migratory restlessness (Zugunruhe), usually associated with captive birds. This migratory internal state may be brought on by developmental processes (such as growth or molt) or by seasonal changes in daylength, temperature, or other bioenvironmental factors. However, the actual migratory movement in a locality is probably induced by the abrupt changes in physical conditions accompanying storms which immediately precede the migrations.

The direction of migration seems characteristic for a given population; for example, files moved southwest at Bimini and northeast at Boca Raton. Each individual in the population may be initially capable of orienting in this heading, as in the case of escape movements by fiddler crabs (8), or directional information may be communicated by the queuing behavior. In any event, individuals other than the leader adopted the original bearing after the foremost lobster was removed.

This question of orientation by a benthic animal is of special interest since the directional movements occurred when the sky was overcast, the sea surface turbulent, and currents and bottom slope variable. Such conditions complicate postulating orientation mechanisms such as a celestial-compass, chemorheotaxy, pressure perception, or sound localization used to explain migrations in birds, fishes, and turtles (9). The single-file migratory behavior may share the navigational capabilities indicated by homing experiments in which 20 percent of released lobsters returned to a home area from distances of up to 2 miles and depths of more than 500 m (10).

The adaptive significance of autumnal mass migrations has not been determined. Available evidence argues against a reproductive function since females do not carry spermatophores until spring (11). Possible functions being examined are: attainment of better feeding grounds (2), attainment of maximum shelter for molting (12), local dispersal, and reduction of population pressure.

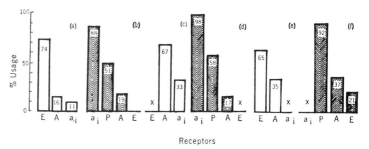

Fig. 2. Histograms showing the proportion of appendicular-receptor usage by captive spiny lobsters during queue formation (unshaded) and maintenance (shaded). (a) Normal lobsters during queue formation (number observed, $n = 19$); (b) maintenance ($n = 87$). (c) Blinded lobsters during queue formation ($n = 18$); (d) maintenance ($n = 60$). (e) Antennule-deprived lobsters during queue formation ($n = 20$); (f) maintenance ($n = 72$). E, Eyes-vision; A, antennae; a_i, antennule inner rami; P, anterior pereiopods; and x, receptor sensitivity eliminated.

References and Notes

1. D. E. Crawford and W. J. J. DeSmidt, *Bull. U.S. Bur. Fish.* **38**, 28, 281 (1922).
2. C. E. Dawson and C. P. Idyll, *Fla. State Board Conserv. Tech. Ser.* No. 2, 39 (1951).
3. R. Bainbridge, in *The Physiology of Crustacea*, T. Waterman, Ed. (Academic Press, New York, 1961), vol. 2, pp. 431–463.
4. W. F. Herrnkind and W. C. Cummings, *Bull. Mar. Sci.* **14**, 123 (1964).
5. I witnessed a mass migration on 22 October 1965, near Hilsboro lighthouse and was reliably informed by individuals from the Bureau of Commercial Fisheries of another occurrence at Boynton Beach on 1 October 1965. Another migration occurred near Boca Raton in the fall of 1966.
6. D. M Maynard and H. Dingle, *Z. Vergl. Physiol.* **46**, 515 (1963).
7. W. Herrnkind, in preparation.
8. ———, *Amer. Zool.* **8**, 585 (1968).
9. R. M. Storm, Ed., *Animal Orientation and Navigation* (Oregon State Univ. Press, Corvallis, 1967).
10. E. P. Creaser and D. Travis, *Science* **112**, 169 (1950).
11. A. B. Williams, *U.S. Dep. Interior Fish. Bull.* **65**, 91 (1965).
12. C. Von Bonde, *Fla. State Board Conserv. Tech. Ser.* No. 2, 29 (1951).
13. I thank Dr. A. A. Myrberg, Jr., P. B. Wright, S. Ha, and J. D. Richard of the Institute of Marine Science, University of Miami, Florida, for aid and facilities. I appreciate the assistance of Dr. R. Mariscal in preparation of the manuscript. Miss L. Spence prepared the illustrations.

4 March 1969; revised 22 April 1969

Part III

WILL MOVEMENTS BE SUCCESSFUL?

Editors' Comments on Papers 17 Through 20

17 JÄRVINEN and VEPSÄLÄINEN
Wing Dimorphism as an Adaptive Strategy in Water-Striders (Gerris)

18 KENNEDY
A Turning Point in the Study of Insect Migration

19 RAINEY
Weather and the Movements of Locust Swarms: A New Hypothesis

20 METZGAR
An Experimental Comparison of Screech Owl Predation on Resident and Transient White-footed Mice (Peromyscus leucopus)

As dispersal strategies evolve, selection will balance the risks of movement against the benefits (see also Table 1 in Part II). In Part II we examined benefits as expressed by motivations to leave home. Now we consider factors that can affect the success of individuals once they initiate dispersal. Success is measured by survival during traveling and the establishment of a new home, either by immigrating into a preexisting population or by establishing a new population. A more rigorous criterion of success would be successful reproduction in the new home, but this seems to be operationally extremely difficult to assess in most cases.

There is no question that movements away from home are inherently dangerous. One general adaptation for coping with high failure rates is the production of large numbers of dispersers by parents. Not all modes of travel are equally dangerous. Salisbury (1975) has explored this question for plant propagules and reports that wind dispersal is the most dangerous while being carried by other organisms (phoresy) is the most reliable. Mitchell (1970) also concludes that for mites phoresy requires the smallest production of dispersers.

INTRINSIC FACTORS

Central to any consideration of dispersal or migratory success are the characteristics of the individual organisms involved. Relevant properties can be subsumed under *viability, vagility,* and behavioral modifications. Viability refers to the condition and durability of the individual to withstand traveling in unfamiliar and often hostile circumstances. In fact, many species have evolved particular life history stages that are specifically adapted for dispersal. Such stages are called "disseminules" and are exemplified in plants by resistant seeds and spores, and in animals by encysted stages, mobile larvae, and vigorous subadult age classes. The rigor of travel for small mammals is reflected in reduced resistance to mortality in live traps (Andrzejewski and Wrocławek, 1961), as well as increased susceptibility to various extrinsic biotic influences (to be discussed later).

Vagility measures the traveling abilities and tendencies of an organism. Mobility may be completely passive such as drifting in air or water currents. In other cases, such as the floating of hydras or ballooning of spiderlings, some active behavior may be involved in initiating the travel. Similarly, disseminules carried by another organism (phoresy) may actively seek out carriers (hummingbird flower mites, Colwell, 1979) or be picked up passively (cockleburrs on fur). Finally, some dispersers may be active, depending only on their own locomotory capabilities.

Aside from morphological and/or physiological differences specifically affecting vagility that vary among species, we often see such variation among individuals or populations within species. These adaptations may be as subtle as slight differences in body size leading to different dispersal capabilities (and viabilities). For example, Dingle et al. (1980) report that dispersal capability appears to be positively correlated with body size in several species and populations of milkweed bugs, *Oncopeltus,* and Roff (1977) found a similar situation in *Drosophila melanogaster.* The possible advantages of increased size to dispersers may range from greater flight capacity and ability to withstand the rigors of dispersal, to increased reproductive potential at the time of colonization (Dingle, 1980). On the other hand, adaptations affecting vagility may specifically involve major alterations of the locomotory apparatus. Some cases will involve adaptations improving dispersal capability; in other cases, where dispersers are at a disadvantage, selection will favor a reduction or loss of vagility. A few examples selected from the literature on insect dispersal will suffice to make this point.

The development and histolysis of flight muscles during adult life in many insects often relates to trade-offs between the need to disperse and the cost of developing and maintaining flight muscles on the one hand, and the energetic demands of reproduction on the other (Young, 1965). In some cases, as in the Colorado potato beetle,

Leptinotarsa decemlineata, development and histolysis of flight muscles are reversible processes so that one individual may switch back and forth between being capable and incapable of flight as environmental conditions dictate (de Kort, 1969). In other insects, however, the development or histolysis of flight muscles is irreversible and is often facultatively triggered by local conditions. For example, the flight muscles in the corrixid, *Sigara scotti*, do not develop and the insects are flightless unless the water in which they develop becomes quite warm. This condition occurs when small pools are about to dry up in the summer. In this situation, functional flight muscles develop; the obvious advantage being that adults can escape from a deteriorating habitat (Young, 1965). In cotton stainer bugs (*Dysdercus* sps.), females histolyze their flight muscles and begin reproducing in response to adequate food and water (Dingle and Arora, 1973; Derr, 1977).

Many insects show polymorphism for wing length, which may considerably affect their ability to disperse. In some species, these differences are relatively minor as in the migratory locust *Schistocercus gregaria*. In this species the ratio of the wing length to hind femur length is usually around 2 in the solitary phase but increases to approximately 2.3 in the gregarious phase (see Uvarov, 1966, 1977). These differences produce smaller wing loading values in the gregarious phase that may increase flight efficiency for these strongly migratory locusts. Many insects, however, demonstrate a much more dramatic wing polymorphism producing winged and wingless (or at least individuals with the wings so reduced that they are flightless) forms. Good examples of this phenomenon may be found in various aphids (Dixon, 1973), plant hoppers (Denno, 1976), carabid beetles (Carter, 1976), and water striders (Vespäläinen, 1978). Generally, the winged morphs are produced in response to a variety of stimuli signaling the approach of unfavorable conditions while the flightless morphs are produced when conditions favor reproduction. In one typical, well-studied case, Denno and Grissell (1979) report that in the saltmarsh plant hopper, *Prokelisia marginata*, the proportion of long- to short-winged morphs is correlated with the proportion of unstable patches of vegetation in the local environment. Along the United States Gulf coast where the saltmarsh is stable, nearly all of the adults are short-winged. On the eastern seaboard where patches of saltmarsh are more likely to be destroyed, there is a larger proportion of long-winged morphs produced. Furthermore, the level of crowding in nymphs is correlated with the proportion of long-winged adults produced; heavy crowding leads to more long-winged morphs. The threshold of this response is set higher in the more stable Gulf coast populations than it is for Atlantic populations, suggesting a genetically controlled developmental switching mechanism responsive to environmental cues. Finally, perhaps one of the best studied examples of wing-length polymorphism, its control, and its relation to environmental variables, is the work by Vepsäläinen and others on the European water striders (Gerridae). Paper 17 by Järvinen and Vepsäläinen (1976) succinctly reviews much of this work.

An additional class of intrinsic adaptations to dispersal is that of behavioral modifications. General cases of this sort would include increased vigilance or activity levels and changing thresholds of response to various environmental stimuli. Still another common behavioral change concerns feeding. Dispersers may change their feeding times, change the kinds of food taken, or cease feeding altogether. In Paper 18 Kennedy summarizes changes in stimuli thresholds, feeding activities, and other "vegetative" behaviors for dispersing insects. Four recent examples illustrate the diversity of behavioral modifications associated with dispersal. Caldwell and Rankin (1974) document the disassociation of dispersal behavior from feeding and reproduction in the milkweed bug, *Oncopeltus fasciatus*. In the intertidal aphid, *Pemphigus trehernei*, young stages are photopositive, increasing the probability of successful dispersal by tidal currents; those successfully finding new host plants become photonegative (Foster, 1978). In the white-footed mouse, *Peromyscus leucopus,* Tardif and Gray (1978) report the interesting finding that dispersers are much more catholic in their food choices than are residents. Migrating Northern Water Thrushes, *Seiurus noveboracensis,* can switch in and out of a territory-holding mode at various stopover sites depending on their fat reserves and local competition (Rappole and Warner, 1976).

EXTRINSIC FACTORS

In the environment of dispersing and migrating individuals factors that can influence their chances of success can be grouped into physical (nonbiotic), habitat, interspecific, and social factors.

The importance of nonbiotic influences such as barriers, weather, distance, navigational cues, and shelter are obvious and will not be discussed in detail. Many such examples are presented by Wolfenbarger (1975) in his review of dispersal among small organisms. An early paper by Rainey (Paper 19) called attention to the fact that African and Middle Eastern plague locust swarms often appear with a particular weather pattern associated with the Intertropical Convergence Zone. This type of weather system produces the rains needed for the locusts to successfully reproduce and develop, and at the same time (because locusts ride the winds to the convergence zone from over a vast area) concentrates their numbers thus facilitating the production of the gregarious phase. The gregarious phase, in turn, is well-adapted to follow shifts in the position of the Intertropical Convergence Zone. Rainey was incorrect in assuming that the direction taken by the swarm was totally passive depending on wind direction, since we now know that locusts may actively orient downwind (Waloff, 1972). The importance of Paper 19 was in suggesting a weather-related dispersal mechanism that allowed locusts to locate areas of rainfall in a relatively nonpredictable and heterogeneous environment.

Heavy abiotic mortality among migrating birds is well known and mainly affects young individuals. Migrating passerine birds that appear lost by virtue of their having accumulated along coastlines or on offshore islands are predominantly young of the year (Greenberg, 1980). Johnson (1973) provides an unusually well-documented case for greater migratory losses among young Western Flycatchers (see Table 1). While all losses reflected in his figures cannot be attributable to migration—and some collecting biases toward young in the fall and males in the breeding season are likely—the decline in percentage of young during the fall migration is so spectacular that much of this differential loss by age can be attributed fairly to this cause. Substantial migratory mortality has also been documented for the Gray Bat, *Myotis grisecens,* by Tuttle and Stevenson (1977).

Habitat features are generally mixtures of abiotic and biotic elements. Discontinuities in habitat represent a nearly ubiquitous factor influencing the success of animal movements. Gaps in suitable habitat are perilous to cross, and their extent will obviously affect the probability of a traveler reaching the next patch of appropriate habitat. Getz et al. (1978) described the effective use of interstate roadsides for range extension by meadow voles, *Microtus pennsylvanicus,* and many similar cases exist. The pattern of habitat patchiness becomes a critical variable in the evolution of dispersal behavior (see Part V).

TABLE 1. Sex and age ratios in the Western Flycatcher, *Empidonax difficilis,* taken from museum specimens at four periods in the annual cycle. Data are from Johnson, 1973 (p. 214).

Season	n	% ♀♀	% young*
Fall migration	121	45.5	81.8
Wintering	152	40.1	42.1
Spring migration	229	36.7	36.2
Breeding	398	29.6	27.4

*less than 1 year old

Interspecific interactions are also critically important in the success of travelers. Some species will provide food and/or shelter, and their presence will be beneficial. Other species acting as predators, parasites, or competitors will reduce chances of successful movements. For species of small mammals, it is well known that dispersers suffer increased predation pressure (Errington, 1946; Pielowski, 1962; Paper 20; Ambrose, 1972) and possibly even increased ectoparasite loads (Janion, 1961). Paper 20 by Metzgar is a particularly good experimental study involving screech owl predation on white-footed mice. On the other hand, predation pressure may be reduced if predators tend to search mainly in their prey's optimal habitat. Competitors can pose significant threats to travelers. In passing through suboptimal habitats, superior competitors in those circumstances may be encountered. Even in cases where inferior competitors are encountered, numerical advantage and resident status may well put the traveler at a disadvantage. Narise (1965) described a situation in two species of *Drosophila* in which numerical advantage reversed the outcome of competition expected based on competition under equal numbers. Conversely, Ayala (1971) documents a case of interspecific competition between two different species of *Drosophila* in which competitive ability among adults is inversely frequency-dependent, a relationship in which travelers would have the advantage when passing through high density patches of a competitor species.

Perhaps the most interesting and least appreciated type of biotic interactions affecting the success of movements is that of intraspecific or social interactions. Conspecifics interact with dispersers and migrants primarily through aggressive spacing behavior and mate choice. Another possibility is that residents are better competitors for resources because of familiarity with their own home range.

Territorial behavior, both individual and group, is widespread among animals. Such behavior tends to exclude nonresidents from establishing in occupied areas and limits their access to resources. Attempting to displace an established resident is a high risk and an expensive operation. In cases of group territories among mammals, immigration is occasionally successful, but such immigrants generally go to the bottom of the social hierarchy and sometimes have only limited access to resources. Two diverse examples of this situation are the house mouse, *Mus musculus,* studied by Andrzejewski et al. (1963) and Mainardi (1964), among others, and the toque monkey, *Macaca sinica,* reported by Dittus (1980). The phenomenon is probably widespread among the more social species of mammals.

Mate choice also can be an important influence on the success of immigrants. Sometimes immigrants are favored, and sometimes they are not. Based on present knowledge, it would be difficult to predict the nature of any particular case because the interaction among potential mates is complex and subject to numerous variables. In a now classic paper, Santibañez and Waddington (1958) showed that in mate choice experiments among six inbred lines of *Drosophila melanogaster,* there was a tendency for males to prefer mates of their own line in four out of six strains. Females showed the same preferences in only two lines. Such positive assortative mating could discriminate against immigrants. Greenwood et al. (1979) describe mate selection in the Great Tit, *Parus major,* and report that residents are more likely to mate with residents and immigrants with immigrants. Moreover, immigrant males generally end up with younger mates than do residents males.

On the other hand, there exists the strange phenomenon of the "rare-male advantage" described in a variety of insects (see White and Grant, 1977, and review by Ayala and Campbell, 1974), guppies, *Poecilia reticulata* (Farr, 1977), and Parasitic Jaegers, *Stercorarius parasiticus* (O'Donald, 1976). Where this condition prevails, immigrant males at least would clearly have a reproductive advantage. Wallace (1970) reports that "migrants" of *Drosophila melanogaster* mate more successfully than "natives."

Mate choice can also be influenced by social status, and there are many examples among mammals. High social rank generally confers mating choice advantages. Thus, to the extent that immigrants tend to have low social rank, this effect would discriminate against them. Positive assortative mating can also arise from imprinting on parents or on those with whom one is reared. Examples come from *Drosophila* (Mainardi, 1968) and domestic pigeons (Warriner et al., 1963). In mammals, there is a general tendency to avoid selecting mates from among parents or siblings, and this tendency is generally thought to represent an antiinbreeding adaptation. For female house mice, however, this inhibition is reversed if they are offered a choice of males of a different subspecies or artificially perfumed males (Mainardi, 1963; Mainardi et al., 1965). Clearly mate choice is a complex and little-understood phenomenon and yet of critical relevance to the potential immigrant.

REFERENCES

Ambrose, H. W., III, 1972, Effect of Habitat Familiarity and Toe-clipping on Rate of Owl Predation in *Microtus pennsylvanicus, J. Mammal.* **53:**909-912.

Andrzejewski, R., K. Petrusewicz, and W. Walkowa, 1963, Absorption of Newcomers by a Population of White Mice, *Ekol. Pol.,* ser. A, **11:**223-240.

Andrzejewski, R., and H. Wrocławek, 1961, Mortality of Small Rodents in Traps as an Indication of the Diminished Resistance of the Migrating Part of a Population, *Acad. Pol. Sci. Bull.*, Cl. 2, **9**:491–492.

Ayala, F. J., 1971, Competition between Species: Frequency Dependence, *Science* **171**:820–824.

Ayala, F. J., and C. A. Campbell, 1974, Frequency-dependent Selection, *Annu. Rev. Ecol. and Syst.* **5**:115–138.

Caldwell, R. L., and M. A. Rankin, 1974, Separation of Migratory from Feeding and Reproductive Behavior in *Oncopeltus fasciatus, J. Comp. Physiol.* **88**:383–394.

Carter, A., 1976, Wing Polymorphism in the Insect Species *Agonum retractum* Leconte (Coleoptera: Carabidae), *Can. J. Zool.* **54**:1375–1382.

Colwell, R. K., 1979, The Geographical Ecology of Hummingbird Flower Mites in Relation to their Host Plants and Carriers, *Recent Adv. Acarol.* **2**:461–468.

de Kort, C. A. D., 1969, *Hormones and the Structural and Biochemical Properties of the Flight Muscles in the Colorado Beetle,* Mededeling no. 159, Laboratorium voor Entomologie, Wageningen, 63p.

Denno, R. F., 1976, Ecological Significance of Wing Polymorphism in Fulgoroidea which Inhabit Tidal Salt Marshes, *Ecol. Entomol.* **1**:257–266.

Denno, R. F., and E. E. Grissell, 1979, The Adaptiveness of Wing-dimorphism in the Salt Marsh-inhabiting Planthopper, *Prokelisia marginata* (Homoptera: Delphacidae), *Ecology* **60**:221–236.

Derr, J. A., 1977, Population Movements of *Dysdercus bimaculatus* (Pyrrhocoridae, Heteroptera) in Relation to Moisture Stress and the Fruiting Cycles of its Different Host Plants, Ph.D. thesis, Washington University, St. Louis.

Dingle, H., 1980, Ecology and Evolution of Migration, in *Animal Migration, Orientation, and Navigation,* S. A. Gauthreaux, Jr., ed., Academic, New York, pp. 1–101.

Dingle, H., and G. Arora, 1973, Experimental Studies of Migration in the Bugs of the Genus *Dysdercus, Oecologia* **12**:119–140.

Dingle, H., N. R. Blakley, and E. R. Miller, 1980, Variation in Body Size and Flight Performance in Milkweed Bugs (*Oncopeltus*), *Evolution* **34**:371–385.

Dittus, W. P. J., 1980, The Social Regulation of Primate Populations: A Synthesis, in *The Macaques: Studies in Ecology, Behavior and Evolution,* D. G. Lindburg, ed., Van Nostrand Reinhold, New York, pp. 263–286.

Dixon, A. F. G., 1973, *Biology of Aphids,* Edward Arnold, London, 58p.

Errington, P. L., 1946, Predation and Vertebrate Populations, *Q. Rev. Biol.* **21**:144–177, 221–245.

Farr, J. A., 1977, Male Rarity or Novelty, Female Choice, Minority Advantages, and Frequency-dependent Selection in the Guppy, *Poecilia reticulata* Peters (Pisces: Poeciliidae), *Evolution* **31**:162–168.

Foster, W. A., 1978, Dispersal Behaviour of an Intertidal Aphid, *J. Anim. Ecol.* **47**:653–659.

Getz, L. L., F. R. Cole, and D. L. Gates, 1978, Interstate Roadsides as Dispersal Routes for *Microtus pennsylvanicus, J. Mammal.* **59**:208–212.

Greenberg, R., 1980, Demographic Aspects of Long-distance Migration, in *Migrant Birds in the Neotropics: Ecology, Behavior, Distribution, and*

Conservation, A. Keast and E. S. Morton, eds., Smithsonian Institute Press, Washington, D. C., pp. 493-504.

Greenwood, P. J., P. H. Harvey, and C. M. Perrins, 1979, Mate Selection in the Great Tit *Parus major* in Relation to Age, Status and Natal Dispersal, *Ornis Fenn.* **56:**75-86.

Janion, S. M., 1961, Studies on the Differentiation of a House Mouse Population According to the Occurrence of Fleas (Aphaniptera), *Acad. Pol. Sci. Bull.*, Cl. II **9:**501-506.

Johnson, N. K., 1973, Spring Migration of the Western Flycatcher, with Notes on Seasonal Changes in Sex and Age Ratios, *Bird-Banding* **44:**205-220.

Mainardi, D., 1963, Speciazione nel topo. Fattori etologici determinanti barriere riproduttive tra *Mus musculus domesticus* e *M. m. bactrianus*, *Ist. Lombardo. Accad. Sci. Lett. Rendiconti, B.* **97:**135-142.

Mainardi, D., 1964, Interazione tra preferenze delle femmine e predominanza sociale dei maschi nel determinismo della selezione sessuale nel topo (*Mus musculus*), *Accad. Naz. Lincei, Cl. Sci. Fis. Mat. e Nat.* **37:**484-490.

Mainardi, D., M. Marsan, and A. Pasquali, 1965, Causation of Sexual Preferences of the House Mouse. The Behaviour of Mice Reared by Parents whose Odour was Artificially Altered, *Soc. Ital. Sci. Nat. e Mus. Civ. Stor. Nat. Milano Atti* **104:**325-338.

Mainardi, M., 1968, Su alcuni fattori etologici determinanti accoppiamenti assortativi in *Drosophila melanogaster, Ist. Lombardo Accad. Sci. Lett. Rendiconti, B,* **102:**160-169.

Mitchell, R., 1970, An Analysis of Dispersal in Mites, *Am. Nat.* **104:**425-431.

Narise, T., 1965, The Effect of Relative Frequency of Species in Competition, *Evolution* **19:**350-354.

O'Donald, P., 1976, Mating Preferences and Their Genetic Effects in Models of Sexual Selection for Colour Phases of the Arctic Skua, in *Population Genetics and Ecology*, S. Karlin and E. Nevo, eds., Academic, New York, pp. 411-430.

Pielowski, Z., 1962, Untersuchungen über die Ökologie der Kreuzotter (*Vipera berus* L.), *Zool. Jahrb. Syst.* **89:**479-500.

Rappole, J. H., and D. W. Warner, 1976, Relationships Between Behavior, Physiology and Weather in Avian Transients at a Migration Stopover Site, *Oecologia* **26:**193-212.

Roff, D. A., 1977, Dispersal in Dipterans: Its Costs and Consequences, *J. Anim. Ecol.* **46:**443-456.

Salisbury, E. 1975, The Survival Value of Modes of Dispersal, *R. Soc. London Proc., B,* **188:**183-188.

Santibañez, S. F., and C. H. Waddington, 1958, The Origin of Sexual Isolation between Different Lines within a Species, *Evolution* **12:**485-493.

Tardif, R. R., and L. Gray, 1978, Feeding Diversity of Resident and Immigrant *Peromyscus leucopus, J. Mammal.* **59:**559-562.

Tuttle, M. D., and D. E. Stevenson, 1977, An Analysis of Migration as a Mortality Factor in the Gray Bat Based on Public Recoveries of Banded Bats, *Am. Midl. Nat.* **97:**235-240.

Uvarov, B., 1966, *Grasshoppers and Locusts,* vol. 1, Cambridge University, London, 481p.

Uvarov, B., 1977, *Grasshoppers and Locusts,* vol. 2, Overseas Centre for Pest Research, London, 613p.

Vepsäläinen, K., 1978, Wing Dimorphism and Diapause in *Gerris:* Determination and Adaptive Significance, in *Evolution of Insect Migration and Diapause,* H. Dingle, ed., Springer-Verlag, New York, pp. 218-253.

Wallace, B., 1970, Observations on the Microdispersion of *Drosophila melanogaster,* in *Essays in Evolution and Genetics,* M. K. Hecht and W. C. Steere, eds., Appleton-Century-Crofts, New York, pp. 381-399.

Waloff, Z., 1972, Orientation of Flying Locusts, *Schistocerca gregaria* (Forsk.), in Migrating Swarms, *Bull. Entomol. Res.* **62:**1-72.

Warriner, C. C., W. B. Lemmon, and T. S. Ray, 1963, Early Experience as a Variable in Mate Selection, *Anim. Behav.* **11:**221-224.

White, H. C., and B. Grant, 1977, Olfactory Cues as a Factor in Frequency-dependent Mate Selection in *Mormoniella vitripennis, Evolution* **31:**829-835.

Wolfenbarger, D. O., 1975, *Factors Affecting Dispersal Distances of Small Organisms,* Exposition, Hicksville, N. Y., 230p.

Young, E. C., 1965, Flight Muscle Polymorphism in British Corixidae: Ecological Observations, *J. Anim. Ecol.* **34:**353-390.

Wing dimorphism as an adaptive strategy in water-striders (*Gerris*)

OLLI JÄRVINEN and KARI VEPSÄLÄINEN

Department of Genetics, University of Helsinki, Finland

> Several hypotheses have been suggested to account for the adaptive significance of the different wing morphs in water-striders (*Gerris*, Heteroptera). Stability and isolation of population sites should favour short-wingedness; increased rates of population extinction should increase the fitness of the long-winged individuals. Further, if the populations are often resource (food) limited, dimorphism may be optimal. Combinations of other selective pressures can also produce local dimorphism, which need not be optimal — dimorphism can result from mixing of individuals from different population sites. The term *morphism cycle* is coined to express a cyclical change: when a region (comprising a great number of population sites) is initially colonized, long-wingedness is favoured, but short-wingedness becomes more advantageous after the colonization phase. However, if the populations become totally short-winged, they probably face a relatively high risk of extinction, and the cycle may begin anew. The ecological genetics of the Finnish water-striders (nine species) is discussed in connection with the numerous predictions suggested by the original cluster of wing-dimorphism hypotheses.

Olli Järvinen, Department of Genetics, University of Helsinki,
P. Rautatiekatu 13, SF-00100 Helsinki 10, Finland

Wing dimorphism is common in Holarctic water-striders (*Gerris* FABR.), a group of heteropters adapted to living on the water surface. The population sites are discrete "islands", some of them very stable and others more or less temporary. Long-winged (LW, macropterous) imagos can fly at least during part of their lives, but the short-winged (SW) cannot. Morphologically distinct, but functionally equivalent SW submorphs (apterous, micropterous, brachypterous) are known, though rarely in the same population or species. Wing morphism is certainly an adaptation to the structure and pattern of the environment — it is simply implausible to explain its occurrence by pleiotropy or by chance.

Because different species and populations live in different environments, patterns of wing or alary morphism vary. The nine Finnish species are univoltine or partially bivoltine, and their wing morphism in Finland can be classified as in Table 1 (see VEPSÄLÄINEN 1974a for data and certain reservations). The situation outside Finland may be quite different: Some Central European populations of *G. najas* are dimorphic (e.g. KRAJEWSKI 1969). Populations of *G. odontogaster*, *G. argentatus* and *G. lacustris* have dimorphic summer generations in some parts of Europe (documented for Hungary by VEPSÄLÄINEN 1974c). The summer generation of *G. thoracicus* is LW in Hungary (VEPSÄLÄINEN 1974c). Certain populations, especially in mountain habitats, support SW *G. rufoscutellatus* individuals seasonally.

Our study is based on the Finnish set of species and mainly on Finnish conditions. The geography of the wing dimorphism in *Gerris* outside Finland may be studied in detail in VEPSÄLÄINEN and KRAJEWSKI (1974; Poland), VEPSÄLÄINEN (1974c; Hungary), MITIS (1937; Austria), VEPSÄLÄINEN and NIESER (unpubl.; the Netherlands), POISSON (1924; France), BRINKHURST (1959; Great Britain) and ANDERSEN (1973; Denmark).

The diapause condition of water-striders is determined by the direction of change in day length (VEPSÄLÄINEN 1971, 1974d). If day length increases during the critical larval stages, the individual begins reproducing after it becomes imago, but if day length shortens, the individual is in diapause after eclosion and overwinters before it reproduces. So the proper column in our table for univoltine populations is that of the diapause generation. In many Finnish populations the diapause condition can be directly connected with long wings, but the correlation does not apply to all species (e.g. *G. lateralis*) or all places (e.g. Hungary).

Table 1. Wing morphism in the nine Finnish water-strider species
LW indicates long-wingedness, and SW indicates short-wingedness. The diapause generation of *G. sphagnetorum* is known only from about 40 overwintered SW individuals captured in May—June; LW individuals occur in Sweden (GAUNITZ 1947). The diapause generation overwinters; it is the diapause generation which occurs in univoltine populations

Species	Summer generation	Diapause generation
G. rufoscutellatus LT.	not known	LW
G. odontogaster (ZETT.)	SW	LW
G. argentatus SCHUMM.	SW	LW
G. thoracicus SCHUMM.	SW	LW
G. paludum FABR.	SW	LW
G. najas (DE G.)	not known	SW
G. sphagnetorum GAUN.	not known	SW
G. lateralis SCHUMM.	not known	dimorphic
G. lacustris (L.)	SW	dimorphic

We shall investigate patterns of wing morphism mainly in univoltine populations. However, the complications due to bivoltinism are studied, although not incorporated in the basic arguments. It is possible that a partial third generation emerges in, e.g. Hungary (VEPSÄLÄINEN 1974c), but the possible existence of a third generation does not affect our conclusions. The determination of wing length has been investigated in detailed studies elsewhere (VEPSÄLÄINEN 1971, 1974b). It is the purpose of this paper to elucidate the adaptive significance of wing dimorphism in *Gerris*.

The hypotheses

1. Seasonal patterns

It has not yet been stated that the probability of SW-ness greatly increases in imagos which moult towards the end of the season (ANDERSEN 1973; VEPSÄLÄINEN 1974a, 1974b). This is true of *G. lacustris*, but possibly also of several other species. The increase can be interpreted as an adaptive response to the hazards of the approaching end of the season: SW individuals do not construct the flight apparatus, and they are thus physiologically ready to overwinter sooner than the LW ones. Not only time may be short, but also food shortage may be severe near the end of the season. As a consequence, the necessity to find surplus energy for the flight muscles can be quite disadvantageous for the LW individuals.

The fitness of SW individuals late in the season might also be increased, because these individuals have been produced in a pond which is likely to be available also the following spring (H. Dingle, in litt.).

Another seasonal pattern can be discerned in our table — if any differences exist, SW-ness is more typical of the summer than the diapause generation. Also this pattern seems best explained by the observation that no flight apparatus has to be constructed by the SW individuals, and they can thus reproduce sooner than the LW individuals (ANDERSEN 1973). The advantages of early reproduction in the summer generation are two-fold: the offspring larvae thus avoid cannibalism by larvae developing earlier, because these do not exist, and they also avoid the hazards of the late season (e.g., early cold spells). Other fitness differences do not seem to account for the tendency of the summer generation to be more characteristically SW than LW. The egg production rates of LW and SW females appear to be similar at least in *G. lateralis* (GUTHRIE 1959) and *G. lacustris* (ANDERSEN 1973; VEPSÄLÄINEN 1974a), despite a considerable difference between the morphs in size. Though not adequately studied, differences in mortality are not likely to occur, as regards the summer generation.

Although the above arguments indicate why SW-ness provides certain advantages in the summer generation or in the late individuals of the diapause generation, they do not suggest that SW-ness would be more favoured than LW-ness. Other selective factors may well tip the balance in favour of LW-ness. Such factors include habitat instability, i.e. drying up of the pond, but its effects will be investigated below. The main principles can nevertheless be easily applied also to bivoltinism — instability is fatal to the SW water-striders. However, opposite selective pressures do not necessarily imply that both morphs are maintained in a population, but it is quite possible that monomorphism is produced.

Our investigation of the adaptive features of the known seasonal patterns amounts to the statement that the occurrence of seasonal variation in *Gerris* wing dimorphism is not a difficult problem. But why is there sometimes monomorphism, sometimes dimorphism in a given generation?

2. Populations in a spatially and temporally varying environment

VEPSÄLÄINEN (1974a) has studied the adaptive significance of *Gerris* wing dimorphism using fitness sets (see LEVINS 1962, 1968). His discussion, including also a review of previous suggestions, has

175

recently been supplemented by JÄRVINEN (1976), who used LEVINS's (1969, 1970) migration-extinction models for an investigation of the same problem. Both of us started from similar general assumptions, and our main conclusions are similar, even if not identical. We have thus concluded that

(1) If the population sites are unstable, LW-ness is favoured.

(2) If the population sites are well-isolated, SW-ness is favoured.

(3) Because poor productivity of the habitats decreases equilibrium population sizes, population extinction is more frequent in less productive environments. Hence, more sites will be available for colonists, and decreased productivity thus favours LW-ness.

The optimum structure of a population is determined by the net effect of these factors on individual fitness values. Our studies suggest that it is impossible to attain dimorphism in *Gerris* populations, unless at least one of the following two conditions prevail:

(1) Dimorphism may result from the mixing of individuals: immigrating LW individuals can make a population dimorphic, even if it would be SW otherwise.

(2) Dimorphism probably results, if different sites, connected with each other by migrants, have varying probabilities of drying up. In other words, habitat heterogeneity based on differences in the degree of stability may give rise to optimal dimorphism in the whole system of sites.

We have previously neglected population dynamics in the study of wing dimorphism in *Gerris*. As a result, our previous view can be summarized briefly as follows. SW-ness is always more advantagous *within* a population site, because the LW individuals may not return after spring and autumn flights. But every population site will sooner or later be "empty" for colonists, owing to population extinction, and LW-ness is thus no doubt *the* strategy for dispersal between the sites.

3. **Resource-limitation in *Gerris***

Several authors have pointed out that water-striders are often cannibalistic in laboratory and in nature (e.g. ESSENBERG 1915; BUENO 1917; RILEY 1922; BRINKHURST 1966; VEPSÄLÄINEN 1971; JAMIESON 1973; MATTHEY 1974). It has been suggested that cannibalism involves nutritional benefits in at least two generally herbivorous beetles, *Tribolium* (HO and DAWSON 1966; MERTZ and ROBERTSON 1970) and *Labidomera clivicollis* (EICKWORT 1973), but *Gerris*

spp. are predators. Their non-cannibalistic food consists of insects, died or still living, which have been captured by the water surface. The only adaptive explanation of cannibalism in *Gerris* seems to be that cannibalism occurs as a response to food shortage. Our interpretation is also supported by our experience from laboratory rearings: if the experimenter causes food shortage, cannibalism may become very intense.

The females lay some tens of egg batches during one or two months. E.g., females of *G. lacustris* laid in laboratory experiments (unpubl.) about 10 eggs/day for several weeks in batches of 1—30 eggs (average size of 173 batches 4.6 eggs). If resources were not limiting, it might be presumed that, as a result of natural selection, the main reproductive effort would be more concentrated in the initial phase of the reproductive season (LEWONTIN 1965).

If the reproductive season is long, the number of overwintering offspring is most efficiently multiplied if the number of generations is increased, i.e. if changes occur in the mechanism which determines diapause (see Introduction). This strategy has not evolved in *Gerris*, but the advantages of a long reproductive season (e.g. in Hungary) are exploited by an extended life-span of the reproducing imagos (VEPSÄLÄINEN 1974a, 1974c). An extended egg-laying period is advantageous, because the diapause offspring are removed to their terrestrial overwintering sites after about three weeks from eclosion. As a result, the total number of diapausers is large compared with the number of water-strider imagos living simultaneously in the population site (pond).

Our next argument is based on population dynamics. It can easily be calculated that realistic birth and death rates, without cannibalism or other density-dependent mortality, very soon produce much higher numbers than those really seen in nature. Parasitism may occasionally be intense (FERNANDO and GALBRAITH 1970; VEPSÄLÄINEN and JÄRVINEN 1974; see also HUNGERFORD 1920; POISSON 1940; LIPA 1968), but its occurrence is sporadic, while predation occurs in most populations, but it appears to be relatively insignificant in most cases (see e.g. ANDERSON 1932; FROST and MACAN 1948; CALLAHAN 1974). Density-dependent mortality (especially cannibalism) thus seems to be the most realistic explanation for the observed stability of numbers. One more argument deserves mention: it is hard to find natural *Gerris* populations which give the impression that food is continuously abundant. Because most of our field work has taken place when conditions have been favourable for insect life, the really critical days when the influx of prey has more or less ceased owing to cold spells, etc., have little influence on our view.

176

4. Resource-limitation and wing dimorphism: Density-dependent selection

We have suggested that many *Gerris* populations are often resource-limited, in other words, near the maximum numbers determined by the food supply. *Gerris* populations cannot "overfish" their food supplies — the carrying capacity cannot be exceeded (except very temporarily through inactivity and consumption of food reserves). We certainly do not believe that the population numbers are continuously near the carrying capacity of the environment, even if the only whole-season study of the numbers of water-striders revealed relatively constant biomasses throughout the season (VEPSÄLÄINEN 1971). Our arguments only presuppose that the populations are often enough affected evolutionarily by the exhaustion of their food supply.

It seems that the simulation models of ROFF (1975) are very pertinent to our problem. He studied discrete "island" populations connected with dispersing individuals. All "islands" had a stochastically varying carrying capacity as well as a stochastic rate of population increase. Density-dependent regulation of the populations was introduced by the assumption that the carrying capacity cannot be exceeded. Dispersion implied a risk of failure in the models. The tendency to disperse was determined by varying mechanisms, genetic and non-genetic. ROFF concluded from his computer simulations that a stable dispersal polymorphism is generated under a wide range of conditions.

The available data suggest that ROFF's general assumptions are valid for *Gerris* populations. We thus add a third possibility of dimorphism in *Gerris* (see above for the two other possibilities): if the populations are often enough limited by the carrying capacity of the environment, a stable dimorphism is evolved. Because ROFF's (1975) models are very general, we give here a simple model devised for *Gerris*.

Fitness of individuals will be measured by the contribution of reproducing individuals to the next generation, taking into account the fact that different offspring individuals may have different chances of leaving offspring. We especially think of the improbable, but potentially great success of a long-winged individual finding an empty, but favourable population site.

Let us assume that N individuals reproduce in a metapopulation of population sites. Each of the animals has b offspring; if the proportion of LW offspring is p, there are bpN LW offspring. The proportion E of the LW individuals emigrate to empty sites in the metapopulation area, while $1-E$ stay in the inhabited sites. As a result, $bpEN$ emigrate. Of the LW individuals, $bp(1-E)N$ stay, along with $b(1-p)N$ SW individuals. If the inhabited population sites do not support more than K individuals, the additional individuals are removed. It is assumed that there is no difference between the morphs when their probability to survive or to perish is assessed. We denote the success of emigrant ($bpEN$ LW individuals) by P, which takes into account both the probability of finding another site and the average fitness there. The average fitness of those individuals that remain is assumed to be 1, i.e. it is assumed that the population size is stable in the inhabited sites.

On these assumptions, it is equally fit to produce LW as SW offspring in the metapopulation if

$$bN(1 - E\hat{p}) = K/P$$

The equilibrium proportion of the LW individuals, $\hat{p}$, is then

$$\hat{p} = \frac{1 - K/PbN}{E}$$

Because values of $\hat{p}$ are meaningful only if $0 \leq \hat{p} \leq 1$, $\hat{p} \leq 0$ implies that production of SW individuals is more advantageous, while $\hat{p} \geq 1$ implies the favourability of LW offspring. Intermediate values imply dimorphism. It can be shown that non-trivial equilibria ($0 < \hat{p} < 1$) are stable: the advantage of producing LW offspring is higher than that of producing SW offspring if $p < \hat{p}$, but the relationship changes if $p > \hat{p}$. It can further be shown that fitness is a function of density, but it is also a function of morph frequency (p), because density depends on p. As CLARKE (1972) points out, density-dependent selection is often difficult to distinguish from frequency-dependent selection.

We conclude that a form of density-dependent selection may account for wing dimorphism in *Gerris* populations. Such dimorphism may well be stable. This conclusion does not exclude the possibilities we have suggested earlier (VEPSÄLÄINEN 1974a; JÄRVINEN 1976). The novel feature of the present hypothesis is that we assume that many *Gerris* populations are relatively often resource-limited.

Discussion

1. Carrying capacity of the environment

Our new model assumes that populations reach the carrying capacity of the environment (K) sufficiently often to be affected evolutionarily. K as such does not have immediate effects on the wing morphism complex, except that lowering of K most probably increases extinction rates, and thus the probability of success in colonization. Consequently, decreased carrying capacities favour LW-ness. Both our previous models and the present one share this conclusion, though the present model may, under certain conditions, predict dimorphism and not monomorphism. In the previous models a decrease of K increased the fitness difference between the two morphs.

2. Emigration

Lowering of K also favours emigration, because LW-ness is favoured. Emigration of the LW waterstriders — with a flight-oogenesis syndrome (see JOHNSON 1969) — implies that the population responds to the selection pressures which arise because of resource-limitation ('K selection'; see MACARTHUR 1962, 1972: p. 226) by an increased production of individuals behaving like K-strategists (KENNEDY 1975).

Regionally low carrying capacities of the environment in combination with efficient isolation may make certain areas inaccessible for water-striders. *Gerris lateralis* occupies mostly isolated and small permanent springs in southern Finland. The species is rare in this region; most populations are dominated by SW individuals (there are even monomorphically SW populations). This fact is compatible with our models, because sufficiently isolated populations should be SW. In northern Finland the habitat range of the species is expanded, presumably owing to competitive release (VEPSÄLÄINEN and JÄRVINEN 1974). As a result, the habitats are less isolated and, on the average, less permanent than the springs the species inhabits in southern Finland. Our models are compatible with the observed northwards increase in the frequency of the LW individuals.

In a resource-limited environment inter-reproductive emigration (i.e., imagos emigrate after they have reproduced for some time) is apparently advantageous. It is thus implied that it is more advantageous to attempt a flight to a potentially empty population site than to remain in the site already occupied. This assumption is generally not valid, but it can be supported here, because the later offspring of the remaining individuals have a very low chance of becoming imagos under the cannibalistic pressure from hundreds of thousands of older larvae. Certain data from natural populations are compatible with the hypothesis, but are hard to explain by any alternative hypotheses. This is especially true of male-dominated mass emigrations (*G. odontogaster*: VEPSÄLÄINEN 1971; *G. lacustris*: VEPSÄLÄINEN 1974a: p. 26) of overwintered imagos before the developing larvae will demand plenty of the resources. Sometimes it may be possible that early senescence evolves by kin selection in resource-limited populations, but this possibility will not be pursued further here.

3. Isolation

In our model, lowering of the success of LW migrants (measured by P) favours SW-ness. Because P is affected by isolation and by the rate of population extinction, we expect that two populations, both of which live in stable environments, differ greatly in their morph frequencies, if one of them is isolated and the other is close to other population sites.

Relatively isolated populations of *G. lacustris* can be found on Hanko peninsula in S Finland. These populations can be compared with the less isolated ones in other parts of S Finland. The frequencies are much higher on Hanko peninsula; one certainly isolated population has even evolved stable monomorphic brachypterism (Tvärminne population 4016, see VEPSÄLÄINEN 1974a). Similarly, *G. lateralis* is predominantly or totally apterous in regions where it is rare — the Netherlands (N. Nieser, pers. comm.), Denmark (LETH 1943; VEPSÄLÄINEN 1974a), and the southernmost parts of Finland (VEPSÄLÄINEN 1974a). All Finnish individuals of *G. sphagnetorum* and *G. najas* which have been examined were SW; these two species occur very sparsely in Finland. LW individuals have been captured in regions where isolation between suitable population sites appears to be less efficient.

4. Winter mortality

Winter mortality is very likely density-independent (i.e. the probability of survival through the winter is not affected by density). If it is heavy, the diapause population must be large if extinction is to be avoided. Interesting hypotheses emerge from this deduction.

Even the northernmost populations of *G. argentatus* produce a partial second generation, at least in favourable summers. The same is apparently true of *G. paludum*. As a hypothesis, we suggest that *G. argentatus* (and *G. paludum*) cannot expand to more northern areas, because the more northern populations would not be able to build up sufficiently large diapause populations. Our argument presupposes that the northernmost populations of *G. argentatus* are *not* resource-limited, even if the more southern populations can be. If the Finnish *G. argentatus* populations were often resource-limited, there would not be a dramatic difference between uni- and bivoltinism, because the population sites do not seem to be particularly unstable (VEPSÄLÄINEN 1973). These assumptions seem restrictive, but, actually, it is sufficient to assume that there is a considerable difference in winter mortality between the Finnish and, say, Hungarian populations.

The diapausing LW *G. lacustris* are probably cold-hardier than the SW ones (VEPSÄLÄINEN 1974a). We thus expect that the nothernmost populations would be more or less LW, because winter mortality is presumably most severe in the north. However, available population sites for *G. lacustris* appear to be more isolated and more permanent in N Finland than in S Finland (VEPSÄLÄINEN 1973, 1974a). As a hypothesis, the sharp northern boundary of this species is suggested to arise from strong selective pressures against both morphs. Because the frequency of SW individuals increases clinally northwards in the region covered by univoltine populations (VEPSÄLÄINEN 1974a), isolation and stability of the habitats seem to be more important selective factors than winter mortality in this case.

5. Voltinism

We have already touched the interesting topic of voltinism. It may appear that the water-striders would put a voluntary end to their reproductive season, because all individuals developing after the summer solstice belong to the diapause generation. Such voluntarism could hardly be naturally selected; in fact, the present hypothesis of resource-limitation offers an adaptive explanation. Recall that water-striders either reproduce or start overwintering after their eclosion, but if they reproduce, they fail to overwinter. It is thus not advantageous to produce more generations than what is allowed by the carrying capacity of the environment, plus the individuals which can emigrate.

An important cause favouring multivoltinism in more southern populations is probably the increased temporarity of the population sites. If the season is long and if the probability that the population sites dry up is comparatively high, multivoltinism may be advatageous, because more individuals are thus able to disperse to new sites.

6. Determination of wing length

Developmental (ontogenetic) switch mechanisms occur widely in the insect world, and insects respond to a great variety of environmental signals so that morphologically distinct morphs are produced. The determination of sex, reproductive type, diapause condition and wing length in aphids is a case in point (see LEES 1966). In earlier studies (VEPSÄLÄINEN 1971, 1974b), it has been observed that the change of day length and temperature are the main environmental stimuli involved in the determination of wing length in *Gerris*. We have now found that resource-limitation, probably food limitation, can maintain wing-length dimorphism in *Gerris* populations. However, we think that crowding or the amount of food are not likely to be involved in the *mechanism* of wing-length determination. In other words, crowding and/or food shortage are probably important ultimate, but not proximate causes for the wing-length dimorphism in *Gerris*. Note that the length of the wings of a water-strider is determined more than two weeks before the individual is able to fly (if it has become a LW individual). Present crowding or food supply are presumably poor predictors of the availability of resources after a period of two or more weeks, because "overfishing" is impossible. Also food crises should be difficult to predict from these variables, because the crises probably occur during occasional cold and rainy periods. Herbivorous animals or many other predators than water-striders are, in principle, able to deplete their resources, and hence crowding now often predicts food shortage tomorrow. Consequently, the type of food consumed may determine whether crowding or food shortage is involved in the determination mechanism of wing length or not!

7. Morphism cycles

Vacant sites are abundant in larger regions (comprising a great number of population sites) after environmental catastrophes (e.g., several cold or hot summers). Thus LW-ness should be favoured after such catastrophes. If the population sites tend to be stable, they are gradually colonized, and the selective advantage involved in flight ability decreases. If the population sites are relatively isolated, it is well possible that monomorphic SW-ness is optimal. As two possible examples, we mention the monomorphically SW species in Finland, *G. najas* and *G. sphagnetorum*. The populations may originally have been LW, but have become SW during a longer period of selection against LW-ness. The population sites of these two species have probably always been relatively isolated, because they are the most extreme habitat specialists of the Finnish water-striders, along with *G. lateralis* in S Finland. As noted above, also the southern populations of *G. lateralis* are SW, or nearly so.

However, SW populations almost certainly face a higher risk of extinction than the dimorphic or LW ones, if any kind of catastrophe occurs, because dispersal ability is most advantageous when the population site inhabited rapidly deteriorates. As a consequence, the populations of *G. najas* and *G. sphagnetorum* may be on their way to extinction in Finland. New colonisations from dimorphic or LW populations could establish the species again, but the cycle — termed here *morphism cycles* — would begin anew. Detailed studies of the geographical distribution of the populations of *G. najas* and *G. sphagnetorum* will probably prove to be very illuminating. LINDROTH (1949) has discussed several factors favouring SW-ness and LW-ness in carabid beetles. As a suggestion, several instances of observed dimorphism could possibly be understood in terms of morphism cycles — i.e., there may be one favoured morph (SW) and one relic morph (LW).

Acknowledgments. — We are very grateful for the comments and criticism given by those who read the manuscript: H. Dingle, J. F. Grassle, I. Hanski, C. H. Lindroth, L. Oksanen, P. Pamilo, R. Ricklefs, B. Svensson, S.-L. Varvio-Aho and C. Wiklund. The work reported above was begun when O. J. was supported by a grant from the Finnish Academy of the Sciences.

Literature cited

ANDERSEN, N. MØLLER 1973. Seasonal polymorphism and developmental changes in organs of flight and reproduction in bivoltine pondskaters (Hem. Gerridae). — *Entomol. Scand.* 4: 1—20

ANDERSON, L. D. 1932. A monograph of the genus *Metrobates*. — *Univ. Kansas Sci. Bull.* 20: 297—311

BRINKHURST, R. O. 1959. Alary polymorphism in the Gerroidea (Hemiptera-Heteroptera). — *J. Animal Ecol.* 28: 211—230

BRINKHURST, R. O. 1966. Population dynamics of the large pond-skater *Gerris najas* DEGEER (Hemiptera-Heteroptera). — *J. Animal Ecol.* 35: 13—25

BUENO, J. R. DE LA TORRE 1917. Life-history and habits of *Gerris remigis* SAY (Hem.). — *Entomol. News* 28: 201—208

CALLAHAN, J. R. 1974. Observations on *Gerris incognitus* and *Gerris gillettei* (Heteroptera: Gerridae). — *Proc. Entomol. Soc. Washington* 76: 15—21

CLARKE, B. C. 1972. Density-dependent selection. — *Am. Natur.* 106: 1—13

EICKWORT, K. R. 1973. Cannibalism and kin selection in *Labidomera clivicollis* (Coleoptera: Chrysomelidae). — *Am. Natur.* 107: 452—453

ESSENBERG, C. 1915. The habits of the water-strider, *Gerris remigis* (ORBA). — *J. Animal Behav.* 5: 397—402

FERNANDO, C. H. and GALBRAITH, D. 1970. A heavy infestation of gerrids (Hemiptera-Heteroptera) by water mites (Acarina-Limnocharidae). — *Can. J. Zool.* 48: 592—594

FROST, W. E. and MACAN, T. T. 1948. Corixidae (Hemiptera) as food of fish. — *J. Animal Ecol.* 17: 174—179

GAUNITZ, D. 1947. *Gerris sphagnetorum* n. sp. (Heteropt.). — *Opuscula Entomol.* 12: 34

GUTHRIE, D. M. 1959. Polymorphism in the surface water bugs (Hemipt.-Heteropt.: Gerroidea). — *J. Animal Ecol.* 28: 141—152

HO, F. K. and DAWSON, P. S. 1966. Egg cannibalism by *Tribolium* larvae. — *Ecology* 47: 318—322

HUNGERFORD, H. B. 1920. The biology and ecology of aquatic and semiaquatic Hemiptera. — *Kansas Univ. Sci. Bull.* 11: 1—328

JAMIESON, G. S. 1973. Coexistence in *Gerris*. — Ph.D. Thesis abstract. Univ. British Columbia, Vancouver, B.C., Canada

JÄRVINEN, O. 1976. Migration, extinction, and alary morphism in water-striders (*Gerris* FABR.). — *Ann. Acad. Sci. Fenn. Ser. A IV* 206: 1—9

JOHNSON, C. G. 1969. Migration and Dispersal of Insects by Flight. — *Methuen, London*

KENNEDY, J. S. 1975. Insect dispersal. — In *Insects, Science and Society* (Ed. D. PIMENTEL), Academic Press, New York, London, p. 103—119

KRAJEWSKI, S. 1969. Wasserwanzen (Heteroptera) des Flusses Grabia und seines Überschwemmungsgebietes. — *Bull. Entomol. Pologne* 39: 465—513 (in Polish)

LEES, A. D. 1966. The control of polymorphism in aphids. — *Advan. Insect Physiol.* 3: 207—227

LETH, K. O. 1943. Die Verbreitung der dänischen Wasserwanzen. — *Entomol. Meddel.* 23: 399—419

LEVINS, R. 1962. Theory of fitness in a heterogeneous environment. I. The fitness set and adaptive function. — *Am. Natur.* 96: 361—378

LEVINS, R. 1968. Evolution in Changing Environments. — *Princeton Univ. Press, Princeton, New Jersey*

LEVINS, R. 1969. Some demographic and genetic consequences of environmental heterogeneity for biological control. — *Entomol. Soc. Am. Bull. (Washington)* 15: 237—240

LEVINS, R. 1970. Extinction. — In *Some Mathematical Questions in Biology, Am. Mathem. Soc.*, p. 75—107

LEWONTIN, R. C. 1965. Selection for colonizing ability. — In *The Genetics of Colonizing Species* (Ed. H. G. BAKER and G. L. STEBBINS), Academic Press, New York, p. 77—94

LINDROTH, C. H. 1949. Die Fennoskandischen Carabidae. Eine Tiergeographische Studie. III. Allgemeiner Teil. — *Meddel. Göteborgs Mus. Zool. Avd.* 122: 1—911

LIPA, J. J. 1968. Some observations on flagellate parasites of hemipterans *Corimelaena, Euschistus, Gerris, Leptocoris* and *Oncopeltus* in the United States. — *Acta Protozool.* 6: 59—68

MACARTHUR, R. H. 1962. Some generalized theorems of natural selection. — *Proc. Nat. Acad. Sci.* 48: 1893—1897

MACARTHUR, R. H. 1972. Geographical Ecology. — *Harper and Row, New York*

MATTHEY, W. 1974. Contribution à l'écologie de *Gerris remigis* SAY sur deux étangs de Montagnes Rocheuses. — *Mitt. Schw. Entomol. Ges.* 47: 85—95

MERTZ, D. B. and ROBERTSON, J. R. 1970. Some developmental consequences of handling, egg-eating, and population density for flour beetle larvae. — *Ecology* 51: 989—998

MITIS, H. VON 1937. Ökologie und Larvenentwicklung der mitteleuropäischen *Gerris*-Arten (Heteroptera). — *Zool. Jahrb. (Abt. Syst.)* 69: 337—372

POISSON, R. 1924. Contributions a l'étude des Hémiptères aqatiques. — *Bull. Biol. France Belg.* 58: 49—305

POISSON, R. 1940. Contributions a l'étude des *Gerris* de la France et de l'Afrique du Nord. — *Bull. Soc. Sci. Bretagne* 17: 140—173

RILEY, C. F. C. 1922. Droughts and cannibalistic responses of the waterstrider, *Gerris marginatus* SAY. — *Bull. Brooklyn Entomol. Soc.* 17: 79—87

ROFF, D. A. 1975. Population stability and the evolution of dispersal in a heterogeneous environment. — *Oecologia 19*: 217—237

VEPSÄLÄINEN, K. 1971. The role of gradually changing daylength in determination of wing length, alary dimorphism and diapause in a *Gerris odontogaster* (ZETT.) population (Gerridae, Heteroptera) in South Finland. — *Ann. Acad. Sci. Fenn. Ser. A IV 183*: 1—25

VEPSÄLÄINEN, K. 1973. The distribution and habitats of *Gerris* FABR. species (Heteroptera, Gerridae) in Finland. — *Ann. Zool. Fenn. 10*: 419—444

VEPSÄLÄINEN, K. 1974a. The life cycles and wing lengths of Finnish *Gerris* FABR. species (Heteroptera, Gerridae). — *Acta Zool. Fenn. 141*: 1—73

VEPSÄLÄINEN, K. 1974b. Determination of wing length and diapause in water-striders (*Gerris* FABR., Heteroptera). *Hereditas 77*: 163—176

VEPSÄLÄINEN, K. 1974c. The wing lengths, reproductive stages and habitats of Hungarian *Gerris* FABR. species (Heteroptera, Gerridae). — *Ann. Acad. Sci. Fenn. Ser. A IV 202*: 1—18

VEPSÄLÄINEN, K. 1974d. Lengthening of illumination period is a factor in averting diapause. — *Nature 247*: 385—386

VEPSÄLÄINEN, K. and JÄRVINEN, O. 1974. Habitat utilization of *Gerris argentatus* (Het. Gerridae). — *Entomol. Scand. 5*: 189—195

VEPSÄLÄINEN, K. and KRAJEWSKI, S. 1974. The life cycle and alary dimorphism of *Gerris lacustris* (L.) (Heteroptera, Gerridae) in Poland. — *Notulae Entomol. 54*: 85—89

A TURNING POINT IN THE STUDY OF INSECT MIGRATION*

By Dr. J. S. KENNEDY

Agricultural Research Council Unit of Insect Physiology, Entomological Field Station,
34A, Storey's Way, Cambridge

C. G. JOHNSON[1] has crystallized a new trend, since the War, in the field of insect migration and dispersal by flight. The trend is inspired largely by work on aphids and locusts, which has been more intensive than that on economically less-important insect migrants such as butterflies. The new knowledge gained cannot be fitted into the previous framework of behavioural ideas due to C. B. Williams[2-4]; therefore Johnson (working on aphids) and Rainey[5] (on locusts) propose shifting the emphasis to the timing and results of migration. This would mean abandoning, along with his particular ideas, Williams's general aim: to characterize migration ecologically, behaviourally and physiologically. The purpose of this article is to suggest that the recent advances, far from giving cause to abandon that aim, have brought it within reach.

Records of migrating insects other than locusts have been collected and collated for many years under the leadership of Williams, and the outstanding achievements of this work are: the demonstration that migration is much commoner than was supposed, even if we count only the long-range and geographically clear-cut cases that Williams considers; and secondly, the detection of apparent return movements in an ever-growing number of species. Lack of evidence of any return in many cases had previously posed a problem of the evolutionary origin of the migratory habit and lent colour to the idea that migration served mainly to reduce high populations[6]. But in all the observations so collated the chief criterion used for distinguishing insects as migrant is the peculiarly persistent, undeviating manner of flight of butterflies, dragonflies, locusts and other large powerful fliers when close to the ground. This gives them a mysteriously purposeful air that rivets the attention of observers and makes teleological interpretation almost irresistible. This in turn proved an impediment to the development of the subject on the physiological and ecological sides alike.

Williams inferred that such migrants continue to fly low, holding to one compass direction for hours and days on end, and in this way directly determine the route of their long-distance movements. This has recently been disproved for locusts and questioned for some butterflies[7], and has not yet been verified for any insect. But taken as fact it posed an intractable problem of the mechanism of orientation. Williams even suggested that migrants had an innate sense of direction independent of the environment. There the matter of direction determination has rested, while other physiological problems of migration have remained in the background.

Furthermore, only insects flying in that self-propelled, self-directed manner were regarded by Williams as migrants. The dispersal of smaller, weaker, more buoyant insects such as winged aphids or moth larvæ that drift on the wind, merely supporting themselves by their wing beating or on silken kites, was not migration. Williams's definitions of migration are to be understood in that exclusive sense: "a change of location which is determined both in distance and direction by the insect itself, and not a passive distribution of individuals by overpowering forces such as strong winds"[2]; or "movement determined by the insect itself; [which] instead of being haphazard in direction, is straightened out so as to result in a change of habitat"[4].

Movements that result in a change of habitat are, of course, of major concern to ecologists[6,8,9]. They, however, regard regular long-range to-and-fro movements between discrete areas as special cases. Hence all the work done on such migration has seemed of doubtful relevance to everyday ecological work. Ecologists use the looser term 'dispersal' for the general phenomenon in which they are interested, and Elton's[8] description of dispersal as "taking place all the time as a result of the normal life of the animals" seems quite the opposite of Williams's basic idea of migration as a distinct biological process. In the absence of more information on dispersal mechanisms, ecologists who do believe there are distinct behavioural adaptations for dispersal', have had difficulty in describing them objectively[9].

Recent Advances

Locusts and aphids have, in the past, served as excellent illustrations of the supposed contrast between active migration and passive dispersal. But it turns out that locusts migrate on the wind no less than winged aphids: and winged aphids are, in their own way, no less specialized as migrants than locusts. Both are active, 'deliberate' migrants; and both achieve their long-range movements 'passively'. This is what has made a broadening of the idea of migration seem pressing[1,5].

The new phase opened in the locust work (mainly on the desert locust, *Schistocerca gregaria* Forsk.) when it was realized that the orientation of fliers and the direction of their long-range migrations were separate problems: the one did not directly determine the other[10,11]. The explanation of their purposeful air when seen flying close to the ground does not lie in any type of orienting stimulus capable of maintaining a fixed course and track for long periods. It could be explained in terms of familiar types of orienting stimuli from the immediate environment. These have a steadying effect on courses and tracks near the ground, but certainly do not fix them. For example, low-flying locusts,

---

* Based on papers contributed by the author to symposia on migration at the eleventh International Congress of Entomology, Vienna, August 18, 1960, and at the Anti-Locust Research Centre, London, September 22, 1960.

like butterflies and aphids as well as locust 'hoppers' migrating on foot, change course away from upstanding objects and so keep in the open where locomotion can be maintained. Non-migrating insects frequently turn toward such objects and are arrested.

But low-level oriented flying achieves relatively little displacement over the ground. The more important way in which the swarming locusts' behaviour ensures that they travel is by their rising actively away from the ground and, moreover, allowing themselves to be lifted by convection, like aphids. In this way, they not only avoid obstacles (including such obstacles as the Atlas Mountains) but also pass out of range of any optomotor orienting effect of the ground pattern and ride the faster-moving upper air. Then, the only obvious orientation reactions are those of locust counteracting the disruptive effects of air turbulence on the swarm. Groups of locusts within it show a common orientation, but this varies from minute to minute and group to group. Cohesion is maintained only because any group that approaches the edge of the swarm, on any side, turns back into it. Meanwhile the swarm as a whole is being bundled across country by the wind[5,11-16].

That this resultantly downwind mode of 'straightened-out' travel can be adaptive was demonstrated by Rainey[16] in an outstanding example of the application of modern synoptic meteorology to insect ecology. The rainfall desert locusts need for breeding occurs in different regions at different seasons, and results from the temporary convergence of winds upon each region, so that the wind-borne swarms always tend to be delivered to the right place at the right time. Thus a more or less regular seasonal to-and-fro pattern of migrations between geographically discrete regions, on the familiar vertebrate model, and by large powerful insects, can be created not by self-directed flights but by 'passive' displacements. It may still be true that the long-range to-and-fro migrations of some butterflies do not depend on such displacement and are actively oriented, but the onus now lies on students of butterflies to prove it.

That the long-range movements of winged aphids are effected by passive transport on the upper air has been very thoroughly confirmed (mainly with the black bean aphid, *Aphis fabae* Scop.)[17,18]. But their embarkment on, remaining in and disembarkment from this transporting vehicle all depend on active locomotor responses[19]. Embarkment depends on a sharp rise in locomotor excitability in the fresh adult which is triggered into flight, and then directed up out of the vegetation into the free-circulating air, by the bright, predominantly short wave-length light from the sky. This oriented 'straightening out' of its movement does not take the aphid far, but again the direct result is a passive, long-range straightening out of its movement by the fast-moving upper air. Disembarkment depends on a change in the reaction to light as flight proceeds, with a weakening of the kinetic effect and of the attraction to sky light and a relative strengthening of the attraction to the longer wave-length light reflected from plants. Before the phototaxis reverses lastingly to negative, there are intermediate phases when the insects fly about actively near the ground, orienting into and making headway against gentle air-currents and avoiding obstacles, like locusts, and eventually making the alightments by which the flora is sampled for suitable host plants. The intensely active nature of the migratory flight is obvious even in this phase, for the great majority of alighters, even on their own host plants, take off again within minutes.

There is, in fact, a definite inhibition in the young adult of the usual settling, feeding and larvi-position responses to visual and other stimuli from plants, together with an arrest of embryo growth in the ovaries. When the inhibition is eventually removed, by locomotion, embryo growth restarts and at the same time the flight muscles degenerate and the fat body enlarges[19,20]. The restriction of the frenetic migratory activity of these aphids to, at most, a few days and usually a few hours out of the individual's otherwise sedentary life, further underlines the differentiation of migration as a biological function.

What all this means, as Johnson justly insists, is that migration is part of the normal development of winged aphids, "an evolved adaptation"[1]. It is not their reaction to crowding, food shortage or other current adversity as Elton[6], for example, supposed most migrations to be. Nevertheless, although obligate in the winged form of aphid, migration is facultative for the species. Production of the winged form is the response of the wingless form to crowding and other signals of impending adversity. With locusts, too, migration of the persistent lifelong type seen in the gregarious members of swarms is not obligatory but facultative for the species and results from a heritable response to crowding. Migration is often facultative without obvious polymorphism, as in the pierid butterfly *Ascia monuste* L., where it seems to depend on both population density and the time of day of emergence[21]. If, as Southwood argues, migration is usually an adaptation to a temporary habitat[22], then whether it is obligate or facultative may depend on whether the habitat typically lasts long enough for only one generation or for more.

Proposed Non-behavioural Definitions of Migration

Rainey[13] claims with justification that recent work on locusts has "dispelled some of the mystery" in Williams's picture of migration. The mystery was primarily behavioural, and in dispelling it neither Rainey nor Johnson puts anything in its place so far as behaviour is concerned, rather shifting the emphasis to some other aspect of migration. Thus Rainey insists that the effects of vertical and horizontal air-currents on insect displacement must be assessed before the "finer effects" of behaviour, "in order to distinguish what the organism is itself doing from what is being done to it by the environment"[13,16]. But Rainey has himself reported that the members of small swarms do not allow themselves to be dispersed upwards, like aphids, by turbulent convection, when the members of large swarms do[15]. This is but one illustration of the need to assess what both the insect and the environment are doing, simultaneously. Rainey suggests that migration is "more usefully characterized by a seasonal displacement of populations than by directionally purposeful behaviour"[5]. The qualification "seasonal" is not generally applicable since aphids and other insects often migrate in response to irregularly arising local conditions. It is not necessary to fall back on a purely ecological description like "population displacement", for "directionally purposeful behaviour" is no longer the only alternative.

Johnson recommends distinguishing migration flights "chronologically in the first place before attempting to distinguish them behaviourally"[1]. Disregarding migration other than by flight, he adduces evidence that migratory flights usually occur as the first flights of new adults. There are a good many exceptions to this post-teneral rule, notably post-diapause migrations[4]. But most of them, although not all, would be covered if migration in adult insects were characterized as pre-reproductive, rather than post-teneral. This is an important new common feature recognized in migration: a pointer to developmental physiology (below). Nevertheless, it is as new adults that insects first fly anyway, so the timing does not distinguish migration. The timing of an event becomes of interest once we know what the event is.

Johnson gives two reasons for believing there is no satisfactory way of characterizing migration behaviourally[1]. He doubts whether there need be any difference of kind between migratory and other flights, and earlier questioned the validity of Berland's division of the atmosphere into "terrestrial" and "plankton" zones for flying insects[17]. But that general idea has now been revived in Taylor's concept of a "boundary layer"[18]; and Johnson[1] concedes there is a clear difference between migratory and other flights in Lepidoptera and mosquitoes. The difference is equally apparent in locusts[10,23] and aphids[19] and other insects[22]. His second and more important reason for omitting behaviour from his general system is that the behavioural qualities of migrants "are so variable that definitions based on them as to what is or is not 'migration' or what is 'active' migration and what is 'passive' dispersal become arbitrary, controversial and unsatisfactory when applied to all the types of insect that are said to migrate"[1]. It is true that those distinctions have now broken down, because they assumed a single mechanism of "straightening out" of movements, when as we have seen there are two, discussed further below. At the level of reactions to particular stimuli there is great variation; but the resultant straightening out remains a common feature.

Without that straightening out (the "klino-kinetic" component) even very persistent locomotor activity is not migration, for example in a stationary swarm of flying midges. But when combined with it, a second common feature is what Andrewartha and Birch[9] give as the main one: a special readiness to move (the "ortho-kinetic" component). Here again various stimuli and reactions may be involved. Persistent employment of the insect's own locomotor organs is most common; but there may be no more locomotion than that involved in leaving the feeding or hibernation site and releasing the foothold or clambering on the body of a winged migrant[22,24]. But again, a common feature remains in the special readiness to move and go on moving.

Negative Features of Migratory Behaviour

If we consider only those reactions that show heightened excitability (low threshold) in migrants, we cannot make any all-embracing generalization about them except in terms of their resultant effect: straightened-out movement. That common behavioural feature is important; but it does not entirely dispose of Johnson's difficulty, for it does not provide a ready means of distinguishing migration from such long unidirectional movements as those of an ant toward the nest, a chafer toward a clump of trees, a mosquito toward a host, or a male moth toward a female[25]. But a further criterion is available in what the migrant does not do.

In a migrant the persistently lowered threshold of responses resulting in straightened-out movement goes with a persistently raised threshold of responses that arrest and deviate movement. With locusts and aphids in mind, these more-or-less inhibited reflexes have been generalized as "vegetative", for lack of a better term, meaning responses promoting growth, of the larva or of the adult's genital products, through the exploitation of present resources[26]. The locomotor ones among them were called "station-keeping" responses. Migrants are distinguished by "a transient accentuation of locomotor functions with depression of vegetative functions, such that the insect now travels, by what method being a secondary matter"[19]. Nielsen and Nielsen[21], Blunck[27] and Provost[28] directed attention to this unresponsiveness in migrant butterflies and mosquitoes, and Andrewartha and Birch[9] made the same general point. Thus the Nielsens listed the reflex patterns of adult *Ascia monuste* "living in a territory" as feeding, mating, egg-laying, basking and cleaning, and reported failure to respond to the appropriate stimuli for any of these when the butterflies were engaged in unidirectional flight elsewhere.

The relative degree to which the diverse "vegetative" reflexes are inhibited varies greatly. But the grouping of them under one label is no mere descriptive convenience. In the aphid they are allied reflexes in the physiological sense, standing together in antagonistic relationship to locomotion[29]. This is so when the insect is not migrating also. But then the vegetative reflexes are readily excitable and only briefly inhibited centrally by a stimulus that provokes locomotion instead. When migration starts they are deeply and lastingly inhibited. A more complete characterization of migratory behaviour is then: persistent, straightened-out movement that is accompanied by and dependent upon the maintenance of an internal inhibition of those "vegetative" reflexes that will, eventually, arrest movement. Conversely, the locomotor reflexes (or other reflexes securing displacement) cannot for some while be inhibited by stimuli that do promptly inhibit them at other times.

The distance moved is indicative but not always a reliable criterion of migration, as Provost[28] and Andrewartha and Birch[9] point out, because an insect that is ready to respond to an arresting stimulus may not receive it for some time. Hunting and foraging movements may be very extensive, as in some mosquitoes, yet not be migration if the insects stop as soon as feeding stimuli are received. On the other hand, when an army ant colony, on the emergence of a new brood of worker adults, enters on a 'nomadic phase' in which the workers shift the larvae to a new bivouac site every night along one of the trails they used for foraging that day, this is migration[30]. In the laboratory, starvation produces a high level of locomotor activity in blowflies; but this is not migration physiologically speaking, since the flies take sugar as soon as it is presented and as a result of filling their crops with it become lastingly less active[31]. On the other hand, when locust 'hoppers' or Mormon crickets[32] launch into 'marching' and are stopped only temporarily by suitable food plants, this is migration when compared with the behaviour of the solitary individuals. Young adult dragonflies which disperse in all directions away from the edge of the water in which they developed may also be showing an elementary form of migratory behaviour, because at this time they are positively repelled by the sight of water which will later attract them[33]. Teneral adult *Calliphora* will burrow for hours while they remain in soil, but this is not migration, for emergence from the soil at any time stops the movement[34]. On the other hand, the adult female of *Anopheles sacharovi* may truly migrate when it enters diapause in the autumn, for, unlike the non-diapausing female of previous generations, its flying is apparently not arrested by the first blood meal and shelter it finds, so that it travels far from the breeding place[35].

Thus there is an objective experimental method of deciding whether a moving insect is migrating or not, by measuring its responsiveness to the stimuli signalling the appropriate 'vegetative' requirement. This test has seldom been applied, so that it is often impossible to draw any firm conclusion from the literature. When the insects are known to travel on the upper air the tendency is to assume this is mere 'vagrancy': an 'involuntary' consequence of non-migratory behaviour under special conditions (Williams's "overpowering forces"). The spruce budworm (*Choristoneura fumiferana*) is a good case in point, for it has been studied extensively[24]. Much is known about the immediate triggering stimuli for flight and the air conditions favouring transport, but little about changes in the responsiveness of the individuals. It appears, however, that both larval and adult dispersal are increased actively when the population rises to very damaging densities. The reduced fecundity of the resulting undersized adults is partially compensated by their readiness to fly upwards and thus be transported to undamaged trees before depositing any eggs. The larger, more fecund females produced at less-damaging densities do not fly at all until after they have deposited one or two egg masses[24]. But whether there is any difference of kind between flights before and after oviposition, and whether there is a temporary internal inhibition of oviposition in the undersized females, are still unexplored questions.

The idea of temporary inhibition of 'vegetative' responses in migrants has often been conveyed in a subjective, teleological way that confuses means with ends and serves as a substitute for, rather than an invitation to, experiment. Migratory behaviour has been distinguished as not searching (for any vegetative requirement: food, mate, oviposition site, etc.), not internally oriented, non-motivated, non-appetitive[9,21,28]. The inference is that any movement with a specifiable end is not migratory (see Provost[28] on pre-hibernation flights of anophelines). But migratory and non-migratory movements both end with readiness to respond to some specific stimulus signalling a vegetative requirement or resting place, and in neither case is the movement directed by its end. In both, the locomotion is, in fact, centrally antagonistic to the reflexes that terminate it, at first inhibiting them and finally being inhibited by them in turn. The only objective sense in which a movement is directed toward an end is that the antagonistic arresting reflexes may be progressively 'primed' (thresholds lowered) as a result of being inhibited by the locomotion (successive induction[29]). The basic difference between migratory and non-migratory behaviour is a difference of relative thresholds. There is no sharp line of demarcation between them; we are concerned rather with degrees of functional differentiation in a particular direction, and our first concern is to identify that direction.

Physiological Differentiation in Migrants

The study of insect migration will remain in its present backward state until we know more of the sensory, nervous, endocrine and metabolic components of the mechanism by which migration is 'primed' (in addition to the immediate stimuli that elicit the primed responses) and likewise of the reverse changes that follow on migratory activity. The best information of this kind concerns aphids[20]; but generally there is little. Something is known about fat deposition in place of egg development in migrants[4] as in diapausing insects, about flight muscle development in place of eggs[36] and the use of fat as the most economical flight fuel for migrants[37].

Ecologically, migration sometimes looks like an alternative to diapause when conditions will no longer permit growth, development or reproduction[9,22]. It is so when the habitat deterioration is merely local so that vegetative activities can be resumed earlier elsewhere than *in situ*. It is not so when the deterioration is synchronized throughout the insect's range, as in a temperate autumn or dry tropical summer. Migration may then be, on the contrary, a condition for diapause, if that is also not possible *in situ*[4,22]. But physiologically, the important point is that migration is in either case at the expense of vegetative functions in the individual. The internal inhibition of certain reflexes is the behavioural component of a whole physiological syndrome.

Fig. 1 illustrates this by comparing the timing and amount of growth or reproduction in migratory and non-migratory (or less migratory) females of the same species in the laboratory, where actual migratory activity is minimal and vegetative requirements are continuously available. Solitary locusts (Fig. 1, right) migrate as young adults between seasonal habitats, but the syndrome is developed to an extreme in gregarious locusts which migrate throughout their lives and will not settle in any place[23,26]. Gregarious locusts may show hastened instead of retarded oviposition when there is a tendency to diapause governed by length of day, but in either case they lay considerably fewer eggs[41,42]. The garden chafer (*Phyllopertha horticola* (L.)) is another interesting variant[43]. All the females are ready to lay on emergence before any flight, and most lay their eggs then and do not migrate. The migrants are females that did not lay the bulk of their eggs on emergence and thereafter seem to have developed an inhibition of the reflexes associated with laying; they lay readily after migrating.

The nature of the physiological syndrome in migrants—accentuation of 'sensori-motor' functions with depression of vegetative ones—makes it easier

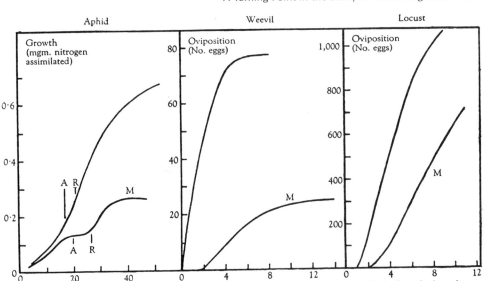

Fig. 1. Examples of the arrest and depression of growth functions in adult migrants (M curves). Left: *Tuberolachnus salignus* winged (M) compared with wingless form, after Mittler (ref. 38). A, final moult to adult; R, start of reproduction (larvi-position) after which the further rise represents embryo growth; Centre: *Callosobruchus maculatus* 'active' (M) compared with normal form (terminology of Caswell (ref. 39)), after Utida (ref. 40); Right: *Locusta migratoria migratorioides* gregarious (M) compared with solitary form, after Norris (ref. 41). All curves cumulative

to understand why migration so commonly occurs in the adult stage and is then usually pre-reproductive[1] or inter-reproductive. For reproduction involves a resumption of the growth functions typical of juveniles, and "an individual is at its most adult during that period of physiological reversal wherein the products of one growth phase are not added to but employed and consumed in ensuring the success of the next"[26]. In this sense migration is physiologically 'super-adult'. In that light it is equally understandable that the next most common stage at which migration differentiates is the first instar larva. This stage is better fitted than later larval instars, because it often has to travel farther than them or at any rate when the eggs are not laid directly on the food, it is smaller and hence more buoyant, and more numerous.

At the same time the negative features of migration represent a differentiation in the direction of diapause. Diapause cannot be characterized by one physiological process in all insects. The one common feature is temporary failure to respond by further growth or development to the conditions that will, eventually, promote those processes[44]. Considered physiologically, migration is an example of this rather than an alternative to it.

During the depression of vegetative functions in both migration and diapause some other process occurs which is a pre-condition for their resumption. In diapause this unidentified other process is called "diapause development"[44]. In the winged aphid the equivalent of diapause development is promoted by an antagonistic process, locomotion, which restarts the arrested vegetative functions by central nervous induction[29]. This distinguishes a *locomotory* type of developmental arrest from other types, such as arrests of larval development that are broken by feeding, arrests of egg development or oviposition broken by copulation, and diapauses proper in which the nature of the response promoting diapause development is still unknown, but the stimuli are known to be low temperature or photoperiod[44]. By no means all the developmental arrests in migrants can be assumed to resemble the aphid's, for some are typical extended diapauses covering also periods of complete inactivity[4]. But in the common situation where the developmental arrest ends with the travelling, it is suggested as a working hypothesis that the two are causally related as in the aphid.

It follows from all this that when, as frequently happens, a deterioration in individual size, fecundity or other vegetative properties is observed under certain conditions, this should not be taken as wholly negative in effect until the possibility of an accompanying increase in migratory capacity has been excluded.

Ecological Functions of Straightened-out Movement

Resultantly downwind displacement independently of orientation was not recognized as a migratory "straightening-out" mechanism by Williams. But it also brings about long-distance travel, instead of confinement within a small area through frequent changes of direction, and does so on a larger scale than does oriented movement on or near the ground; and it results from the active, specialized behaviour of the insects. In locusts, aphids and many other insects[1,22] this consists in rising actively away from the ground and so out of what Taylor[18] calls the "boundary layer" within which non-migrants actively confine their movements. Such resultantly downwind travel cannot therefore be dismissed as the "passive distribution of individuals by overpowering forces", which Williams[2] excluded from migration. It can be assimilated readily into his definitions of migration cited in the first section above, provided we do not read into them his restrictions.

Moreover, these two mechanisms of straightening out, actively oriented and passively wind-borne, can be observed at different times in the same migrant, powerful locust and weak aphid alike. Migrants cannot therefore be separated into two classes according to which mechanism they exhibit, although they may differ in the relative importance of the two mechanisms in their migration. After their oriented ascent, aphids use the wind-borne mechanism in the earlier part and the low oriented one toward the end of a migration[19]. Garden chafer migrants seem to use the two more equally from the start[43]. According to the available accounts[2-4,21], many butterfly migrants keep close to the ground more persistently than aphids, chafers or locusts, and may thus use the oriented mechanism during a greater proportion of their time, although this does not necessarily mean for the greater part of the distance they travel.

Hitherto, it has been solely in the context of seasonal to-and-fro migrations between geographically discrete regions that the ecological function of straightening out seems to have been considered. Williams's hypothesis that straightening out is directionally adaptive, serving to direct the migrants to the right place at the right time, provided the motive for his successful search for evidence of return movements; and his hypothesis has now been confirmed for the seasonal movements of locusts by Rainey[16]. But the continued migratory movements of locust swarms through the mosaic of soils and vegetation within a wind convergence zone where suitable food plants and oviposition sites occur in patches are not directionally adaptive. Some long-range seasonal movements of aphids and other small plant-feeding insects seem to deliver them also to suitable breeding areas[45]. But again this is not true of their movements within very considerable regions wherein the host plant is not concentrated in one place but patchily distributed. Many sorts of habitat, hibernation site, etc., are similarly scattered.

In that situation migrants are no more likely to find what they require in one direction than in any other, or at one distance more than another. The straightening out of their movements has then no directionally adaptive significance. They travel in many directions by both straightening-out mechanisms; and this is adaptive inasmuch as some of them find the requirements in all directions and at all distances. Given enough individuals, the two mechanisms together make an efficient 'scanning' device that works on a coarse and a fine scale by means of long wind-borne, and short, low-level, oriented traverses, respectively.

Straightening out here serves the simpler and, presumably, more primitive function of displacement *per se*: the opposite of 'station keeping'. This is not, of course, an efficient procedure so far as the individual is concerned. The positive advantage of having large numbers of individuals if the migratory habit is to develop[19,26] is a point that has been little considered in discussions of the association between density and migration. Enormous losses of individuals *en route* certainly occur, and even entirely abortive movements by enormous numbers of individuals together, as in Elton's famous case of the lachnids on Spitsbergen. But these losses the species can evidently afford; and if migration tends to be associated with temporary habitats[22] this may mean that such habitats tend to permit higher multiplication-rates than permanent ones. The losses do not call in question the adaptive value of the behaviour pattern that brings them about because the same behaviour brings about biologically successful movements as well, with the added advantages of escape from overcrowding, enemies, etc. What the procedure does achieve is a very efficient coverage of the available habitats over a large area and avoidance of 'too many eggs in one basket'[19].

Given multi-directional movements between places arranged in mosaic the evolutionary problem of return movements disappears, as Wiltshire[38] pointed out in connexion with butterfly migrations. Many species of aphid alternate seasonally between woody and herbaceous plants. It has never been supposed that the descendants of a given individual necessarily returned to the same tree the same year. Nevertheless, untidy, constantly reshuffled returns are assumed, and no problem has been raised of how the migratory habit could persist in the population. The same applies to any insects that repeatedly disperse and colonize new habitats available at the time, new or old, so that no particular movement can be singled out as a return.

Movements of this kind are so common as to be of most interest to ecologists, but are included under dispersal rather than migration[9] because they are not directionally adaptive. Since that is a secondary matter it seems best to apply the term migration, as Johnson[1] and Southwood[22] are now in effect doing, to all cases where a change of location can be shown to be an active 'evolved adaptation' along the lines distinguished here, regardless of exactly how or at what stage in the life-history it occurs. The continued use of dispersal instead of migration for the particular mechanism of transport by wind or water can cause only confusion. Dispersal will, no doubt, continue to be used in the purely spatial sense of population scattering, especially when a loose term is needed because there is not yet enough information to judge whether the movement is migratory or not. If it is used in any more precise sense this should be specified. Thus dispersal is likely to be used for non-migratory movements involving no general inhibition of vegetative reflexes, but merely a change in them, as when the adult's food is different from the larva's; or for movements that may involve some temporary inhibition of them, as when new-hatched caterpillars scatter over a uniformly palatable leaf, but are not persistent or straightened out enough to rate as migration.

To sum up: recent work has undermined C. B. Williams's behavioural assumptions about insect migration while confirming his idea that it is a differentiated biological process. Migration is a persistent, straightened-out movement with some internal inhibition of the responses that will eventually arrest it. It may be effected by the insect's own locomotory exertions or by its active embarkment on some transporting vehicle. These are the behavioural aspects of a physiological syndrome which is 'super-adult' in character but may develop at any stage of the life-history, and in common with diapause involves a phase of depression of 'vegetative' (growth-promoting) functions as a condition for their resumption.

Grateful acknowledgment is made to Drs. A. D. Lees, P. E. Madge, T. R. E. Southwood and B. P. Uvarov, and to Prof. V. B. Wigglesworth, for their criticisms of this article in draft.

[1] Johnson, C. G., *Nature*, **186**, 348 (1960); *Rep. Seventh Commonwealth Ent. Conf.*, 140 (1960); *Proc. Eleventh Int. Ent. Congr., Vienna*, 1960 (in the press).
[2] Williams, C. B., *The Migration of Butterflies* (Oliver and Boyd, Edinburgh and London, 1930); *Trans. Roy. Ent. Soc. London*, **92**, 101 (1942); *Proc. Roy. Ent. Soc. Lond.*, C, **13**, 70 (1949).
[3] Williams, C. B., *Ann. Rev. Ent.*, **2**, 163 (1957).
[4] Williams, C. B., *Insect Migration* (Collins, London, 1958).
[5] Rainey, R. C., *Proc. Eleventh Int. Ent. Congr., Vienna*, 1960 (in the press).
[6] Elton, C., *Animal Ecology and Evolution* (Clarendon Press, Oxford, 1930).
[7] Roer, H., *Z. angew. Ent.*, **44**, 272 (1959); *Proc. Eleventh Int. Ent. Congr., Vienna*, 1960 (in the press).
[8] Elton, C., *Animal Ecology* (Sidgwick and Jackson, London, 1927).
[9] Andrewartha, H. G., and Birch, L. C., *The Distribution and Abundance of Animals* (Chicago Univ. Press, 1954).
[10] Kennedy, J. S., *Phil. Trans. Roy. Soc. Lond.*, B, **235**, 169 (1951).
[11] Waloff, Z., and Rainey, R. C., *Anti-Locust Bull.*, **9**, 1 (1951). Rainey, R. C., and Waloff, Z., *ibid.*, 51.
[12] Rainey, R. C., and Sayer, H. J., *Nature*, **172**, 224 (1953).
[13] Rainey, R. C., *Int. J. Bioclim. Biomet.*, **2** (3B) (1957).
[14] Waloff, Z., *Proc. Tenth Int. Ent. Congr., Montreal*, 1956, **2**, 567 (1958).
[15] Rainey, R. C., *Quart. J. Roy. Met. Soc.*, **84**, 334 (1958).
[16] Rainey, R. C., *Nature*, **168**, 1057 (1951).
[17] Johnson, C. G., and Penman, H. L. *Nature*, **168**, 337 (1951). Johnson, C. G., *Biol. Revs.*, **29**, 87 (1954); *Quart. J. Roy. Met. Soc.*, **83**, 194 (1956); *J. Anim. Ecol.*, **26**, 479 (1957).
[18] Taylor, L. R., *Proc. Linn. Soc. Lond.*, **169**, 67 (1958).
[19] Kennedy, J. S., and Stroyan, H. L. G., *Ann. Rev. Ent.*, **4**, 139 (1959). Kennedy, J. S., Booth, C. O., and Kershaw, W. J. S., *Ann. App. Biol.*, **47**, 410, 424 (1960); **48** (1961) (in the press).
[20] Johnson, B., *Nature*, **172**, 813 (1953); *J. Inst. Physiol.*, **1**, 248 (1957); **3**, 367 (1959).
[21] Nielsen, E. T., and Nielsen, A. T., *Amer. Mus. Novit.*, 1471 (1950).
[22] Southwood, T. R. E., *Proc. Eleventh Int. Ent. Congr., Vienna*, 1960 (in the press); *Biol. Revs.* (in the press).
[23] Kennedy, J. S., *Trans. Roy. Ent. Soc. London*, **89**, 385 (1939).
[24] Wellington, W. G., and Henson, W. R., *Canad. Ent.*, **83**, 240 (1951). Blais, J. R., *ibid.*, **85**, 446 (1953). Greenbank, D. O., *Canad. J. Zool.*, **35**, 385 (1957). Morris, R. F., Miller, C. A., Greenbank, D. O., and Mott, D. G., *Proc. Tenth Int. Ent. Congr. Montreal*, 1956, **4**, 137 (1958). Henson, W. R., *Proc. Eleventh Int. Ent. Congr. Vienna*, 1960 (in the press).
[25] Schneider, F., *Mitt. schweiz. ent. Ges.*, **25**, 269 (1952). Hocking, B., *Proc. Eleventh Int. Ent. Congr., Vienna*, 1960 (in the press). Ivanova, L., *ibid.* Schwink, I., *Proc. Tenth. Int. Ent. Congr., Montreal*, 1956, **2**, 577 (1958).
[26] Kennedy, J. S., *Biol. Revs.*, **31**, 349 (1956).
[27] Blunck, H., *Beitr. Ent.*, **4**, 485 (1954).
[28] Provost, M. W., *Mosquito News*, **12**, 174 (1952).
[29] Kennedy, J. S., *Proc. Tenth Int. Ent. Congr., Montreal*, 1956, **2**, 397 (1958). Ibbotson, A., and Kennedy, J. S., *J. Exp. Biol.*, **36**, 377 (1959).
[30] Schneirla, T. C., *Biol. Bull.*, **88**, 166 (1945); *Smithsonian Rep.* 1955, 379 (1956).
[31] Browne, L. B., and Evans, D. R., *J. Inst. Physiol.*, **4**, 27 (1960).
[32] Johnson, S. A., *Bull. Agr. Exp. Sta. Colorado*, 101 (1905). Cowan, F. T., Shipman, H. J., and Wakeland, C., *U.S. Dep. Agr. Farmers' Bull.*, 1928 (1943).
[33] Corbet, P. S., *J. Anim. Ecol.*, **26**, 1 (1957). Corbet, P. S., Longfield, C., and Moore, N. W., *Dragonflies* (Collins, London, 1960).
[34] Fraenkel, G., *Proc. Zool. Soc. Lond.*, 893 (1935).
[35] Kligler, I. J., and Mer, G., *Riv. Malariol.*, **9**, 363 (1930).
[36] Chapman, J. A., *Nature*, **177**, 1183 (1956); *Proc. Tenth Int. Ent. Congr. Montreal*, 1956, **4**, 375 (1958). Chapman, J. A., and Kinghorn, J. M., *Canad. Ent.*, **90**, 362 (1958).
[37] Weis-Fogh, T., *Trans. Ninth Int. Ent. Congr., Amsterdam*, 1951, **1**, 341 (1952). Urquhart, F. A., *The Monarch Butterfly* (Toronto Univ. Press, 1960).
[38] Mittler, T. E., *J. Exp. Biol.*, **35**, 626 (1958).
[39] Caswell, G. H., *Bull. Ent. Res.*, **50**, 671 (1960).
[40] Utida, S., *Res. Pop. Ecol. (Kyoto)*, **3**, 93 (1956).
[41] Norris, M. J., *Anti-Locust Bull.*, **6** (1950).
[42] Norris, M. J., *Anti-Locust Bull.*, **13** (1952); **18** (1954); **28** (1957); **36** (1959). Albrecht, F. O., Verdier, M., and Blackith, R. E., *Bull. Biol. Fr. Belg.*, **92**, 349 (1958); *Nature*, **184**, 103 (1959). Albrecht, F. O., *Bull. Biol. Fr. Belg.*, **93**, 414 (1959).
[43] Milne, A., *Bull. Ent. Res.*, **51**, 353 (1960).
[44] Lees, A. D., *The Physiology of Diapause in Arthropods* (Cambridge Univ. Press, 1955).
[45] Medler, J. T., *J. Econ. Ent.*, **50**, 493 (1957); *Proc. Eleventh Int. Ent. Congr., Vienna*, 1960 (in the press).

WEATHER AND THE MOVEMENTS OF LOCUST SWARMS: A NEW HYPOTHESIS

Dr. R. C. Rainey

Desert Locust Survey Headquarters, Nairobi

MIGRATING locust swarms have attracted attention since the earliest times, and recent evidence has fully established the spectacular nature of some of these movements. Thus, following the heavy locust breeding experienced on the Red Sea coast of Saudi Arabia between February and April 1950, the first wave of desert locust (*Schistocerca gregaria* Forsk.) swarms to reach the Nile valley during the present plague appeared near Atbara on May 17, 1950, and were reported in Darfur, some seven hundred miles away to the west-south-west, fifteen days later. A detailed examination of the evidence leaves little doubt that these swarms had, in fact, originated on the Arabian coast, and that it was some of the same swarms which afterwards reached the Ouaddai area of French Equatorial Africa in mid-June, representing a total net displacement of some 1,300 miles in about a month. An illustration of the distance which may at times be covered in a single flight is provided by two well-documented records of numbers of desert locusts coming on board ships in the Atlantic some 1,200 miles from the nearest land[1,2].

The importance of the wind in the dispersal of airborne insects in general has often been stressed (ref. 3, etc.), and the part played by the wind in such swarm movements as those mentioned is indicated by the off-shore wind, which reached gale force at times, associated with the only one of these ocean records for which adequate weather data are available[4], and by the fact that certainly the greater part and probably the whole of the movement across the Sudan during May–June 1950 took place down the north-east monsoon. Moreover, while prolonged flight has been regarded[5,6] as characteristic of the behaviour of gregarious adult locusts, field[7,8] and laboratory[9,10] determinations of the air-speed of flying locusts over a considerable range of conditions have given mean values ranging from 7 to 11 m.p.h., which at once set very modest upper limits to the wind-speeds against which swarms may be expected to be capable of making headway.

Furthermore, while there are a number of records of locusts landing into the wind, and turning out of winds in excess of flying speed (probably both optomotor reactions[6]), and of taking off into wind (probably involving the aerodynamic sense-organ[7,11]), field observations have so far failed to find any fixed and consistent orientation of flying locusts to potential orientating factors such as wind and sun. Thus, in winds of less than the probable air-speed of the locusts, up-wind, down-wind and cross-wind orientations are all observed[6,7,12]. Moreover, a swarm failing to maintain any persistent orientation, either circling (a form of swarm movement not infrequently seen), or even changing course sufficiently frequently at random, would also be displaced down-wind. On this evidence, swarm movements in directions other than down-wind would appear likely to be of limited extent, of the order of a few miles a day, and to occur only in areas and seasons of light winds.

This is in general agreement with most of the regular major seasonal movements of swarms so far described in, for example, the Indo-Pakistan area[13] and eastern Africa[5], which take place down the corresponding prevailing winds, and change direction with them. Furthermore, certainly some of the exceptions to this generalization, and possibly all, may well represent movements down non-prevailing winds. Thus perhaps the most striking of these exceptions is the northward movement of swarms in spring to the countries of North Africa and the Middle East[6,14,15], against the direction of the corresponding prevailing northerly winds. At this season, however, all the territories concerned, from Morocco to Baluchistan, are intermittently affected by depressions, moving eastwards and characterized by the temporary interruption of the northerlies by warmer winds from a southerly quarter[16], the subsequent re-establishment of the cooler northerlies often taking place at a well-marked cold front[17,18,19]. An association between the arrival of swarms in Egypt and the passage of depressions has long been recognized[20], and an example of a similar association was provided by the heavy invasion of Tripolitania in early 1951; the main wave of swarms arrived during January 29–30, with the strong south-westerly winds that developed ahead of the primary cold front of one of the deepest and most active depressions of the winter. Rao[21] has similarly attributed northerly and north-easterly swarm movements in spring in Baluchistan to down-wind displacement in the warm sectors of passing depressions. The higher temperatures associated with the southerly winds in the warm sectors of such depressions there appear to operate as a 'rectifying factor'[6], favouring displacement in a northerly direction. A further example of a frequently recorded swarm movement which probably takes place down a non-prevailing wind is provided by the swarms landing on the north coast of the Somali peninsula during the south-west monsoon. Such swarms have been shown[22] to be probably mainly of local origin, returning with the well-marked afternoon sea-breeze characteristic of this area and season, after having been swept out to sea by the strong morning monsoon.

It has accordingly been found useful to examine some of the meteorological implications of down-wind movement. A synoptic chart, recording simultaneous observations of weather over a wide area, demonstrates areas of convergence, across the boundaries of which surface winds show a net excess of inflowing air over outflow, and within which the overall vertical motion of the air is one of ascent; and areas of divergence, across the boundaries of which outflow exceeds inflow, and within which the overall vertical motion of the air is one of descent. Convergence is an essential factor in the production of widespread heavy precipitation, while divergence is generally associated with fair weather; and the recognition and characterization of such areas is accordingly an essential step in the process of weather forecasting. Examples of areas of convergence affecting regions invaded by the desert locust are the Intertropical Convergence Zone (refs. 23, 24, etc.), between trade-wind and monsoon currents originating on

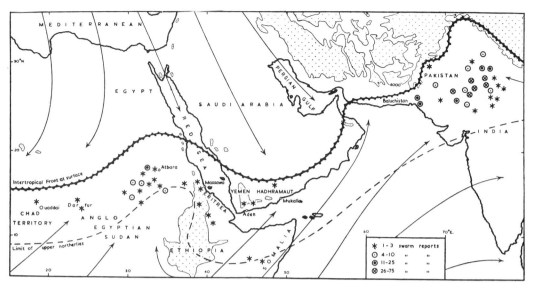

Fig. 1. The Intertropical Convergence Zone and the distribution of desert locust swarms, July 12-31, 1950

opposite sides of the equator, associated with the 'short', 'long' and 'monsoon' rains of most of Africa and India; the westerly depressions already mentioned which give the winter and spring rains of the Mediterranean and Persian Gulf (ref. 16, etc.); and the semi-permanent front, between north-westerly and south-easterly winds, associated with the winter rains of the central and southern Red Sea (refs. 17, 25). Now, in general, winds within a few thousand feet of the earth's surface may be regarded ultimately as blowing from areas of divergence to areas of convergence; and displacement down-wind may therefore be expected to result in movement sooner or later into areas of convergence. It is therefore suggested that the major displacements of locust swarms take place towards areas of convergence, and that swarms may in general be expected to collect in the vicinity of such areas.

This hypothesis may readily be tested by a comparison of swarm records with the corresponding current synoptic charts. In dealing with records from thinly peopled regions, such a comparison is most conveniently made over a period of a number of days during which the synoptic situation is showing little change, in order to be able to consider a number of swarm reports sufficiently large to be representative of the current swarm distribution. The first such period examined was July 12-31, 1950, when the Intertropical Convergence Zone remained relatively stationary in the position indicated in Fig. 1, which also gives the distribution of all desert locust swarms reported during this period, together with an indication of the corresponding surface winds. It will be noted that all the swarms reported were in the vicinity of the Intertropical Convergence Zone, from Chad Territory in the west, through the Anglo-Egyptian Sudan, Eritrea, Ethiopia, Yemen, the Somalilands, the Aden Protectorates and Pakistan to northern India. Practically all recorded swarms were, in fact, between the position of the air-mass boundary at the surface (the Intertropical Front) and the southern limit of the over-running upper northerlies (the 'limit of deep monsoon air'[23], or 'nose of the Intertropical Front'[26]), which at this season is usually closer to the main zone of precipitation than is the surface Intertropical Front[23,26,27]. During this particular season, both the Intertropical Convergence Zone and locust breeding were incidentally recorded as extending considerably farther north than usual in the Anglo-Egyptian Sudan, and the Intertropical Convergence Zone was, as indicated in Fig. 1, probably also farther north than usual over Arabia.

Such evidence of an association of swarm distribution with the Intertropical Convergence Zone encouraged the hope that the movements of the latter might prove to be associated with some of the corresponding swarm movements. The first opportunity of investigating this possibility was provided by the swarm movements in eastern Africa during September and October, 1950, including the first invasion of the British East African territories during the present plague (Fig. 2).

During most of September, 1950, the only desert locust swarms reported south of the Gulf of Aden were in the western end of the Somaliland Protectorate, with the adjacent parts of Ethiopia, where breeding had occurred during the previous month, and near Erigavo in north-eastern Somaliland, where swarms had been present since early August.

During the fourth week of September the Intertropical Convergence Zone began to move southwards from the Gulf of Aden across the Somalilands, with surface north-easterlies established for the greater part of each day at Djibouti from September 22, at Berbera from September 23, and at Hargeisa from September 26; on September 27 a large swarm was reported in south-eastern Somaliland at Las Anod, where 2 in. of rain had been recorded the previous day. On September 29 and 30 rain was recorded in the neighbouring parts of Somalia and of the Ogaden; and on October 1 immature swarms were seen in this area, near Gardo. Away to the west, swarms were reported on September 30 to have left Durdur, where breeding had been particularly heavy, in a southerly direction; and on October 1 a swarm was seen near

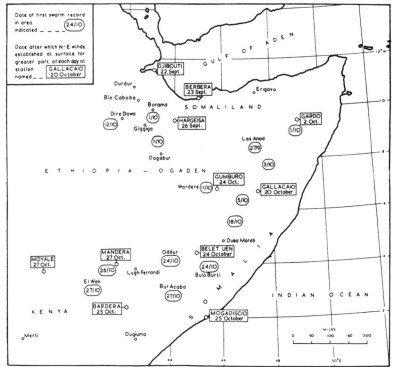

Fig. 2. Advance of the Intertropical Front and the movement of desert locust swarms in eastern Africa during September and October 1950

area, between Gallacaio and Belet Uen, and during October 20–25 widespread egg-laying took place in this area, where heavy rains also fell at about this period.

The Intertropical Convergence Zone continued to move southwards, with the establishment of northerlies at Belet Uen and of north-easterlies at Gumburo on October 24, and of north-easterlies at Mogadiscio and easterlies at Bardera by October 25; on October 24 flying swarms were reported at Oddur, between Belet Uen and Bardera, and near Bulo Burti, between Belet Uen and Mogadiscio.

On the night of October 25–26, scattered locusts flew over Lugh Ferrandi; on October 26 there were the strongest north-easterlies of the month at Belet Uen and Mogadiscio, widespread showers occurred in Kenya, and swarms arrived at three points in the Mandera area. On October 27, easterlies had become established at Mandera, and swarms were reported near El Wak and at Bur Acaba, again very close to the limit of the surface north-easterlies in these areas.

Giggiga, to the south of the Borama-Bio Caboba area, in which a number of young swarms had been reported during the latter half of September. By October 2 northerlies were established at Gardo, and for the next few days the Intertropical Front remained between this station and Gallacaio, still in south-westerlies. On October 3 an immature swarm was seen and heavy rains fell between Gardo and Gallacaio; there was further rain at Gallacaio on October 4; and on October 6, following freshening north-easterlies at Gardo and Hargeisa, swarms were seen to the south-west of Gallacaio.

Following a temporary recession of the Intertropical Front, with south-westerlies re-established at Gardo between October 7 and 11 and at Hargeisa between October 8 and 10, its southward movement was resumed, with surface north-easterlies on the afternoon of October 11, with gusts up to 20 knots at Gardo and reaching 15 knots at Hargeisa. On this date flying locusts were seen from an aircraft at 4,500 ft. over Wardere, and an immature swarm was reported between Giggiga and Dagabur; on October 12 there were two further reports of immature swarms, south and south-west respectively of Dire Dawa.

The south-westerlies receded from Gallacaio and Gumburo (I am indebted to the Sinclair Petroleum Co. for the Gumburo records) after October 18 and from Belet Uen after October 19; easterlies were established at Gallacaio on October 20, with rain, continuing on October 21, when showers were also recorded at Belet Uen. Easterlies were temporarily established at Gumburo during October 19–21, with 2·2 in. of rain on October 20. On October 18 a heavy invasion of mature swarms began in the Dusa Mareb

The Intertropical Convergence Zone became relatively stationary and generally inactive during most of November, with no further sustained southward movement until early December. There were signs of increased activity during mid-November, with rains on November 11 and 12 in south-central Somalia and north-east Kenya; and on these dates swarms reached the Merti (Kenya) and Dugiuma (Somalia) areas, the most southerly points known to have been attained by this generation; both areas were in the immediate vicinity of the Intertropical Front at the time.

The north-easterlies usually over-run the retreating south-west monsoon in East Africa[24], as elsewhere in the region under consideration[26–28], and were, for example, established at 7,000 ft. above sea-level over Kenya as far south as the equator when they had only reached 5° N. in Somalia at the surface. Plateau-level north-easterlies and rains were, in fact, experienced at times well in advance of the southern limit of the surface north-easterlies of the lowlands. The leading swarms, however, remained in the immediate vicinity of this latter limit (the Intertropical Front) throughout the migration; the beginning of the southerly swarm movement from Somaliland, for example, and the arrival of the first swarms in Kenya, five hundred miles away and a month later, both appear to have taken place within a day of the passage of the Intertropical Front across the areas concerned. Over this period the speed as well as the general direction of displacement of the leading swarms thus appear to have been determined by the movement of the Front.

The hypothesis that major swarm movements take place towards, and with, zones of convergence thus appears to merit further consideration as a potentially useful means of supplementing the present essentially historical method of forecasting swarm movements, since forecasts of the movements of the Intertropical Convergence Zone over central Africa, for example, based on the pressure distribution to the north and south of the continent, are issued by the East African Meteorological Department up to a week ahead. The hypothesis may well indicate a need for further information on limiting conditions for flight, which are likely to be important, for example, in determining the speed of swarm displacements approaching zones of convergence from a distance. It may, in addition, help to provide a background against which the finer effects of behaviour on displacement may be more effectively studied, by enabling appropriate allowance first to be made for the part played by the purely transporting effect of the wind in particular recorded swarm displacements.

Convergence also provides a mechanism for the frequently recorded association of exceptional rains with the arrival of locusts; during November–December 1949, to quote a single example, the first swarm record anywhere in Arabia after a lapse of four months, at Mukalla, with the first full-scale gregarious breeding in the Hadhramaut during the present plague, followed a rainfall of 7 in. in 36 hr. at this station, for which the mean annual rainfall, over the whole nine years for which records are now available, is 2·5 in. A similar effect (I am indebted to Mr. P. R. Stephenson for this suggestion) may perhaps also contribute at times to the process of gregarization, by producing a measure of 'real concentration'[29] of adult *solitaria* from long distances into limited areas which provide suitable conditions of moisture and vegetation for breeding. Thus in late November 1950, for example, scattered locusts, at densities ranging from 1 to 100 per 100 yd. of traverse, were encountered in uniformly dry conditions at a series of points extending along 58 miles of the Eritrean coast near Massawa, while on a second reconnaissance over the same route a fortnight later locusts were found, at densities of 400–600 per 100 yd., only within the eight-mile stretch to which rains in the meantime (totalling 5·6 in. at Massawa) had been restricted[30]. This mechanism may well be of vital importance to the continued existence of species such as the desert locust, inhabiting regions of scanty and erratic rainfall with an egg-stage almost aquatic in its moisture requirements[31], and moreover suggests a possible survival value of flight without consistent orientation for other insects.

The data used in this work have been made available by the co-operation of the meteorological services and of the locust control and research organizations of the whole area considered, from French Equatorial Africa to India. I am furthermore particularly indebted to members of the forecast staffs of the meteorological offices at Aden, Asmara, Cairo, Djibouti, Ismailia, Karachi, Khartoum, Mogadiscio and, particularly, Nairobi, for the benefit of their synoptic experience; to Dr. B. P. Uvarov, of the Anti-Locust Research Centre, and Mr. P. R. Stephenson, of the Desert Locust Survey, for their support and active interest in this investigation; to Miss Z. Waloff and Messrs. Y. R. Ramchandra Rao, K. F. Sawyer, W. J. Stower, T. Weis-Fogh and N. W. Wootten, for the use of unpublished data; and to members of the Nairobi Scientific and Philosophical Society, for their comments on the hypothesis at the Society's meeting of November 21, 1950.

[1] Sélys Longchamps, E. de., *Ann. Soc. ent. Belg.*, *C.R.*, v–viii (1878).
[2] Howard, L. O., *Proc. Ent. Soc. Wash.*, **19**, 77 (1917).
[3] Felt, E. P., *Bull. N.Y. State Mus.*, **274**, 59 (1928).
[4] Meteorological Office, Air Ministry, unpublished data of the Marine Branch (M.O. 1).
[5] Waloff, Z., Anti-Locust Mem., London, No. 1 (1946).
[6] Kennedy, J. S., *Phil. Trans. Roy. Soc.*, B, **235**, 163 (1951).
[7] Waloff, Z., and Rainey, R. C., Anti-Locust Bulletin, London, No. 9 (1951).
[8] Waloff, Z., field data in course of preparation for publication.
[9] Weis-Fogh, T., report on locust experiments, January to July 1950 (unpublished).
[10] Sawyer, K. F., and Wootten, N. W. (unpublished laboratory data).
[11] Weis-Fogh, T., *Nature*, **164**, 873 (1949).
[12] Gunn, D. L., Perry, F. C., *et al.*, Anti-Locust Bull., London, No. 3 (1948).
[13] Ramchandra Rao, Y., *Bull. Ent. Res.*, **33**, 241 (1942).
[14] Donnelly, U., Anti-Locust Mem., London, No. 3 (1947).
[15] Davies, D. E., Anti-Locust Mem., London, No. 4 (in the press).
[16] Kendrew, W. G., "The Climates of the Continents" (3rd edit., Oxford, 1937).
[17] El-Fandy, M. G., *Quart. J. Roy. Meteor. Soc.*, **66**, 323 (1940).
[18] Meteorological Office. Weather in the Indian Ocean, II, 3. The Persian Gulf and Gulf of Oman, etc. Air Ministry (M.O. 451*b* (3)) (London: H.M.S.O., 1941).
[19] Lunson, E. A., Prof. Notes, Meteor. Office, London, **7**, No. 102 (1950).
[20] Mackillop, A. T., and Gough, L. H., in Ballard, E., Mistikawi, A. M., and El Zoheiry, M. S., *Bull. Min. Agric. Egypt*, Cairo, No. 110 (1932).
[21] Ramchandra Rao, Y. (in preparation).
[22] Rainey, R. C., and Waloff, Z., *J. Anim. Ecol.*, **17**, 101 (1948).
[23] Memorandum on the Intertropical Front, A. 941848/47 (Meteorological Office; 1947).
[24] Forsdyke, A. G., Geophys. Mem. Meteor. Office, London, **10**, No. 82 (1949).
[25] Hydrographic Department, Admiralty, Red Sea and Gulf of Aden Pilot (9th edit., 1944).
[26] Sawyer, J. S., *Quart. J. Roy. Meteor. Soc.*, **73**, 346 (1947).
[27] Solot, S. B., Research Report, 19th Weather Region U.S.A.A.F.
[28] Meteorological Office, Weather in the Indian Ocean, II, 2 (M.O. 451*b* (2)) (London: H.M.S.O., 1944).
[29] Kennedy, J. S., *Trans. Roy. Ent. Soc.*, **89**, 385 (1939).
[30] Stower, W. J. (unpublished field data).
[31] Needham, J., "Biochemistry and Morphogenesis" (Cambridge, 1942).

ERRATA

Page 1058, left hand column, lines 10 and 11 should read: "feet of the earth's surface may be regarded as ultimately blowing from areas of divergence to . . . ."

Page 1059, left hand column, line 18 should read: "noon of October 11, gusting up to 20 knots . . . ."

AN EXPERIMENTAL COMPARISON OF SCREECH OWL PREDATION ON RESIDENT AND TRANSIENT WHITE-FOOTED MICE (*PEROMYSCUS LEUCOPUS*)

Lee H. Metzgar

The cause of mortality in small mammal populations is frequently unknown, although aggressive encounters, disease, starvation, temperature stress, and predation are considered potentially important (Blair, 1953; Burt, 1940; Chitty, 1958, 1960; Christian, 1961; Errington, 1946; Howard, 1949; Pearson, 1964; Pitelka, 1957). This study deals with the predator-prey relationship between the screech owl (*Otus asio*) and the white-footed mouse (*Peromyscus leucopus*).

Live-trapping studies of white-footed mice have indicated that while most mice know the terrain well and will rapidly take refuge when released, a few others move less adeptly (Burt, 1940; Metzgar, unpublished data). If the former are residents with established home ranges, and the latter are wandering or transient mice, transients may be preyed upon more readily than residents. With the elimination of transients, the survivors would become progressively more difficult to capture, and predators would leave the established breeding population relatively untouched. The following laboratory experiment was designed to test the relative rates of predation on white-footed mice conditioned in such a way as to be analogous to residents and transients.

I am indebted to Dr. F. C. Evans for his encouragement and numerous suggestions. He and Dr. Stephen T. Emlen read the manuscript. Mr. David Ligon kindly donated his screech owl, and Mr. Richard Hill gave much valuable statistical advice. The work was done at The University of Michigan Museum of Zoology where facilities of the Bird Division were made available by Dr. H. B. Tordoff.

Materials and Methods

Two adult white-footed mice, only one of which was familiar with the test area, were simultaneously exposed to a gray-phase screech owl that had been hand raised from a nestling. The mice were wild-caught in southern Michigan, toe-clipped or ear-punched for identification, and kept in the laboratory for 1 to 12 weeks in 20 × 30 cm box cages. The owl was maintained in a 1 × 1 × 2 m cage and was denied food for 24 hrs before each test.

The test area was a windowless room that was 2.8 × 1.4 m and 3.1 m high. Two owl perches were placed across the room at right angles to each other 1.4 m and 2.8 m above the floor. Beet pulp litter and leaves covered the floor, and several branching sticks 0.3 to 1.0 m long and 2 to 8 cm thick were also present. Abundant food and water were continually available in the room. A broad band of cardboard surrounded the room 0.6 m above the floor to prevent mice from climbing the cement block walls. Several fluorescent ceiling lights provided the 12-hr "day." Three General Electric glow-plate night-lights grouped at the ceiling provided a light intensity at "night" which, judged subjectively, was equivalent to a clear night with a quarter moon. All tests were carried out at "night."

Prior to each experiment, a bisexual pair of "resident" mice spent a conditioning period of 2.5 to 10.5 nights in the test room while two "transients" spent an equal amount of time in a 36 × 25 × 20 cm wire cage furnished with litter like that of the test room. At the time of the test, the two pairs were removed from their quarters and kept separately in box cages for 0.5 to 3 hrs, while the owl was placed in the test room. The mice then were released in the presence of the owl by placing the residents and transients in opposite sides of a two-cell container near the doorway of the room, quickly removing the top and sides of the container, and then shutting the door of the room. In tests 1, 2, and 3 all four animals were released together and left for 30 min. Because this density of mice might have interfered with normal behavior patterns, the two pairs of mice (2 males or 2 females) were used separately in tests 4 to 17. Each test was terminated when the owl was heard to make a capture or after 30 min, at which time both mice were removed. The owl was then allowed to rest for ½ hr and the second pair (2 females or 2 males) was then introduced and left for 30 min. No mouse was used in more than one test.

In addition to knowledge of the terrain, the results could have been influenced by the fact that the residents spent the conditioning period in a room while the transients spent it in a medium-sized cage. As an experimental control for this variable, mice were conditioned as described above and held in box cages immediately before each test. While the mice were in these cages, the litter and sticks were removed from the test room, the room was thoroughly vacuumed, and different litter was spread in the room. Thus, of the two mice subsequently exposed to the owl, one had been conditioned in a room and the other in a cage but neither had had any previous experience with the terrain in the test environment. Consequently, they differed only in the type of place in which they spent the conditioning period.

Residents and transients were selected randomly with respect to such factors as method of marking (toe-clip or ear-punch) and time in captivity so that these factors would not influence the results.

Two conditions had to be met in order to justify accepting the hypothesis that transients are more vulnerable to predation than residents. The experimental ratio of $\frac{\text{residents captured}}{\text{transients captured}}$ had to be less than 1.0 and also had to differ significantly ($P \leq 0.05$) from the control ratio of $\frac{\text{room-conditioned mice captured}}{\text{cage-conditioned mice captured}}$. Significance was tested by the one-tailed Fisher Exact Probability Test (Siegel, 1956).

RESULTS

Forty mice were exposed to the owl in 17 experimental tests. On the 13 experimental occasions when the owl captured a mouse, two residents and 11 transients were taken (Table 1). In the 18 control tests (Table 2), the owl captured eight room-conditioned mice and five cage-conditioned mice. The experimental ratio (2/11) differs significantly ($P = 0.021$) from the control ratio (8/5), demonstrating that, under the test conditions, transient white-footed mice are more vulnerable to owl predation than residents. Furthermore, since

TABLE 1.—*Results from 17 tests in which equal numbers of white-footed mice which were acquainted (Residents) and unacquainted (Transients) with the test area were exposed to an owl.*

	No. animals used	No. animals killed
Residents	20	2
Transients	20	11

8/5 does not differ significantly from 6.5/6.5, differences in the size of the environment during conditioning apparently did not contribute to the greater transient mortality. It is concluded that the greater death rate of transients resulted from their lack of familiarity with the terrain.

Discussion

The results of this experiment suggest that a white-footed mouse in familiar terrain is much less vulnerable to avian predation than one in an unfamiliar area. My experience with these mice in other situations suggests three possible bases for this advantage. First, resident mice may become aware of danger more readily. If a mouse with no prior experience in an area is released, it begins to explore, even in the presence of a person. An animal familiar with the area seeks cover in a similar situation. Second, animals that know the terrain may escape more effectively. Mice that have been free in the test room for several days are more difficult for a person to catch than those that have spent only a few hours there. Also, if a mouse caught in the woods is released at the point of capture, it escapes rapidly, but if taken to another woods and released, it can easily be recaptured by hand. Third, a transient may be more active and hence more exposed to predation than a resident. When two mice are released in a large cage ($1.0 \times 1.0 \times 0.5$ m) that is familiar to only one of them, the mouse that is unacquainted with the cage is more active, apparently exploring the cage. Any of these factors could apply to this experimental situation or to field conditions where the screech owl is known to prey on white-footed mice (Bent, 1937), and all of them would result in transient mice being more conspicuous, both visually and audibly. Since owls seek prey by sight (Cushing, 1937; Dice, 1945) and by hearing (Bent, 1937; Payne, 1962), transients would experience greater danger.

Young white-footed mice are weaned at about 23 days of age (Nicholson,

TABLE 2.—*Results from 18 control tests in which pairs of white-footed mice were exposed to an owl. One member of each pair had spent the conditioning period in a room and the other had spent it in a cage.*

	No. animals used	No. animals killed
Room-conditioned	18	8
Cage-conditioned	18	5

1941; Snyder, 1956), when they may emigrate, possibly being forced to move by the residents (Burt, 1949; Nicholson, 1941; Svihla, 1932). Periods of emigration in small mammals have been held by some authors to be times of heavy predation (Blair, 1953; Errington, 1964). Pearson and Pearson (1947) found that shrews captured by owls were (on the average) younger than those taken from the same area at the same time by trapping. One possible explanation is that the younger shrews were individuals that did not have established home ranges and were therefore more subject to predation.

Most adult white-footed mice possess home ranges that they seem to know well (Burt, 1940), and this knowledge supposedly confers several advantages such as the ability to avoid predators (Blair, 1953). The present study demonstrates such an advantage in a laboratory situation, and the results imply that wandering individuals will bear a disproportionate amount of pressure from predation. From these ideas, it follows that the age classes and sex classes as well as the numbers of white-footed mice taken by avian predators will depend in large part upon social interactions within the population, which cause certain animals to wander. Errington (1937, 1956, 1964) and McCabe and Blanchard (1950) have previously drawn similar conclusions from their field studies.

Literature Cited

Bent, A. C. 1937. Life histories of North American birds of prey (part 2). Bull. U.S. Nat. Mus., 170: viii + 1–482.

Blair, W. F. 1953. Population dynamics of rodents and other small mammals. Advances in genetics (M. Demerec, ed.), Academic Press, New York, 5: 1–41.

Burt, W. H. 1940. Territorial behavior and populations of some small mammals in southern Michigan. Misc. Publ. Mus. Zool., Univ. Michigan, 45: 1–58.

———. 1949. Territoriality. J. Mamm., 30: 25–27.

Chitty, D. 1958. Self-regulation of numbers through changes in viability. Cold Spring Harbor Symp. Quant. Biol., 22: 277–280.

———. 1960. Population processes in the vole and their relevance to general theory. Canadian J. Zool., 38: 99–113.

Christian, J. J. 1961. Phenomena associated with population density. Proc. Nat. Acad. Sci., 47: 428–449.

Cushing, J. E. 1939. The relation of some observations upon predation to theories of protective coloration. Condor, 41: 100–111.

Dice, L. R. 1945. Minimum intensities of illumination under which owls can find dead prey by sight. Amer. Nat., 79: 385–416.

Errington, P. L. 1937. What is the meaning of predation? Ann. Rep. Smithsonian Inst. for 1936, pp. 243–252.

———. 1946. Predation and vertebrate populations. Quart. Rev. Biol., 21: 145–177; 221–245.

———. 1956. Factors limiting higher vertebrate populations. Science, 124: 304–307.

———. 1964. The phenomenon of predation. Ann. Rep. Smithsonian Inst., pp. 507–519.

Howard, W. E. 1949. Dispersal, amount of inbreeding, and longevity in a local population of prairie deermice on the George Reserve, southern Michigan. Contr. Lab. Vert. Biol. Univ. Michigan, 43: 1–50.

McCabe, T. T., and B. D. Blanchard. 1950. Three species of *Peromyscus*. Rood Associates, Santa Barbara, California, 136 pp.

NICHOLSON, A. J. 1941. The homes and social habits of the woodmouse, (*Peromyscus leucopus noveboracensis*) in southern Michigan. Amer. Midland Nat., 25: 196–223.

PAYNE, R. S. 1962. How the barn owl locates prey by hearing. Living Bird, 1: 151–159.

PEARSON, O. P. 1962. Carnivore-mouse predation: an example of its intensity and bioenergetics. J. Mamm., 45: 177–188.

PEARSON, O. P., AND A. K. PEARSON. 1947. Owl predation in Pennsylvania with notes on the small mammals of Delaware County. J. Mamm., 28: 137–147.

PITELKA, F. A. 1957. Some characteristics of microtine cycles in the arctic. Pp. 73–88, *in* Arctic biology (H. P. Hansen, ed.), Biology Colloquium, Oregon State College, 1957.

SIEGEL, S. 1956. Nonparametric statistics for the behavioral sciences. McGraw Hill, New York, 312 pp.

SNYDER, D. P. 1956. Survival rates, longevity, and population fluctuations in the whitefooted mouse, *Peromyscus leucopus*, in southeastern Michigan. Misc. Publ. Mus. Zool., Univ. Michigan, 95: 1–33.

SVIHLA, A. 1932. A comparative life history study of the mice of the genus *Peromyscus*. Misc. Publ. Mus. Zool., Univ. Michigan, 24: 1–39.

Laboratory of Vertebrate Biology, Department of Zoology, The University of Michigan, Ann Arbor 48104. Accepted 19 February 1967.

Part IV

WHAT ARE THE CONSEQUENCES OF DISPERSAL AND MIGRATION?

Editors' Comments
on Papers 21 Through 26

21 GILLESPIE
The Role of Migration in the Genetic Structure of Populations in Temporally and Spatially Varying Environments. I. Conditions for Polymorphism

22 DINGLE
Migration Strategies of Insects

23 ŁOMNICKI
Individual Differences between Animals and the Natural Regulation of Their Numbers

24 TAYLOR and TAYLOR
Aggregation, Migration and Population Mechanics

25 DETHIER and MacARTHUR
A Field's Capacity to Support a Butterfly Population

26 BAILEY
Immigration and Emigration as Contributory Regulators of Populations through Social Disruption

The consequences of dispersal must be considered from the point of view of the population that supplies the emigrant and of the population that receives the immigrant. The effects on both populations are pervasive and complex and, for our purposes, can be divided into genetic and demographic components. Part IV provides a basis for considering both aspects.

GENETIC

Unless the dispersers differ on the average in their genetic constitutions from residents, no genetic consequences will accrue from their activity. If they do differ, as seems often to be the case, the genetic impact will be a direct function of the degree of genetic difference between the residents and dispersers and the rate of gene flow (that is, the extent to which dispersers integrate into the new gene pool). Population geneticists call this *migration pressure*.

If a species carries genetic variation for dispersal tendencies, then populations that are net exporters of emigrants will tend to lose dispersal genotypes and any other traits behaviorally or genetically linked to dispersal. Where immigration rates are near zero, such as on islands, dispersal tendencies can be completely eliminated in this way. On the other hand, populations that are net importers of dispersers will tend to accumulate dispersal genotypes, eventually increasing their export rates until they come into balance with imports.

An extremely difficult and important question concerns the effects of dispersal on population fitness. To the extent that emigration acts to purge populations of genotypes relatively poorly adapted to local conditions, it functions like differential mortality and serves to improve local adaptation. Saturation types of emigration would be expected to have this sort of effect. It is less clear whether presaturation emigration would also produce this result since, in this case, some locally fit individuals would be lost to the population as well. In either case, emigration in excess of immigration would tend to reduce population variability. Before the classic publications of Lewontin and Hubby (1966) and Harris (1966) and the plethora of insights that followed, such losses of variability would have been likely viewed as reductions in "load" and hence an increase in population fitness. Now we realize that genetic variability may in itself be an adaptation conferring long-term fitness relative to environmental change. Whether such an adaptive advantage will apply in specific cases will depend on whether the environment varies more spatially or temporally and the rate of gene flow among demes (Levins, 1968; Gillespie, 1976, 1981, and Paper 21; see also Part V). The reverse argument applies to populations receiving a net input of immigrants and hence a tendency to increase its genetic variability. Paper 21 by Gillespie is an example of this approach.

Other potential effects on fitness result from changing inbreeding coefficients. High levels of inbreeding often seem to reduce fitness (for example Roberts, 1960). Where this condition applies, populations receiving immigrants will tend to avoid such negative consequences, while those inhibiting immigration will increase such risks. Similarly, increased genetic variation will generally lead to increases in heterozygosity levels, which in turn could have heterotic benefits. Populations with low variability lose this possible advantage (unless heterozygosity is fixed in some way).

Finally, several improbable but dramatic effects can be imagined. Immigrants joining gene pools unlike those from which they originated may, upon breeding with local population members, produce offspring with new and advantageous recombinations of genetic material. Such offspring could be significantly more fit than other young in the population. On the other hand, immigrants have the negative potential for breaking down locally adapted homeostatic systems and destroying some advantageous group property, either for their own offspring or possibly even for the whole population. Such dramatic effects may be unlikely, but even if rare could have fundamental significance for the evolution of species' properties.

DEMOGRAPHIC

The demographic impacts of dispersal is an area of population biology only beginning to receive the attention that its importance requires. For far too long ecologists considered the consequences of dispersal to be inconsequential or at best no more than an additional source of mortality losses. This view fails to take into account the magnitude of the dispersal process, the fact that dispersers do not always die (especially in pre-saturation dispersals), the fact that emigrants often become immigrants, and the fact that dispersers may represent a biased sample of a population with respect to such demographically important parameters as age, sex, health, and reproductive condition. Perhaps most importantly, this earlier view failed to recognize that dispersal behavior could evolve by natural selection and hence be disassociated from its traditional role as a signaler of saturated resource conditions. More than half a century ago, Charles Elton (1930, p. 63) pointed out the right direction with a characteristically prophetic statement when he concluded: "I do not mean to imply that the study of migration will instantly solve the problem of the regulation of numbers among animals. I only wish to emphasize the fact that no solution can be arrived at without including migration as an important factor . . . ."

The emancipation of dispersal from being considered a minor mortality factor closely associated with saturated habitats has permitted us to appreciate dispersal as a much more significant demographic factor. An early review of the ecological consequences of dispersal by Andrzejewski et al. (1963) has generally been overlooked. Dingle, in Paper 22, reviews the interaction of life history traits and dispersal strategies in insects. Lidicker (Paper 11) reviewed the role of dispersal in the demography of small mammals (pages 108-117 and 120-125 are relevant here), covering such topics as mortality rates, age structure, sex ratio, population growth rates, social structure, and reproduction. Most of these topics can be extrapolated beyond small mammals to animals in general. It becomes clear that population dynamics can be profoundly affected by dispersal both into and out of populations.

Editors' Comments on Papers 21 Through 26

With all these demographic impacts, it is only natural to consider dispersal in the context of regulation of numbers. Lidicker (Paper 11, pages 120-125) proposed three categories of involvement that represent increasing levels of impact on the population:

1. Contributing factor—dispersal accounts for some losses helping to maintain a population at its carrying capacity. It is a nonessential component because if dispersal is prevented, there is no change in density levels.
2. Key factor—in this capacity, dispersal accounts for a major fraction of losses at carrying capacity. If dispersal is stopped or reduced, density increases, overshooting carrying capacity.
3. Regulating factor—three models are proposed in which dispersal acts to temporarily or permanently stop population growth below carrying capacity. Two of these models require large dispersal sinks, that is, empty or marginal habitats into which emigrants can move, and one model involves the absence of a sink leading to frustrated dispersal.

An independently derived model in which emigration can regulate numbers below carrying capacity is proposed by Łomnicki (Paper 23). This model is based on the unequal partitioning of resources among individuals within a given population superimposed on spatial and temporal heterogeneity of the environment. Under this model, population outbreaks are the consequence of a failure of regulation by emigration which in turn is brought about by a temporary loss of spatial heterogeneity in the environment. In most respects this model is analogous to Model II of Lidicker (Paper 11) and also to one proposed by Taylor and Taylor in Paper 24, an especially important contribution. Two empirical studies that seem to support dispersal-mediated regulation of numbers below carrying capacity are those of Dethier and MacArthur (Paper 25) on checkerspot butterflies and Sanders and Knight (1968) on an aphid.

Two other views of density regulation through dispersal mechanisms invoke social disruption and spacing behavior as the proximate promoters of dispersal behavior. Among mammals social disruption by immigrants can stimulate dispersal (Calhoun, 1948; Davis and Christian, 1956; Healey, 1967; Ramsey and Briese, 1971), increase levels of stress (Paper 26), reduce population levels (Davis and Christian, 1956), modify homing behavior (Terman, 1962), and have profound effects on reproduction. Reproductive responses initiated by immigrants are summarized for small mammals in Paper 11. One paper not mentioned there is that by Bailey (Paper 26) who demonstrated a reduction in testicular development and seminal vesicle size in laboratory mice subjected to simulated immigration and emigration. It should be noted that stimulatory as well as inhibitory effects are possible. Petrusewicz (1963) has shown stimulatory effects on captive populations of house mice, *Mus musculus*, and DeLong (1967) reports evidence that this mechanism also works in the field. Even off-season breeding stimulated by immigrants has been reported by Sheppe (1965) for *Peromyscus leucopus*. The importance of dispersal in encouraging reproduction in humans is implied in an interesting paper by Hammel et al. (1979). They argue that negative impacts on human reproduction are to be expected where incest taboos exist and populations are effectively small, that is, where outcrossing opportunities are limited.

Dispersal can also be negatively correlated with density and produce an "anti-regulating" effect (see Lidicker, 1978). Gilbert and Singer (1973) report that larger colonies of the butterfly *Euphydryas editha* are more cohesive than smaller colonies. Similar results have been reported by Mykytowycz and Gambale (1965) for the European rabbit, *Oryctolagus cuniculus*.

Spacing behavior has been postulated to stimulate dispersal among excess individuals and thus stabilize population density. Densities reached may be at or below carrying capacity, depending on local conditions. Spacing behavior as a general theoretical construct for viewing population regulation is supported by Krebs (1978, 1979). A good example of a field population in which spacing of adult females is reported to stabilize breeding densities is that of Bujalska (1970) on *Clethrionomys glareolus*. Similar results have been reported for the congener, *C. rufocanus*, by Saitoh (1981). Territorial behavior among breeding birds represents a well-known example of such forced dispersal regulating numbers. What generally is not clear in most cases, however, is whether or not breeding densities are adjusted to spatial and temporal variations in carrying capacity, or whether spacing merely reflects a density that is appropriate for average conditions. An interesting variation on this common pattern among breeding birds occurs in *Passer montanus*. According to Pinowski (1971), autumnal territorial displays, which may include nest building, pair formation, and copulations, strongly influence over-winter survival and subsequent breeding season residency.

Little interest has been directed toward the demographic effects of migration, except to assume that migration itself extracts a heavy toll in the form of mortality. This mortality selects against the inexperienced and poorly conditioned individuals. Greenberg (1980) discussed demographic aspects of migration in birds that migrate between temperate or arctic regions and the tropics. His model explores the payoffs and debits associated with the quality of wintering habitat, migrational losses, and adjustments in the breeding season required by the distance of migration and seasonality of breeding habitat. Dingle and his co-workers have pioneered the exploration of some life history implications of migratory behavior in insects (Paper 22; Dingle, 1974; Dingle et al., 1977; Derr et al., 1981).

REFERENCES

Andrzejewski, R., A. Kajak, and E. Pieczyńska, 1963, Efekty migracji [Effects of Migration], *Ekol. Pol.,* ser. B, **9:**161-172.

Bujalska, G., 1970, Reproductive Stabilizing Elements in an Island Population of *Clethrionomys glareolus* (Schreber, 1780), *Acta Theriologica* **15:**381-412.

Calhoun, J. B., 1948, Mortality and Movement of Brown Rats (*Rattus norvegicus*) in Artificially Super-saturated Populations, *J. Wildl. Manage.* **12:**167-172.

Davis, D. E. and J. J. Christian, 1956, Changes in Norway Rat Populations Induced by Introduction of Rats, *J. Wildl. Manage.* **20:**378-383.

DeLong, K. T., 1967, Population Ecology of Feral House Mice, *Ecology* **48:**611-634.

Derr, J. A., B. Alden, and H. Dingle, 1981, Insect Life Histories in Relation to Migration, Body Size, and Host Plant Array: A Comparative Study of *Dysdercus, J. Anim. Ecol.* **50:**181-193.

Dingle, H., 1974, The Experimental Analysis of Migration and Life-history Strategies in Insects, in *Experimental Analysis of Insect Behavior,* L. Barton Browne, ed., Springer-Verlag, New York, pp. 329-342.

Dingle, H., C. K. Brown, and J. P. Hegmann, 1977, The Nature of Genetic Variance Influencing Photoperiodic Diapause in a Migrant Insect, *Oncopeltus fasciatus, Am. Nat.* **111:**1047-1059.

Elton, C., 1930, *Animal Ecology and Evolution,* Oxford, London, 96p.

Gilbert, L. E. and M. C. Singer, 1973, Dispersal and Gene Flow in a Butterfly Species, *Am. Nat.* **107:**58-72.

Gillespie, J. H., 1976, The Role of Migration in the Genetic Structure of Populations in Temporally and Spatially Varying Environments. II. Island Models, *Theor. Pop. Biol.* **10:**227-238.

Gillespie, J. H., 1981, The Role of Migration in the Genetic Structure of Populations in Temporally and Spatially Varying Environments. III. Migration Modification, *Am. Nat.* **117:**223-233.

Greenberg, R., 1980, Demographic Aspects of Long-distance Migration, in *Migrant Birds in the Neotropics: Ecology, Behavior, Distribution, and Conservation,* A. Keast and E. S. Morton, eds., Smithsonian Institute Press, Washington, D.C., pp. 493-504.

Hammel, E. A., C. K. McDaniel, and K. W. Wachter, 1979, Demographic Consequences of Incest Tabus: A Microsimulation Analysis, *Science* **205:**972-977.

Harris, H., 1966, Enzyme Polymorphism in Man, *R. Soc. London B, Proc.,* **164:**298-310.

Healey, M. C., 1967, Aggression and Self-regulation of Population Size in Deermice, *Ecology* **48:**377-392.

Krebs, C. J., 1978, A Review of the Chitty Hypothesis of Population Regulation, *Can. J. Zool.* **56:**2463-2480.

Krebs, C. J., 1979, Dispersal, Spacing Behaviour, and Genetics in Relation to Population Fluctuations in the Vole *Microtus townsendii, Fortschr. Zool.* **25:**61-77.

Levins, R., 1968, *Evolution in Changing Environments,* Princeton University Press, Princeton, N.J., 120p.

Lewontin, R. C., and J. L. Hubby, 1966, A Molecular Approach to the Study of Genetic Heterozygosity in Natural Populations, *Genetics* **54:**595-609.

Lidicker, W. Z., Jr., 1978, Regulation of Numbers in Small Mammal Populations—Historical Reflections and a Synthesis, in *Populations of Small Mammals under Natural Conditions,* D. P. Synder, ed., Pymatuning Lab. Ecol., University of Pittsburgh, Pittsburgh, pp. 122-141.

Mykytowycz, R., and S. Gambale, 1965, A Study of the Inter-warren Activities and Dispersal of Wild Rabbits, *Oryctolagus cuniculus* (L.), Living in a 45-ac. Paddock, *CSIRO Wildl. Res.* **10:**111-123.

Petrusewicz, K., 1963, Population Growth Induced by Disturbance in the Ecological Structure of the Population, *Ekol. Pol.* ser. A, **11:**87-125.

Pinowski, J., 1971, Dispersal, Habitat Preferences and the Regulation of Population Numbers in Tree Sparrows, *Passer m. montanus* (L.), *Int. Studies on Sparrows* **5:**21-39.

Ramsey, P. R., and L. A. Briese, 1971, Effects of Immigrants on the Spatial Structure of a Small Mammal Community, *Acta Theriologica* **16:**191-202.

Roberts, R. C., 1960, The Effects on Litter Size of Crossing Lines of Mice Inbred without Selection, *Genet. Res.* **1:**239-252.

Saitoh, T., 1981, Control of Female Maturation in High Density Populations of the Red-backed Vole, *Clethrionomys rufocanus bedfordiae, J. Anim. Ecol.* **50:**79-87.

Sanders, C. J., and F. B. Knight, 1968, Natural Regulation of the Aphid *Pterocomma populifoliae* on Bigtooth Aspen in Northern Lower Michigan, *Ecology* **49:**234-244.

Sheppe, W., 1965, Island Populations and Gene Flow in the Deer Mouse, *Peromyscus leucopus, Evolution* **19:**480-495.

Terman, C. R., 1962, Spatial and Homing Consequences of the Introduction of Aliens into Semi-natural Populations of Prairie Deermice, *Ecology* **43:**216-223.

21

Copyright ©1975 by The University of Chicago
Reprinted from *Am. Nat.* **109**:127–135 (1975), by permission of The University of Chicago Press

THE ROLE OF MIGRATION IN THE GENETIC STRUCTURE OF POPULATIONS IN TEMPORARILY AND SPATIALLY VARYING ENVIRONMENTS
I. CONDITIONS FOR POLYMORPHISM

JOHN H. GILLESPIE

Department of Biology, University of Pennsylvania,
Philadelphia, Pennsylvania 19174

[*Editors' Note:* In the above title, "Temporarily" is a misprint for "Temporally." The correct title is given on the cover of the journal in which the article appears.]

Natural populations are continually faced with evolutionary problems presented by spatial and temporal changes in the environment. These changes can cause polymorphism, which, in turn, will buffer the effects of the changes on the species. The conditions which permit the maintenance of the variation in changing environments have been investigated in three recent papers (Gillespie 1973*a*, 1974; Gillespie and Langley 1974). The first of these papers deals with diploid populations subjected only to temporal changes and shows that polymorphism will occur if the geometric mean fitness of the heterozygote exceeds both homozygotes. It had previously been shown (Gillespie 1972, 1973*b*) that polymorphism never occurs in haploid populations subjected only to random temporal changes, thus pointing out an inherent advantage to diploidy for dealing with temporally changing environments. The third paper deals with purely temporal changes and purely spatial changes. By examining a variety of models, it shows very generally that polymorphism will occur whenever the variance in the environment is sufficient to override mean advantages which may be possessed by particular genotypes. Finally, the second paper deals with the combined effects of spatial and temporal variation. In the model described in that paper, an extension of the model of Levene (1953), the environment is subdivided into a set of patches, each one experiencing the same environmental process through time, although within any generation the patches will be different from each other. The species is assumed to distribute itself at random over these patches each generation, thus removing the possibility of significant local adaptations.

In the present series of papers I extend this model of temporal and spatial changes to include both the effects of migration between patches and the role of different processes in the different patches. In this first paper I examine the conditions for polymorphism in the special case of two patches. Future papers in this series will concentrate on the existence of optimal migration rates and on statistical techniques for using electrophoretic data to discover the relevance of changing environments to natural populations.

PRELIMINARIES

Consider a haploid species with two genotypes, A_1 and A_2, which occupy an environment which is spatially separated into two patches. Assume that the fitnesses of the two genotypes in the ith patch, $i = 1, 2$, in the tth generation, are $1 + U_i(t)$ and $1 + V_i(t)$, respectively. Let $x_i(t)$ be the frequency of allele A_1 in the ith patch in the tth generation, and assume that these frequencies are first changed by selection and then by migration. If migration consists of exchanging a certain fraction of the individuals, m, between the two patches, we can write

$$x_1(t+1) = (1-m)[x_1(t) + \Delta_s x_1(t)] + m[x_2(t) + \Delta_s x_2(t)],$$
$$x_2(t+1) = m[x_1(t) + \Delta_s x_1(t)] + (1-m)[x_2(t) + \Delta_s x_2(t)], \quad (1)$$

where

$$\Delta_s x_i(t) = \frac{x_i(t)[1 - x_i(t)][U_i(t) - V_i(t)]}{1 + x_i(t)U_i(t) + [1 - x_i(t)]V_i(t)}.$$

The corresponding difference operators are

$$\Delta x_1(t) = \Delta_s x_1 + m(x_2 - x_1) - m(\Delta_s x_1 - \Delta_s x_2),$$
$$\Delta x_2(t) = \Delta_s x_2 + m(x_1 - x_2) - m(\Delta_s x_2 - \Delta_s x_1). \quad (1b)$$

The stochastic element to this system is introduced by assuming that the vectors $[U_i(t), V_i(t)]$ are white-noise processes. The following parameters will be required of these processes:

$$\Delta \Gamma_i = (EU_i - \tfrac{1}{2} \operatorname{Var} U_i) - (EV_i - \tfrac{1}{2} \operatorname{Var} V_i);$$
$$\sigma_i^2 = \operatorname{Var}(U_i - V_i);$$
$$\sigma_{12} = \operatorname{Cov}(U_1 - V_1, U_2 - V_2); \quad (2)$$
$$\rho = \frac{\sigma_{12}}{\sigma_1 \sigma_2}.$$

The system (1) is intractable in its present form, but by approximating its behavior with a diffusion process, its major features may be described. In order to arrive at this approximation, the parameters must be shrunk in appropriate ways along with the time scale. The resulting diffusion processes will differ according to the relative rates of shrinking of the genetic versus the migration parameters.

Let τ be an arbitrary parameter which we will shrink to zero. If time is to be measured in units of τ, then all the parameters in (2) must also be of the same order as τ in order to obtain a nontrivial limiting process. We also require that the third- and higher-order moments of the difference operators are of order τ^2 or smaller. To make this precise, we will introduce new parameters as follows:

$$\tau \Delta \Gamma_i^* = \Delta \Gamma_i,$$
$$\tau \sigma_i^{2*} = \sigma_i^2, \quad (3)$$
$$\tau \sigma_{12}^* = \sigma_{12},$$

where the starred parameters will be those appearing in the final diffusion equation. Notice that (3) implies that EU_i, EV_i, Var U_i, and Var V_i are all of the same order as τ or of a smaller order. There are no restrictions on the numeric value of the starred parameters. As $\tau \to 0$, it is easily shown that

$$E\,\Delta_s x_i = \tau M_i(x_i) + O(\tau^2),$$
$$\text{Var}\,\Delta_s x_i = \tau V_i(x_i) + O(\tau^2),$$
$$\text{Cov}\,(\Delta_s x_1, \Delta_s x_2) = \tau C(x_1, x_2) + O(\tau^2),$$

where

$$M_i(x_i) = x_i(1 - x_i)[\Delta\Gamma_i^* + \sigma_i^{2*}(\tfrac{1}{2} - x_i)],$$
$$V_i(x_i) = x_i^2(1 - x_i)^2 \sigma_i^{2*}, \tag{4}$$
$$C(x_1, x_2) = x_1 x_2(1 - x_1)(1 - x_2)\sigma_{12}^*.$$

For the diffusion approximations we require the first two moments of the difference operators (1b). The limiting behavior of these moments depends critically on the relationship of m to τ. In what follows we will examine three such relationships:

Case 1: $m = O(\tau)$.—As in (3), introduce a new parameter $\tau m^* = m$. Under this parameterization

$$E\,\Delta x_1 = \tau M_1(x_1) + \tau m^*(x_2 - x_1) + O(\tau^2),$$
$$E\,\Delta x_2 = \tau M_2(x_2) + \tau m^*(x_1 - x_2) + O(\tau^2),$$
$$\text{Var}\,(\Delta x_i) = \tau V_i(x_i) + O(\tau^2), \quad i = 1, 2, \tag{5}$$
$$\text{Cov}\,(\Delta x_1, \Delta x_2) = \tau C(x_1, x_2) + O(\tau^2).$$

Since time is also measured in units of τ, and since third- and higher-order moments of the operators will be of order τ^2 or smaller, the limiting process obtained by allowing $\tau \to 0$ is a diffusion process with drift coefficients

$$M_1(x_1) + m^*(x_2 - x_1)$$
$$M_2(x_2) + m^*(x_1 - x_2) \tag{6a}$$

and diffusion coefficients

$$V_i(x_i), \quad i = 1, 2$$
$$C(x_1, x_2). \tag{6b}$$

The major portion of this paper will be concerned with the stochastic stability properties of this diffusion process.

Case 2: $m = O(\tau^2)$.—In this case the migration rate approaches zero faster than the genetic parameters. To be precise, set $m^*\tau^2 = m$. The second-order moments for the difference operators may be written as in (5), but the first-order moments will be $E\,\Delta x_i = \tau M_i(x_i) + O(\tau^2)$, $i = 1, 2$. For this model the limiting diffusion process has drift coefficients $M_i(x_i)$, $i = 1, 2$, and diffusion coefficients $V_i(x_i)$, $i = 1, 2$ and $C(x_i, x_2)$.

Case 3: m a fixed constant.—For many cases of biological importance it is unrealistic to assume that m is of the same order of magnitude as or smaller than the genetic parameters. In these cases a limiting diffusion model may be obtained which turns out to be a one-dimensional process. This may be seen most easily if we transform $[x_1(t), x_2(t)]$ into a new process with state variables $y = \frac{1}{2}(x_1 + x_2)$ and $z = \frac{1}{2}(x_1 - x_2)$. From (1b) we get

$$\Delta y = \tfrac{1}{2}(\Delta_s x_1 + \Delta_s x_2),$$
$$\Delta z = (\tfrac{1}{2} - m)(\Delta_s x_1 - \Delta_s x_2) + m(x_2 - x_1).$$

Proceeding as before, it is easily shown that

$$E\,\Delta y = \tfrac{1}{2}[M_1(x_1) + M_2(x_2)]\tau + O(\tau^2),$$
$$\mathrm{Var}\,\Delta y = \tfrac{1}{2}[V_1(x_1) + V_2(x_2) + 2\,\mathrm{Cov}\,(x_1, x_2)]\tau + O(\tau^2),$$
$$E\,\Delta z = (\tfrac{1}{2} - m)[M_1(x_1) - M_2(x_2)]\tau + m(x_2 - x_1) + O(\tau^2),$$
$$\mathrm{Var}\,(\Delta z) = O(\tau),$$
$$\mathrm{Cov}\,(\Delta z, \Delta y) = O(\tau).$$

As $\tau \to 0$, $E\,\Delta z = m(x_2 - x_1) = -2mz$. Thus, $E\,\Delta z$ does not approach zero as $\tau \to 0$, although Var Δz does approach zero. At the limit, as $\tau \to 0$, z becomes a deterministic process. Since time is measured in units of τ, at any fixed time, t, the process has undergone t/τ generations. If z were purely deterministic (as it is at the limit), it would go from its initial value to $z(t) = (1 - 2m)^{t/\tau} z(0)$ in this period of time. As $\tau \to 0$, $z(t) \to 0$, and this state variable becomes identically zero in the limiting diffusion process. This is not surprising when we consider that by shrinking the genetic parameters but not the migration rate we are allowing migration to move on a much shorter time scale than selection. Since migration works to equalize the allele frequencies in the two patches, at the limit it is infinitely better at equalizing allele frequencies than selection is at separating them.

Since $z = 0$ in the limiting process, the drift and diffusion coefficients for the one-dimensional process are $\frac{1}{2}[M_1(y) + M_2(y)]$ and $\frac{1}{4}[V_1(y) + V_2(y) + 2C(y,y)]$, respectively. If we set $\sigma_1^{2*} = \sigma_2^{2*} = \sigma^{2*}$ and $\Delta\Gamma_1^* = \Delta\Gamma_2^* = \Delta\Gamma^*$, we recover the process described by Gillespie (1974) under the initial assumption that $m = \frac{1}{2}$. This particular model has drift and diffusion coefficients

$$y(1 - y)[\Delta\Gamma^* + \sigma^{2*}(\tfrac{1}{2} - y)]$$

and

$$y^2(1 - y)^2 \frac{\sigma^{2*}(1 + \rho)}{2}.$$

In order for a polymorphism to occur,

$$|\Delta\Gamma^*| < \frac{\sigma^{2*}}{4}(1 - \rho) \tag{7}$$

must hold. For the more general case, the same methods as used in Gillespie (1974) yield the following condition for polymorphism:

$$|\overline{\Delta\Gamma^*}| = |\tfrac{1}{2}(\Delta\Gamma_1^* + \Delta\Gamma_2^*)| < \frac{\sigma_1^{2*} + \sigma_2^{2*} - 2\sigma_{12}^*}{8}. \tag{8}$$

The three cases just described correspond to three different levels of migration. Cases 2 and 3 have already been described sufficiently in the literature. The next section will be concerned only with case 1. It will emerge, however, that cases 2 and 3 are actually incorporated into case 1 if appropriate values of m^* are used.

STABILITY ANALYSIS

This section covers the conditions on the starred parameters which will lead to genetic polymorphism. There are two states which will be termed monomorphic states: $(x_1, x_2) = (0, 0)$ and $(x_1, x_2) = (1, 1)$. All other states will be called polymorphic. If both of the monomorphic states are unstable, both alleles will be found in the population.

The stability properties of the monomorphic states may be described by first linearizing (6) in the region of a monomorphic state and then examining the stochastic stability of the linear system. In the region of (0, 0), for example, linearization of (6) yields the linear *Ito* equation with drift coefficients

$$x_1\left(\Delta\Gamma_1^* + \frac{\sigma_1^{2*}}{2} - m^*\right) + x_2 m^*$$

and

$$x_1 m^* + x_2\left(\Delta\Gamma_2^* + \frac{\sigma_2^{2*}}{2} - m^*\right)$$

and diffusion coefficients $x_1^2\sigma_1^{2*}$, $x_2^2\sigma_2^{2*}$, and $x_1 x_2 \sigma_{12}^*$.

The stability conditions may be obtained by using a theorem of Khaz'minskii (1967). The theorem, as applicable to the present use, may be stated as follows. Let $\|x\|$ be a suitable norm for the process $x = (x_1, x_2)$. Define a new process,

$$[p(t), q(t)] = \left[\frac{x_1(t)}{\|x\|}, \frac{x_2(t)}{\|x\|}\right].$$

This process is realized on the edge of a unit ball around the origin. Under suitable conditions this new process will have an asymptotic density, $\varphi(p)$. Using this density, Khaz'minskii showed that the origin will be asymptotically stable if and only if

$$I_1 = \int_0^1 E d \ln \|x\| \varphi(p)\, dp < 0.$$

Note that $E d \ln \|x\|$ depends only on p and q. In the present context, since $x_1(t)$ and $x_2(t)$ are always positive, a convenient norm is $\|x\| = x_1(t) + x_2(t)$. This norm will be used throughout this section.

Before considering the general case, consider the case where both patches have the same moments for the genetic parameters: $\Delta\Gamma_1^* = \Delta\Gamma_2^* = \Delta\Gamma^*$; $\sigma_1^{2*} = \sigma_2^{2*} = \sigma^{2*}$. These patches will be referred to as *equivalent patches*. For this case it is easily shown that the normalized process

$$p(t) = \frac{x_1(t)}{x_1(t) + x_2(t)}$$

satisfies a one-dimensional diffusion process with drift coefficient
$$2p(1-p)\sigma^{2*}(1-\rho) + m^*(1-2p)$$
and diffusion coefficient
$$2p^2(1-p)^2\sigma^{2*}(1-\rho).$$
This process has natural boundaries and the asymptotic density.
$$\varphi(p) = C\,\frac{\exp\{-\lambda/[p(1-p)]\}}{p(1-p)},$$
where
$$C^{-1} = \int_0^1 \frac{\exp\{-\lambda/[p(1-p)]\}}{p(1-p)}\,dp,\quad \lambda = \frac{m^*}{\sigma^{2*}(1-\rho)}.$$
To evaluate the integral I_1, we need $Ed\ln\|x\| = \Delta\Gamma^* + \sigma^{2*}(1-\rho)p(1-p)$. With this,
$$I_1 = \Delta\Gamma^* + \sigma^{2*}(1-\rho)C\int_0^1 \exp\left[-\frac{\lambda}{p(1-p)}\right]dp.$$
Define
$$\zeta(\lambda) = \left[\int_0^1 \exp\left[-\frac{\lambda}{p(1-p)}\right]dp\right]\Big/\left[\int_0^1 \frac{\exp\{-\lambda/[p(1-p)]\}}{p(1-p)}\,dp\right].$$
The origin will be unstable if
$$\Delta\Gamma^* + \sigma^{2*}(1-p)\zeta\left[\frac{m^*}{\sigma^{2*}(1-\rho)}\right] > 0.$$
A similar analysis near $(1, 1)$ yields
$$-\Delta\Gamma^* + \sigma^{2*}(1-\rho)\zeta\left[\frac{m^*}{\sigma^{2*}(1-\rho)}\right] > 0$$
for instability. Thus, for polymorphism—that is, for the instability of both $(0, 0)$ and $(1, 1)$—we require
$$|\Delta\Gamma^*| < \sigma^{2*}(1-\rho)\zeta\left[\frac{m^*}{\sigma^{2*}(1-\rho)}\right]. \tag{8}$$

This condition is analagous to those arrived at by Gillespie and Langley (1974) and Gillespie (1974) in that the difference of the geometric means may be isolated from the variance of the selection differential.

The function $\zeta(\lambda)$ has the following properties:

$$\lim_{\lambda\to 0}\zeta(\lambda) = 0, \tag{i}$$

$$\lim_{\lambda\to 0}\zeta'(\lambda) = \infty, \tag{ii}$$

$$\zeta'(\lambda) > 0, \tag{iii}$$

$$\lim_{\lambda\to\infty}\zeta(\lambda) = \tfrac{1}{4}. \tag{iv}$$

Using these properties, we can say the following about this case. (a) Increasing the rate of migration makes polymorphism more likely. (b) As $m^* \to \infty$, (7) becomes

$$|\Delta\Gamma^*| < \frac{\sigma^{2*}(1-\rho)}{4}. \tag{9}$$

This is the same as condition (7). (c) If $m^* = 0$, polymorphism is impossible. In this case both populations will become quasi-fixed for the allele with the highest geometric mean fitness. This case corresponds to case 2 of the previous section. (d) Property (ii) has the interesting implication that when migration is very low, increasing migration only a little bit makes polymorphism much more likely. In some ways this property is analagous to the effect of migration in spacially structured finite population models, in that very little migration makes many aspects of the population similar to those in populations with very large amounts of migration. (e) If $\rho = 1$, the two patches behave as one and polymorphism is impossible.

The most general situation occurs if we allow the moments of the two patches to differ. The theorem of Khaz'minskii (1967) is equally applicable to this case. Following through as before, the asymptotic density of the normalized process is

$$\varphi(p) = Cp^{(\alpha/\beta)-1}(1-p)^{-(\alpha/\beta)-1} \exp\left[-\frac{2m^*}{\beta}\frac{1}{p(1-p)}\right],$$

where $\alpha = \Delta\Gamma_1^* - \Delta\Gamma_2^*$, and $\beta = \sigma_1^{2*} + \sigma_2^{2*} - 2\rho\sigma_1^*\sigma_2^*$. Again, C is to assure that

$$\int_0^1 \varphi(p)\,dp = 1.$$

We also require $Ed \ln \|x\|$, which is given by

$$p\,\Delta\Gamma_1^* + (1-p)\,\Delta\Gamma_2^* + \tfrac{1}{2}p(1-p)\beta.$$

The stability properties will now depend on

$$I_1 = \int_0^1 Ed \ln \|x\|\varphi(p)\,dp.$$

Unfortunately, the integrals are intractable, and very little in general can be said about this integral. Certain cases are, however, worth examining.

a) As $m^* \to \infty$, $\varphi(p)$ becomes concentrated at one-half. Examination of the linearized systems at both endpoints yields the following conditions for polymorphism:

$$|\overline{\Delta\Gamma}^*| < \frac{\sigma_1^{2*} + \sigma_2^{2*} - 2\sigma_{12}^*}{8}.$$

This is the same as condition (8).

b) If $\Delta\Gamma_1^* = \Delta\Gamma_2^* = 0$, that is, if the geometric mean fitnesses of both alleles are the same in both environments, polymorphism always occurs.

c) If $\Delta\Gamma_1^* = -\Delta\Gamma_2^*$, polymorphism always occurs.

d) In analogy to the previous case, we can write the conditions for polymorphism as

$$|\overline{\overline{\Delta\Gamma}}^*| < \frac{\beta}{2}\zeta^*(m^*, \alpha, \beta),$$

where

$$\overline{\overline{\Delta\Gamma}}^* = E[p\,\Delta\Gamma_1^* + (1-p)\,\Delta\Gamma_2^*],$$
$$\zeta^*(m^*, \alpha, \beta) = Ep(1-p).$$

e) As a final case let the temporal variation in the environment be zero. If we write $\tau\mu_1^* = EU_1$ and $\tau\mu_2^* = EU_2$, then setting all variance terms equal to zero converts the diffusion equation of case 1 into a pair of ordinary differential equations:

$$\dot{x}_1 = \mu_1^* x_1(1-x_1) + m^*(x_2 - x_1);$$
$$\dot{x}_2 = \mu_2^* x_2(1-x_2) + m^*(x_1 - x_2).$$

These equations are a continuous version of the difference model discussed by Moran (1962, p. 173). Elementary considerations yield the following conditions for polymorphism based on these equations:

$$\mu_1^*\mu_2^* < 0;$$
$$|\mu_1^*\mu_2^*| > m^*|\mu_1^* + \mu_2^*|. \tag{10}$$

Notice that higher migration makes polymorphism less likely, unlike the case for equivalent patches with temporal variation.

DISCUSSION

The most important property to come out of this analysis concerns the role of migration in the maintenance of genetic variation. It is clear from the results on equivalent patches with temporal fluctuations that *increasing* migration makes polymorphism more likely (property [iii] of ζ), while for spatially differentiated patches with no temporal variation, *reducing* migration makes polymorphism more likely. This suggests that there may exist an optimum migration rate which will be larger when temporal fluctuations are larger, and smaller when spatial fluctuations are larger. While this is intuitively appealing, its proof requires, in the first place, a notion of the mean fitness of the population which will be meaningful as a quantity for maximization. As has been pointed out before (Gillespie 1973*a*, 1973*b*), the arithmetic mean fitness of the population can decrease under the action of natural selection in a temporally fluctuating environment. The role played by the geometric mean fitness of a genotype in the theory of random environments suggests that the geometric mean fitness of the population would be preferable as a measure of population fitness. This is particularly appealing in light of the multiplicative nature of population growth models. In the second paper in this series the action of selection on the geometric mean fitness will be examined and the criterion for the existence of optimal migration rates will be established.

A second interesting result of the analysis concerns the order-of-magnitude relationships between migration and the genetic parameters. Case 3 and the infinite slope of ζ at the origin together argue that as long as migration rates are larger than the magnitude of the genetic parameters, the population behaves as a single panmictic unit with respect to the qualitative properties concerning the existence of polymorphism. It is probably the case that the selective pressures acting on those loci which are presently being described in electrophoretic studies are extremely small. If this is so, then even very low migration rates could be sufficient to explain both the existence of the variation and the fact that there is very little geographic differentiation in allele frequencies.

As a final point, notice that all of the conditions for the maintenance of variation were written in terms of the starred parameters. Using their definition (3), it is obvious that they could also have been written without the stars, in terms of the original parameters. This is done with the implicit assumption that the original parameters are very small. For example, values of $\Delta\Gamma$ of the order of 10^{-3} would possibly be appropriate for enzyme loci, and would also assure excellent agreement between the original model (1) and the diffusion approximation.

LITERATURE CITED

Gillespie, J. H. 1972. The effects of stochastic environments on allele frequencies in natural populations. Theoret. Pop. Biol. 3:241–248.

———. 1973a. Polymorphism in random environments. Theoret. Pop. Biol. 4:193–195.

———. 1973b. Natural selection with varying selection coefficients—a haploid model. Genet. Res. 21:115–120.

———. 1974. Polymorphism in patchy environments. Amer. Natur. 108:145–151.

Gillespie, J. H., and C. Langley. 1974. A general model to account for enzyme variation in natural populations. Genetics 76:837–848.

Khaz'minskii, R. Z. 1967. Necessary and sufficient conditions for the asymptotic stability of linear stochastic systems. Theory Probability Appl. 1:144.

Levene, H. 1953. Genetic equilibrium when more than one ecological niche is available. Amer. Natur. 87:311–313.

Moran, P. A. P. 1962. The statistical processes of evolutionary theory. Clarendon, Oxford.

Copyright ©1972 by the American Association for the Advancement of Science
Reprinted from *Science* 175:1327-1335 (1972)

Migration Strategies of Insects

Migration is an environmentally modified physiological syndrome adapted for dispersal and colonization.

Hugh Dingle

In recent years there has been a crystallization of new ideas concerning the nature of insect migration. These ideas have come from several sources and include physiology and behavior on the one hand and ecology, especially as it relates to strategies of colonization, on the other. On the face of it, these approaches seem complementary, but there have been until now no comprehensive attempts either to test their combined validity for any one species or to formulate a general theory of insect migration in which both an insect's physiology and behavior, and its population and life history statistics, are taken into account. In this article I have gathered data from studies of a variety of insect migrants; these data suggest that the holistic view arising from a synthesis of the various concepts does in fact provide a sound basis for future migration research, not only in insects but in other migrants, such as birds and fish, as well.

Migration in insects is a distinct behavioral and physiological syndrome (*1–3*). Kennedy defines it in terms of enhanced locomotory behavior which results in persistent, straightened-out movement (which need not be in any specific direction). At the same time "vegetative functions" associated with maintenance and growth, for example, feeding and reproduction, are inhibited. Following long-distance flight the thresholds for vegetative activity are lowered and further migration is inhibited (*1, 4, 5*). Kennedy has demonstrated (*2*) that in the aphid *Aphis fabae* migration involves a complex interaction between flight and feeding and depositing larvae. His definition of migration does not dis-

The author is an associate professor in the department of zoology, University of Iowa, Iowa City 52240.

tinguish between active migration and passive dispersal by wind, considered separate phenomena by many older entomologists (*6*), because the latter usually requires active flight to attain altitudes permitting wind displacement and then sufficient wing beating to remain airborne.

Studies of female insects, summarized by Johnson (*3*), indicate that migration involves an "oogenesis-flight syndrome." The available evidence suggests that most migration takes place prior to egg development (oogenesis) and reproduction although some may be interreproductive. What is important is that the development of the reproductive system is minimized at a time when that of the flight system, including wings, wing muscles, and associated biochemistry, is maximized. Migration therefore occurs chiefly in young adults, and its ontogeny shows a distinct pattern. Its chief functions are to allow escape from unfavorable environments and dispersal over and colonization of all available habitats. This is in marked contrast to the older view that it represented a desperate attempt to relieve population pressure. Insect migrants are thus colonizers and not refugees.

It is the colonization aspect of migration that has received considerable attention from ecologists. Some time ago both Brown and Jackson (*7*) pointed out that there was a relation between migration and habitat; specifically, aquatic insect migrants were likely to be species inhabiting small ponds or ponds subject to periodic drying. More recently, Southwood (*8*) has explored the question further and has summarized an abundance of evidence indicating that migration occurs most often in denizens of "temporary" habitats or, in other words, habitats in early stages of ecological succession such as old fields, roadsides, waste areas, and ephemeral ponds. As old habitats deteriorate, new ones are colonized, although it should be emphasized that considerable migration takes place before the deterioration of the old habitat because of the tendency to exodus before reproductive maturation. Migration may also be abetted by responses to proximate stimuli such as photoperiod, crowding, or alterations in the physiology of a food plant; its function in adaptive dispersal is thus further indicated (*9*).

If a migrant insect is to be a successful colonizer upon arriving in a new habitat, it must of course be able to reproduce and leave descendants. In this situation evolution favors high productivity (*10*); in order to attain this, the migrant would be expected to modify its life history statistics, under the influence of either its environment or its genes, to produce a high rate of population increase. It can do this either by increasing fecundity or, more effectively, by reproducing earlier (*11*). Further, an individual should have a high "reproductive value" when it migrates. This value was defined by Fisher (*12*) as the expected contribution of an individual to population growth. Its computation is analogous to the computation of compound interest on a bank account (see below). Since young adults have their reproductive life ahead of them and at the same time have already survived the causes of juvenile mortality, their reproductive values are usually high relative to juveniles or older adults, indicating a high colonization potential. We would therefore expect evolution to favor dispersal in most species just prior to reproduction (*10*), and the evidence indicates that in insects this is exactly what happens (*3*).

Field Evidence for Insect Migration

Older studies on insect migration tended to concentrate on movements presumed to be long distance and to and fro (*4, 6*), and most information was gained from large insects, such as butterflies, often seen flying across the countryside in sizable swarms. By compiling observations of such swarms, Williams (*6*) succeeded in demonstrating that many insect species were migrants, and that migration was an important behavioral and ecological phenomenon. As data have continued to accumulate, however, it has become evident that many small species are also migrants whose movements are neither very far nor particularly spectacular (*3*). In fact, unless specifically looked for, their movements usually go undetected. Close study, however, reveals that in favorable weather there is a veritable rain of insects descending to the earth following migratory flight. Most of these insects, whether large or small, are transported across the countryside by the wind (*3*).

The role of meteorological factors in the transport of insect migrants was first clearly described by Rainey working on the African migratory locust *Schistocerca gregaria* (*13*). The gregarious phase of this species forms large migratory swarms traveling on winds which converge at the intertropical convergence zone. This zone moves back and forth across the equator once each year causing seasonal rains in much of tropical Africa. These rains promote greening and growth of the vegetation, and the locust swarms are thus deposited in areas suitable for breeding.

In North America several species migrate north in the spring, especially up the Mississippi Valley where they are aided by the southerly winds which blow consistently at that time. Three of the best studied species in this respect are the leafhoppers *Macrosteles fascifrons* and *Empoasca fabae* (Homoptera) and the harlequin bug *Murgantia histrionica* (Heteroptera) (*14*). Influxes of these species are correlated with the appropriate winds and weather fronts. Migrations into the valleys of California by another leafhopper *Circulifer tenellus* are also correlated with the winds (*15*). Perhaps the best known North American migrant is the monarch butterfly, *Danaus plexippus*, which has been shown from tagging experiments to migrate up to 1900 miles (1 mile = 1.6 kilometers) in the fall (*16*). The monarch may migrate independently of the wind although proof is lacking; the Florida salt marsh butterfly *Ascia monuste*, a short-distance migrant, almost certainly does so (*17*). Thus some insects may migrate without eolian transport, but present evidence suggests that they are the exception.

If they arrive in a suitable area, migrants begin a phase of rapid population growth. A good example is provided by studies of field populations of the milkweed bug *Oncopeltus fasciatus* (Heteroptera). This is a wide-ranging species occurring from Canada to Brazil (*18, 19*). It is a migrant that arrives in the northern parts of its range between spring and early summer, depending on latitude and temperature. The arriving migrants settle in patches of milkweed, mate, and lay eggs on the milkweed plant usually close to the developing seed pod. The egg is succeeded by five nymphal instars and after a few weeks by the adult. The population increases rapidly thereafter until it reaches its maximum in late summer. Numbers then decline because, with shortening days, the maturing adults leave. They presumably migrate south on the generally northerly winds of autumn, thus avoiding the oncoming winter. Four or more generations may occur in the southern United States during the course of a summer, while there may be only one in the north. If the weather is favorable, some nymphs and young adults can still be found in early November at the latitude of Iowa. Cold kills those unable to complete the adult molt before the first severe frost. The pattern of summer population growth and decline is given in Fig. 1 which indicates the numbers of *Oncopeltus fasciatus* in an Iowa field in the very cool summer of 1967 and the more unusual summer of 1968.

Age and Migration

A sufficient number of laboratory studies of migrants is now available to permit the general conclusion that migration occurs for the most part early in the life of the adult insect (*3, 20–23, 28*). Flight duration is measured with individuals tethered and either flying in place or on a movable flight mill. The results are the same whichever technique is used; the longest

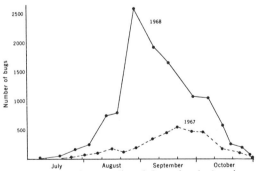

Fig. 1 (left). Growth and decline of *Oncopeltus fasciatus* in an Iowa field (approximately 1 acre) in the summers of 1967 and 1968. In 1967 (cool) there was one major generation with a few offspring from a second generation in early October. In 1968 (warm) there were three generations; the young of generation 2 constitute the sudden rise in late August, and the young of generation 3 account for the break in the population decline occurring in early October. Fig. 2 (right). Duration of tethered flight as a function of age in (A) the frit fly *Oscinella frit* (*20*), (B) the fruit fly *Drosophila funebris* (*21*), and (C) the milkweed bug *Oncopeltus fasciatus* (*22*). In (C) the dotted line indicates the flight of females tested only once during their lifetimes, and the solid line indicates females which flew for at least 30 minutes on day 8 and were flight tested every 2 days thereafter.

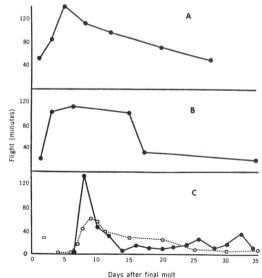

flights occur soon after the molt to adulthood with decreasing flight durations as the insects get older. These effects have been demonstrated for various species of bugs, flies, mosquitoes, aphids, moths, and grasshoppers, and there are some data for beetles which are suggestive of similar relationships (*3*). The decline in flight occurs concurrently with an increase in reproduction (the oogenesis-flight syndrome). Examples of the relation between age and flight activity for three species are given in Fig. 2.

In very young adults flight activity may be either nonexistent or very low. This is because flight cannot occur until the cuticle has hardened; the period during which hardening occurs is known as the teneral period, and flight is thus postteneral. In most insects (flies, mosquitoes, butterflies, and moths are apparent exceptions) the teneral period can be objectively defined by counting cuticular growth rings. These are laid down in the cuticle during the first days of adult life. The structure of the cuticle laid down during the day differs from that deposited at night; when viewed through a polarizing microscope, it is birefringent with one dark and one light ring for each 24 hours of growth (*24*). At the end of the growth period, which is a function of temperature (*25*), no further rings are deposited. In *O. fasciatus*, eight rings on the average are formed at 23°C; the teneral period is thus 8 days long, and the major migration flight occurs at the end of it (Fig. 2C). At 27°C six rings, broader than those at 23°C, are laid down, and migration begins 6 days after the adult molt rather than 8, and is again postteneral (*25–27*).

As insects age, a variety of biochemical changes occur. Included in these are declines in the amount of fuel used for flight and in the amount of enzymatic activity which makes the fuel available to the wing muscles (*21, 28*). Furthermore, because insects are flown to exhaustion while tethered, fuel consumption is commensurate with flight duration (*29*). As a result, there has been a tendency to assume that fuel limits migration. In the ultimate sense this may be so, but there are a variety of proximate stimuli, both internal and external, which trigger the changeover from migratory to other activities long before all fuel is consumed or age reduces enzymatic activity. Therefore statements that fuel limits migration, even though this may ultimately be true, are not particularly helpful (*2, 30*). In fact biochemical changes may themselves be initiated by the same stimuli that inhibit further migration (*2, 31*).

Migratory as Opposed to Vegetative Behavior

Kennedy's definition of migration, cited earlier, offers an objective and experimental test of whether or not flight is migratory (*1*); if it is, the insect should not be responsive to stimuli triggering vegetative behavior (for example, food or oviposition sites). The test has, unfortunately, seldom been applied, and in fact there is little information on the relation between locomotory and vegetative behavior in migrant insects. A notable exception is Kennedy's own work on the black bean aphid, *Aphis fabae* (*2*). In this species there is a reciprocal interaction between migratory flight and the postflight settling responses which include landing on a suitable bean leaf, probing with the mouthparts, and eventually feeding and depositing young. Stimuli which evoke flight inhibit settling, and conversely, stimuli which evoke settling inhibit flight. The two behaviors also have aftereffects on each other which can be either priming or depressing, depending on the respective excitatory states and on the strength of the incoming stimuli.

The whole migration syndrome thus involves a complex integration of stimuli and responses.

The migration-settling sequence in aphids takes place in at most a few hours. A longer term interaction between migratory and vegetative behavior occurs in *O. fasciatus*. Oviposition in this species begins 13 to 15 days after the adult molt when the bugs are maintained on 16 hours of light and 8 hours of dark (LD 16:8) at 23°C; since peak flight occurs at 8 to 10 days under these conditions (Fig. 2), migration is largely prereproductive. The overlap between migratory flight and oviposition is therefore minimized. This is also interesting in another context. Both flight and oviposition show a daily periodicity with a peak at approximately midday (*27, 32*) as shown in Fig. 3. The two do not, however, come into conflict since they are temporally segregated in the life history. Mating also shows a circadian periodicity with a peak at the end of the daily light period (Fig. 3). In this instance, it is separated from flight not only in the life history but also in the day.

Analysis of daily periodicities reveals a separation of flight and feeding also. Females begin feeding a few hours after the adult molt. They continue at a high rate with little or no evidence of a circadian rhythm until day 7 when a rhythm becomes apparent; this rhythm is fully developed at day 8

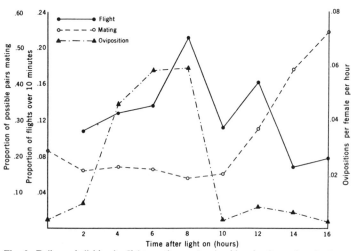

Fig. 3. Daily periodicities in flight, mating, and oviposition in *Oncopeltus fasciatus*. Flights lasting over 10 minutes were used to give a sufficient population size (*N*) at each interval (range, 36 at hour 12 to 304 at hour 4). Data for mating and oviposition are mean values for 40 pairs observed for 10 days. Temperature, 23°C. [After Caldwell and Dingle (*32*) and Dingle (*27*)]

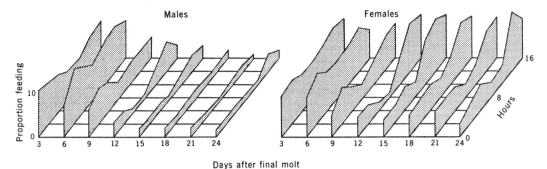

Fig. 4. Daily periodicities in feeding in *Oncopeltus fasciatus* as a function of age; data are averaged for each 3-day interval for 46 bugs of each sex. Hour 0 represents light on and hour 16, light off. Note that from day 9 feeding is low at midday when first flight and then oviposition are maximal. [After Caldwell and Rankin (*33*)]

when migration occurs (*33*). The peak of feeding activity, like mating, occurs at the end of the day (Fig. 3) so that at the time of longest flights, flight and feeding are temporally segregated by their daily periodicities. The maintenance of the rhythm from this point on in the life cycle also insures segregation of oviposition and feeding since the rhythm of the former peaks at midday. Mating proceeds simultaneously with feeding (*32*). In males, the overall rate of feeding falls markedly after day 6 and persists at a low rate throughout the remainder of life. Therefore, at the time of migration there is little feeding going on and indeed there is also more flight activity in older males than in older females (*22*). The relation between the feeding rhythms and the age of the insects is illustrated in Fig. 4.

It is evident from the foregoing that peak migration in the milkweed bug occurs at a specific time in the life history before reproduction occurs; these two activities, one locomotory and one vegetative, therefore do not conflict. Furthermore, there are also timing devices, based on daily periodicities, to assure nonoverlap between flight and feeding. The entire syndrome thus satisfies the physiological criteria defining insect migration.

Environmental Stimuli for Migration

Environmental factors can act directly to trigger migratory takeoffs or landings or can have longer term physiological effects, especially on reproduction, which determine the role of migration in the life history. Takeoff on a migratory flight requires the appropriate conditions of temperature, sunshine, and wind; populations can build up on the ground until conditions are favorable, whereupon a mass exodus may occur (*3*). Aphids provide good examples of this phenomenon as do the various migratory locusts. Locusts, the milkweed bugs *Lygaeus kalmii* and *Oncopeltus fasciatus*, and undoubtedly other migrants bask in the sun with bodies appropriately oriented to increase temperature to a point where flight can occur (*23, 34*); similarly, many butterflies, moths, and beetles increase body temperature by vibrating their wings or "pumping" the thorax (*35*). As flight proceeds, the insect becomes more responsive to stimuli which promote landing and subsequent vegetative behavior. Some aphids, for example, respond to blue wavelengths (of sky origin) on takeoff, but switch over to yellow wavelengths (from leaf surfaces) after a period of flight (*36*). Our knowledge of landing responses in other species is extremely meager.

Temperature can also have longer term influences on migration. At higher temperatures reproductive maturation is speeded, and the period of the life history available for migratory flight is correspondingly reduced (since migration is largely prereproductive). Temperature may also exert a more direct effect. In *O. fasciatus*, raising the temperature from 23° to 27°C results in a significantly smaller proportion of the population exhibiting tethered flights of 30 minutes or longer, the operational criterion for migration. This temperature is about optimal for population growth in *O. fasciatus*. Extrapolated to field situations, this result implies that once bugs find themselves in a thermally favorable environment, they will be inclined to stay there (*27, 37, 38*). Similarly, high temperatures tend to suppress the production of alate (migrant) forms in aphids (*39*).

In temperate regions photoperiod may exert a major influence on migration. Many temperate species of insect use the length of day as a cue to entering a state of suppressed development and metabolism known as diapause (*40*). This suspended state allows survival over the winter or any other set of unfavorable conditions. Diapause occurs in various species in all stages of the life history; when it occurs in the adult, it results in the arrest of reproductive development.

Since migration is largely a prereproductive phenomenon, prolonging the prereproductive period would be expected to prolong migration. This expectation was tested on laboratory populations of *O. fasciatus* and found to hold. At LD 12:12 and 23°C egg laying is delayed in this species until at least 45 days following the adult molt; at LD 16:8 it begins at around 15 days. Adult females 25 days past the adult molt were tested for duration of tethered flight. The LD 16:8 bugs were laying eggs and generally flew for only a few minutes or less; the LD 12:12 bugs showed no signs of reproductive development, and their performance was typical of prereproductive migrants (*27*). One of the possible consequences of this would be that in the autumn, females would be capable of migrating for much longer periods and therefore might have an improved chance of escaping the oncoming winter. Short photoperiods also suppress reproductive development in other known migrants such as the diamond-

219

back moth (*Plutella maculipennis*) and the red locust (*Nomadacris septemfasciata*), but the exact relationships remain to be defined (*41*). As these data indicate, diapause and migration are physiologically similar in that in both, growth and development are suppressed (*1*). They also both involve escape, the one in time and the other in space, and in many species they probably evolved together as parts of the same syndrome.

In addition to prolonging migration by suppressing oogenesis, short photoperiods increase the proportion of migrants in *O. fasciatus*. When LD 12 : 12 bugs were repeatedly tested between 8 and 30 days after the adult molt, a greater proportion of both males and females flew for a long period (at least 30 minutes, usually 2 to 3 hours) when compared to LD 16 : 8 bugs tested over the same interval (*42*). Especially interesting here are the results for the males which are not subject to the constraints of ovarian development. Photoperiod thus seems to have a direct effect on migration as well as an indirect one acting via the reproductive system.

In regions where it fluctuates sufficiently, photoperiod is an excellent proximate environmental cue signaling future conditions before they occur. Many species, however, especially those whose ranges span the equator where photoperiods are essentially constant, must rely directly on ultimate environmental factors. In particular they respond to the availability of food. When starved, the general pattern is for reproductive development to be delayed and migration prolonged; this is true of certain moths, for example (*43*). Three African cotton stainer bugs (*Dysdercus*) are especially interesting in this regard. Females of these bugs histolyze the wing muscles and produce eggs when fed; two species, *D. nigrofasciatus* and *D. superstitiosus*, show some flight activity before histolysis, but the third, *D. fasciatus*, never flies. When starved, there is no histolysis and no egg production, but rather migration in all three species. With food, flight ceases and egg production and muscle histolysis are initiated (*31*). Locusts (*Schistocerca*) delay reproduction when fed senescent leaves, but undergo reproductive development when fed fresh leaves. The active substance is apparently the plant growth hormone gibberellin A_3. They also mature more rapidly when in the presence of terpenoids from the young leaves of various aromatic shrubs. Since these grow young leaves after the rains, the locusts are more likely to mature reproductively in favorable habitats (*44*). At least some photoperiodically sensitive migrants also respond to food shortage. In *O. fasciatus*, for instance, starvation promotes takeoff in young adults, and prolongs migration while suspending reproduction in older individuals (*27*).

Stimuli from other individuals of the species can also be important environmental factors affecting migration. Outstanding examples are the various migratory locusts of which *Schistocerca gregaria*, *Locusta migratoria*, and *Nomadacris septemfasciata* are the best known. Here high population densities result in the development of a gregarious phase whose behavior is characterized by extreme restlessness and extraordinary migratory capability. It is also distinguishable morphologically, and in *Schistocerca* possibly also ecologically, with the gregarious form able, through its powers of dispersal, to exploit widely separated and only marginally suitable habitats (*45*). In the aphids *Megoura viciae* and *Aphis craccivora* crowding results in the production of winged individuals although the effect is modified by temperature and photoperiod (*39*). There is also some suggestion that other insects respond to crowding with increased migration, but the evidence, though tantalizing, is incomplete (*3*). In the milkweed bug *Lygaeus kalmii*, it is a stimulus from the opposite sex that is important, for following mating the amount of long-distance flight is significantly reduced. Mating apparently has a role in terminating the spring dispersal flights (*23*).

Heredity and Migration

Like all characteristics of organisms, migration has a genetic component. But the precise role of hereditary influences, especially their interaction with the environment, has received very little study. In locusts (*Schistocerca*) four generations of selection produces two lines differing significantly with respect to morphological characters defining the gregarious phase, but a single generation of crowding or isolation produces the same or greater differences. Gregariousness is also transmitted by a kind of maternal inheritance since crowding of females produces effects over more than one succeeding generation. The same is true of isolation although the magnitude and rates of change differ. The mechanism is not therefore a simple one (*45, 46*). When 30 minutes of tethered flight is used as the criterion for migration, about 25 percent of an Iowa population of *O. fasciatus* are migrants. Strong selection could increase this to just over 60 percent in one generation (*27*). Migratory capability in this species can thus be altered rapidly by selection.

Natural selection rapidly modifies populations of spruce budworms, *Choristoneura fumiferana*, and tent caterpillars, *Malacosoma disstria* and *M. pluviale*. These species are polymorphic for size. Smaller individuals mature more rapidly, lay fewer but larger eggs, are migrants, and are tolerant of environmental fluctuations. Larger individuals mature slowly, lay many small eggs, are generally sedentary, and survive only in favorable environments. Depending on both past history and current conditions, populations vary in the proportions of the two kinds of individuals. The polymorphism allows these species to deal with fluctuating environments; the large inactive forms exploit favorable conditions while the small migrants disperse and colonize new habitats. A model which assumes that the characters are transmitted by the X chromosome largely accounts for both experimental and field observations (*47*).

Migratory capability, however measured, is a continuous variable and can therefore be analyzed with the techniques of quantitative genetics where methods and theory are well developed (*48*). In only one case, however, have they been applied; this was in estimates of the heritability of the flight duration in the milkweed bug *Lygaeus kalmii* (*49*). Heritability is the proportion of the total variance in a trait resulting from the so-called "additive genetic variance," the variance contributing specifically to parent-offspring resemblance (in human height, for example) as opposed to the resemblance between all members of a species (as in the number of fingers, for example). Heritability is important because in combination with selection differential, it determines response to selection; since it is the ratio of additive genetic to total variance, it varies from zero to one (*50*). In *Lygaeus* the values were 0.20 when estimated from the offspring–male parent regression and 0.41 when estimated from offspring–female parent regression, indicating a possible maternal effect. There is thus sufficient additive genetic variance for selection to operate. In view of the results for *L. kalmii* and the behavioral differences between various migrants, quantitative genetic analysis could be a useful tool in studying dispersal strategies.

Migration and Populations

It is in the analysis of life history and population statistics with respect to migration that the complementarity of physiological and ecological parameters becomes evident. Since migration includes both escape from deteriorating habitats and the colonization of new ones, it is necessary to consider the physiological strategies which facilitate both. Such strategies involve varying reproduction either by varying the age at which egg laying starts or by altering fecundity itself (that is, the total number of eggs laid). As discussed above, many species prolong the prereproductive period with the result that the period in the life history given over to migration is also prolonged. Aphids are particularly interesting in this regard. The parthenogenetic apterous (vegetative) females have shortened the prereproductive period so much that larval development begins before the mother is born (that is, in the body of the grandmother) and a female begins giving birth to young almost as soon as she herself is born. The production of alate migrants (for example, by crowding) results in delaying births until after flight has taken place (*2, 39*). In some insects fecundity may also be reduced by migration, especially if there is a distinguishable migratory phase. This is true for the locust, *Locusta migratoria*, for example, where the gregarious phase lays fewer eggs and lays them later than the solitary phase. The same relationship holds for the "active" and "normal" forms of the beetle *Callasobruchus maculosus* (*1, 51*).

An interesting evolutionary outcome of the interactions between migration and reproduction occurs in females of three species of African cotton stainer bugs (*Dysdercus*) (*31*). These species appear to have modified a basic *Dysdercus* migration-reproduction strategy in which a short period of flight soon after the adult molt is followed by feeding, wing muscle breakdown, and egg development. In all three species, when no food is available, there is no breakdown of the wing muscles and no egg development; instead, flight continues until food is again encountered. Each species has modified the basic pattern to meet its own ecological requirements. *Dysdercus fasciatus* is the most opportunistic. It feeds on a few fruits which are abundant at certain times of the year, but absent at others. Females of this species do not fly at all in the presence of food; they reproduce early, and fecundity is highest of the three. *Dysdercus fasciatus* thus maximizes the dichotomy between locomotory and vegetative behavior. *Dysdercus nigrofasciatus* has adopted an intermediate strategy. It feeds on the seeds of a variety of annuals and perennials so that food is apt to be scattered but generally available. Females do fly for a short period before the wing muscles break down; reproduction is thus delayed for 2 to 3 days relative to *D. fasciatus*, but fecundity is almost as high. Finally, *D. superstitiosus* is a species which feeds on a far greater variety of host plants than either of the other two and thus faces fewer risks from the environment. Its females also undertake some flight before reproduction which begins at about the same time as in *D. nigrofasciatus*. Its fecundity, however, is considerably reduced indicating far less emphasis on reproductive opportunism, as is characteristic of species in more stable environments (*10, 11*).

Factors which affect reproduction also profoundly influence the rate at which a population grows (*11*). An effective measure of the rate of population growth is r, the so-called intrinsic rate of increase, which is usually expressed as growth per unit of time. It combines survivorship, fecundity, and developmental rate in a biologically meaningful way to estimate the growth potential of a population with unlimited resources, and has the particular advantage that it can be compared from one species to another without regard to generation times (*52*). A population of given size, N, growing in an unlimited universe will approach a stable age distribution and expand according to the relation $dN/dt = rN$. To determine r, one must solve for it in the equation

$$\int_0^\infty l_x m_x e^{-rx} dx = 1$$

where e is the base of natural logarithms, x is age, l_x is the proportion of individuals surviving to age x, and m_x is the birthrate defined as the average number of female offspring produced per female in the age interval $x - 0.5$ to $x + 0.5$. In practice r is approximated by trial-and-error substitution in the equation

$$\sum_\alpha^\omega l_x m_x e^{-rx} = 1$$

where α and ω are, respectively, the ages at first and last reproduction. Note that survival (l_x), fecundity (m_x), and rate of development (α) all enter into the solution of r.

The relation between r, life history statistics, and migration is illustrated by results from the milkweed bug *O. fasciatus* (*37*). In this species high temperatures and long photoperiods advance the age at which reproduction first occurs while low temperatures and short photoperiods delay it. The values of r for *O. fasciatus* at a density of 20 pairs per container under four different experimental conditions are given in Table 1 which also indicates the time it would take for the respective populations to double. On a long day (LD 16:8) at a high temperature (27°C) the bugs begin reproducing at 47 days after birth, and r, expressed as the rate of increase per individual per day, is 0.0810; this leads to a doubling of the population in 8.90 days. In contrast, a short day (LD 12:12) and a lower temperature (23°C) results in first reproduction at day 95, an r of 0.0369, and a doubling time of 19.13 days, about twice as long. Long days and high temperatures therefore produce much more rapid population growth; they do so by promoting earlier and more rapid reproduction. This is in fact the most effective way of increasing the value of r (*11*).

A high value of r (or more correctly the ratio r/m where m is the per capita birthrate) is of distinct advantage to a colonizing species (*10, 11*), for it allows such a species to increase its population rapidly when it invades a new environment. In the North American spring and summer *O. fasciatus* enters a previously empty universe, an environment which favors high productivity (*10*). The fact that long days and high temperatures promote rapid population growth means that this species can take maximum advantage of the habitat. There is thus a phenotypic modification of *r* which maximizes colonizing ability. The reduction in migration caused by high temperatures is apparently a still further adaptation to colonization since the bugs will tend to remain and reproduce where conditions are favorable.

In contrast the short days and cool temperatures of autumn delay reproduction and depress the value of *r*. In delaying reproduction, however, they lengthen the period during which migration can take place. Thus *O. fasciatus* under these conditions sacrifices the colonizing advantages of rapid population growth, but for alternative gain, namely, a longer time in which to migrate to a more favorable habitat. The time is apparently necessary because the winds of autumn, on which the bugs presumably travel, are less reliable than those of spring and because it is necessary to travel farther to avoid being overtaken by the oncoming winter. These insects, since they have their reproductive life ahead of them, are still potential colonizers; the extent to which they are actual colonizers is not known. Lack of knowledge of the reproductive fate of any southward migrating insect is one of the largest lacunae in our understanding of the whole migration process.

Maximization of growth potential under optimal conditions and some sacrifice of this potential under conditions promoting migration would seem to be a general strategy of migrant insects. The data from aphids, locusts, *Dysdercus*, and *O. fasciatus* all suggest that this is so.

Delay in reproduction is also interesting in relation to the other major escape mechanism of insects, diapause, in which reproductive delay occurs in adults or delayed growth occurs in immature stages. In both instances vegetative functions are suppressed, and diapause and migration are thus similar (*1*). As diapause allows escape in time, so migration allows escape in space. In many insects reproductive delay or diapause allows migration to a new environment and an escape from unfavorable conditions. In fact adult reproductive diapause may have evolved through delays in reproduction which allowed migration to new areas. The intimacy of the relationships between diapause and migration in adult insects is thus further emphasized.

The relation between life history statistics and colonizing ability is also

Table 1. Rates of increase for *Oncopeltus fasciatus*.

Photoperiod (light : dark)	Temperature (°C)	Age at first reproduction (days)	Increase per day (r)	Doubling time (days)
16 : 8	27	47	0.0810	8.90
12 : 12	27	61	.0593	12.03
16 : 8	23	63	.0499	14.24
12 : 12	23	95	.0369	19.13

brought out in the concept of reproductive value (*11, 12*). Because this is defined as the expected contribution of an individual of specified age to future population growth, it indicates at what age it is most likely to found a successful colony. The reproductive value, v_x, of an individual of age x in a growing population is defined mathematically relative to this value at birth, v_0. Thus where t is time and other terms are as previously defined

$$\frac{v_x}{v_0} = \frac{e^{rx}}{l_x} \int_x^\infty l_t m_t e^{-rt} dt$$

Since v_x is relative to v_0, we can set $v_0 = 1$. Rate of increase, survivorship, and fecundity all enter into the calculation of v_x and integration is from x to ∞ so that only future births are included.

What matters here is that since reproductive value indicates the chances of successful colonization, it is an important factor to consider when discussing a migrant insect. Accordingly, reproductive values were computed for *O. fasciatus* reared at LD 16 : 8 (high r) and at LD 12 : 12 (low r), both at 23°C, and for *Aphis fabae* (*22, 27*). In *O. fasciatus* at LD 16 : 8, v_x rises rapidly after the adult molt to reach a peak at day 20; since flight is maximal at 8 to 10 days, a migrant will reach its maximum v_x shortly after arriving in a new location and thus stands an excellent chance of founding a colony. At LD 12 : 12 the situation is more complex. In this case reproductive value does not reach a maximum until day 100 at the earliest or day 160 at the latest depending on density. Reproduction starts occasionally as early as 45 days after the adult molt, but more usually around days 60 to 70 which can thus be taken as the time most migration ceases. Bugs migrating under these conditions are therefore not so likely to be successful colonizers, for there is a time lag between maximum migration and peak reproductive value. Here the chances of successful colonization, although reduced, are still better than the alternative, which is death through failure to escape the approaching winter. Both high r and high reproductive value are sacrificed for the alternative gain of escape from unfavorable conditions. In *Aphis fabae* the adult molt occurs 4 to 5 days after birth, and migration follows within 24 hours; there is then a burst of reproductive activity. The latter is reflected in the reproductive value which peaks at day 5 indicating that the aphids have a high colonizing potential. In fact, physiological and ecological events are so tightly linked in these aphids that ordinarily flight is a prerequisite for settling and deposition of young (*2*).

Conclusions

Physiological and ecological results from a variety of species are consistent with what seem to be valid general statements concerning insect migration. These are as follows: (i) During migration locomotory functions are enhanced and vegetative functions such as feeding and reproduction are suppressed. (ii) Migration usually occurs prereproductively in the life of the adult insect (the oogenesis-flight syndrome). (iii) Since migrant individuals are usually prereproductive, their reproductive values, and hence colonizing abilities, are at or near maximum. (iv) Migrants usually reside in temporary habitats. (v) Migrants have a high potential for population increase, r, which is also advantageous for colonizers. (vi) Both the physiological and ecological parameters of migration are modifiable by environmental factors (that is, phenotypically modifiable) to suit the prevailing conditions. Taken together, these criteria establish a comprehensive theory and adumbrate the basic strategy for migrant insects. This basic strategy is modified to suit the ecological requirements of individual species. Comparative studies of these modifications are of considerable theoretical and practical interest, the more so since most economically important insects are migrants.

No satisfactory general statements can as yet be made with respect to the genotype and migration. Certainly we expect colonizing populations to possess genotypes favoring a high r, but genotypic variation in r depends on the heritabilities of life table statistics, and such measurements are yet to be made (*10, 53*). The fact that flight duration can be increased by appropriate selection in *Oncopeltus fasciatus*, and the demonstration of additive genetic variance for this trait in *Lygaeus kalmii*, suggest that heritability studies of migratory behavior would also be worth pursuing. Most interesting, of course, will be possible genetic correlations between migration and life history parameters. Also, migration often transports genotypes across long distances with considerable mixing of populations. An understanding of its operation therefore carries with it implications for population genetics, zoogeography, and evolutionary theory.

Finally, at least parts of the above general theory would seem to be applicable to forms other than insects. Bird and insect migrations, for example, are in many respects ecologically and physiologically similar. Birds, like insects, emphasize locomotory as opposed to vegetative functions during long-distance flight; the well-known *Zugunruhe* or migratory restlessness is a case in point. Further, many birds migrate at night at a time when they would ordinarily roost (vegetative activity). Because their life spans exceed single seasons, bird migrants are not prereproductive in the same sense that insect migrants are, and hence reproductive values do not have the same meaning (but note that some insects are also interreproductive migrants). The situation is complicated further by the fact that in many birds adult survivorship is virtually independent of age so that colonizing ability tends to be also (*10, 54*). Nevertheless, birds arrive on their nesting grounds in reproductive condition with the result that migration is a colonizing episode. It is also phenotypically modifiable by environmental factors, some of which, for example, photoperiod, influence insects as well (*55*). The similarities between birds and insects thus seem sufficient to indicate, at least provisionally, that the theory developed for insects applies also to birds with appropriate modifications for longer life span and more complex social behavior; comparisons between insects and fish (*56*) lead to the same conclusion. In birds especially, and also in other forms, various functions accessory to migration such as reproductive endocrinology, energy budgets, and orientation mechanisms have been studied extensively (*55, 56*). But there is need in vertebrates for more data and theory on the ecology and physiology of migratory behavior per se in order to better understand its evolution and its role in ecosystem function (*5, 57*). Migration in any animal cannot be understood until viewed in its entirety as a physiological, behavioral, and ecological syndrome.

References and Notes

1. J. S. Kennedy, *Nature* **189**, 785 (1961).
2. ———, *Proc. Int. Congr. Entomol. 10th Montreal* **2**, 397 (1958); ——— and C. O. Booth, *J. Exp. Biol.* **40**, 67 (1963); *ibid.*, p. 351; *ibid.* **41**, 805 (1964); J. S. Kennedy, *ibid.* **43**, 489 (1965); *ibid.* **45**, 215 (1966).
3. C. G. Johnson, *Nature* **198**, 423 (1963); *Annu. Rev. Entomol.* **11**, 233 (1966); *Migration and Dispersal of Insects by Flight* (Methuen, London, 1969).
4. Definitions of migration have long bedeviled students of behavior. Anthropologists, demographers, and entomologists have tended to use the term in the dictionary sense of moving from one place to another. Vertebrate biologists, on the other hand, largely as a result of a preoccupation with birds and some fish and mammals, have adopted a somewhat ornithocentric view which establishes return movements as the criterion for "true migration" [W. Heape, *Emigration, Migration, and Nomadism* (Heffner, Cambridge, 1931), p. 16; A. L. Thompson, *Bird Migration* (Witherby, London, 1949), p. 24; J. Dorst, *The Migration of Birds* (Heinemann, London, 1962), p. xii]. But return movements in birds and other vertebrates are necessitated by the fact that life spans are in most cases longer than single breeding seasons, whereas this is not true for most insects. The concept of migration as a to and fro journey dies hard, but this criterion now seems unnecessary even for birds [where there is also the possibility of some migrations within the tropics occurring without return movements—see Moreau (*5*)]. The physiological and ecological conditions of migration are so analogous across taxa, regardless of whether or not return movements occur, that something akin to Kennedy's definition herein cited would seem more useful. Migration is thus recognized as a distinct behavioral and physiological syndrome whose result is adaptive dispersal. Dispersal and migration are sometimes equated, but note that dispersal can take place via mechanisms other than migration, for example, complex social behavior (see my conclusions).
5. R. E. Moreau, *The Bird Faunas of Africa and Its Islands* (Academic Press, New York, 1966).
6. C. B. Williams, *Annu. Rev. Entomol.* **2**, 163 (1957); *Insect Migration* (Collins, London, 1958).
7. E. S. Brown, *J. Zool.* **121**, 539 (1951); D. J. Jackson, *Proc. Roy. Entomol. Soc. London Ser. A* **27**, 57 (1952).
8. T. R. E. Southwood, *Biol. Rev. Cambridge Phil. Soc.* **37**, 171 (1962).
9. For clarity I have discussed the contributions of Johnson, Kennedy, and Southwood as though they were separate and distinct with each advancing a particular viewpoint. In fact, these three workers have developed an essentially integrated view with each emphasizing a different aspect.
10. R. H. MacArthur and E. O. Wilson, *The Theory of Island Biogeography* (Princeton Univ. Press, Princeton, N.J., 1967).
11. L. C. Cole, *Quart. Rev. Biol.* **29**, 103 (1954); R. C. Lewontin, in *The Genetics of Colonizing Species*, H. G. Baker and G. L. Stebbins, Eds. (Academic Press, New York, 1965), p. 77.
12. R. A. Fisher, *The Genetical Theory of Natural Selection* (Dover, New York, 1958).
13. R. C. Rainey, *Nature* **168**, 1057 (1951); *Proc. Int. Congr. Entomol. 11th Vienna* **3**, 47 (1960); *World Meteorol. Organ. Tech. Note No. 54* (1963).
14. G. W. Bruehl, *Amer. Phytopathol. Soc. Monogr. No. 1* (1961); L. N. Chiykowski and R. K. Chapman, *Wis. Agr. Exp. Sta. Res. Bull.* **261**, 21 (1965); A. C. Hodson and E. F. Cook, *J. Econ. Entomol.* **53**, 604 (1960); F. H. Huff, *J. Appl. Meteorol.* **2**, 39 (1963); J. T. Medler, *Proc. Int. Congr. Entomol. 11th Vienna* **3**, 30 (1960); W. Nichiporick, *Int. J. Bioclim. Biometeorol.* **9**, 219 (1965); R. L. Pienkowski and J. T. Medler, *Ann. Entomol. Soc. Amer.* **57**, 588 (1964).
15. H. H. P. Severin, *Hilgardia* **7**, 281 (1933); W. C. Cook, *U.S. Dep. Agr. Tech. Bull. No. 1365* (1926).
16. F. A. Urquhart, *The Monarch Butterfly* (Univ. of Toronto Press, Toronto, 1960).
17. E. J. Nielsen, *Biol. Medd. Kgl. Dan. Vidensk. Selsk.* **23**, 1 (1961).
18. J. A. Slater, *A Catalogue of the Lygaeidae of the World* (Univ. of Connecticut Press, Storrs, 1964), p. 95.
19. The evidence for the northward migration of *O. fasciatus* comes from an extremely scattered literature and from observations by myself and others. I particularly thank Dr. R. B. McGhee for making available his notes on *O. fasciatus* in the southern United States, the West Indies, and Mexico.
20. T. D. Rygg, *Entomol. Exp. Appl.* **9**, 74 (1966).
21. C. M. Williams, L. A. Barness, W. H. Sawyer, *Biol. Bull.* **84**, 263 (1943).
22. H. Dingle, *J. Exp. Biol.* **42**, 269 (1965); *ibid.* **44**, 335 (1966).
23. R. L. Caldwell, thesis, Univ. of Iowa (1969).
24. A. C. Neville, *Advan. Insect Physiol.* **4**, 213 (1967).
25. H. Dingle, R. L. Caldwell, J. B. Haskell, *J. Insect Physiol.* **15**, 373 (1969).
26. The term "teneral," from the Latin *tener* meaning soft or delicate, was first used to describe the insect imago before the cuticle had hardened sufficiently for flight. In *O. fasciatus* this coincides with cuticle deposition which thus fairly precisely defines the teneral period. In other insects the coincidence is not so great; locusts, for example, can migrate before cuticle deposition ceases (although hardening has occurred). The relationships between cuticle deposition and flight need to be worked out in more species before any definitive general statement can be made regarding the teneral period. In the meantime it is not helpful to view postteneral migration too rigidly (*3, 24*).
27. H. Dingle, *J. Exp. Biol.* **48**, 175 (1968).
28. A. M. Clark and R. Rockstein, in *The Physiology of Insecta*, M. Rockstein, Ed. (Academic Press, New York, 1964), p. 227; M. Rockstein and P. L. Bhatnagar, *J. Insect Physiol.* **11**, 481 (1965); W. A. Rowley and C. L. Graham, *ibid.* **14**, 719 (1968).
29. V. B. Wigglesworth, *J. Exp. Biol.* **26**, 150 (1949); T. Weis-Fogh, *Phil. Trans. Roy. Soc. London Ser. B* **237**, 1 (1952); B. Hocking, *Trans. Roy. Entomol. Soc. London* **104**, 223 (1953); A. J. Cockbain, *J. Exp. Biol.* **38**, 163 (1961); W. J. Yurkiewicz, *J. Insect Physiol.* **14**, 335 (1968).
30. J. S. Kennedy, in *Insects and Physiology*, J. W. L. Beament and J. E. Treherne, Eds. (Oliver & Boyd, Edinburgh, 1967); Inaugural Lecture, Imperial College of Science and Technology, London, 11 November (1969).
31. H. Dingle and G. Arora, in preparation.
32. R. L. Caldwell and H. Dingle, *Biol. Bull.* **133**, 510 (1967).
33. R. L. Caldwell and M. A. Rankin, *J. Insect Physiol.*, in press.
34. G. Fraenkel, *Z. Vergl. Physiol.* **13**, 300 (1930).
35. A. Krogh and E. Zeuthen, *J. Exp. Biol.* **18**, 1 (1941); D. A. Dorsett, *ibid.* **39**, 579 (1962); P. A. Adams and J. E. Heath, *Nature* **201**, 20 (1964).
36. J. S. Kennedy, C. O. Booth, W. J. S. Kershaw, *Ann. Appl. Biol.* **49**, 1 (1961).

223

37. H. Dingle and R. L. Caldwell, *Ann. Entomol. Soc. Amer.* **64**, 1171 (1971).
38. H. Dingle, *Amer. Natur.* **102**, 149 (1968).
39. J. S. Kennedy and H. L. G. Stroyan, *Annu. Rev. Entomol.* **4**, 139 (1959); A. D. Lees, *J. Insect Physiol.* **3**, 92 (1959); *Advan. Insect Physiol.* **3**, 207 (1966); *J. Insect Physiol.* **13**, 289 (1967); B. Johnson, *Entomol. Exp. Appl.* **9**, 301 (1966).
40. A. S. Danilevskii, *Photoperiodism and Seasonal Development of Insects* (Oliver & Boyd, Edinburgh, 1965); S. D. Beck, *Insect Photoperiodism* (Academic Press, New York, 1968).
41. M. J. Norris, *Symp. Roy. Entomol. Soc. London* **2**, 56 (1964); D. G. Harcourt and M. Cass, *Nature* **210**, 217 (1966).
42. R. L. Caldwell, in preparation.
43. M. Koch, *Biol. Zentralbl.* **85**, 345 (1966).
44. D. B. Carlisle, P. E. Ellis, E. Betts, *J. Insect Physiol.* **11**, 1541 (1965); P. E. Ellis, D. B. Carlisle, D. J. Osborne, *Science* **149**, 546 (1965).
45. F. O. Albrecht, *Polymorphisme Phasaire et Biologie des Acridiens Migrateurs* (Masson, Paris, 1967); J. P. Dempster, *Biol. Rev. Cambridge Phil. Soc.* **38**, 490 (1963); J. S. Kennedy, *Symp. Roy. Entomol. Soc. London* **1**, 80 (1961); *Colloq. Int. Cent. Nat. Rech. Sci.* **114**, 283 (1962); B. P. Uvarov, *Proc. Roy. Entomol. Soc. London Ser. C* **25**, 52 (1961).
46. F. O. Albrecht and R. E. Blackith, *Evolution* **11**, 165 (1957); P. Hunter-Jones, *Anti-Locust Bull.* No. 29 (1958); W. J. Stower, D. E. Davies, I. B. Jones, *J. Anim. Ecol.* **29**, 309 (1960).
47. I. M. Campbell, *Can. J. Genet. Cytol.* **4**, 272 (1962); in *Breeding Pest-Resistant Trees*, H. Gerhold, Ed. (Pergamon Press, New York, 1966); W. G. Wellington, *Can. Entomol.* **96**, 436 (1964).
48. D. S. Falconer, *Introduction to Quantitative Genetics* (Oliver & Boyd, Edinburgh, 1960).
49. R. L. Caldwell and J. P. Hegmann, *Nature* **223**, 91 (1969).
50. The basic models of quantitative genetics assume that individual differences in continuously varying traits are the result of both genetic and environmental differences. The total (phenotypic) variance for any trait is expressed by: $V_P = V_A + V_N + V_E$ where V_A is the additive genetic variance, V_N is the nonadditive genetic variance (due to dominance and epistasis), and V_E is the environmental variance. Phenotypic variance can be partitioned into these components and heritability estimates (h^2) can be obtained from this partitioning where $h^2 = V_A/V_P$. Heritability is, thus the proportion of the total variance accounted for by variance in additive genetic influences.
51. G. H. Caswell, *Bull. Entomol. Res.* **50**, 671 (1960); M. J. Norris, *Anti-Locust Bull.* No. 6 (1950); S. Utida, *Res. Pop. Ecol. Kyoto* **3**, 93 (1956).
52. L. C. Birch, *J. Anim. Ecol.* **17**, 8 (1948); L. B. Slobodkin, *Growth and Regulation of Animal Populations* (Holt, Rinehart & Winston, New York, 1961).
53. K. P. V. Sammeta and R. Levins, *Annu. Rev. Genet.* **4**, 469 (1970).
54. D. Lack, *The Natural Regulation of Animal Numbers* (Oxford Univ. Press, London, 1954).
55. D. S. Farner, *Amer. Sci.* **52**, 137 (1964); B. Lofts and R. K. Murton, *J. Zool.* **155**, 327 (1968); G. V. T. Matthews, *Bird Navigation* (Cambridge Univ. Press, Cambridge, ed. 2, 1968).
56. F. R. Harden-Jones, *Fish Migration* (Arnold, London, 1968).
57. R. H. MacArthur, *Auk* **76**, 318; G. W. Cox, *Evolution* **22**, 180 (1968).
58. I thank R. L. Caldwell, J. P. Hegmann, S. B. Kater, J. S. Kennedy, T. R. Odhiambo, E. Reese, and R. Strathman for comments on various drafts of this manuscript. Supported by NSF grants GB-2949, GB-6444, and GB-8705 and a special fellowship from the PHS.

23

Copyright ©1978 by the British Ecological Society
Reprinted from *J. Anim. Ecol.* **47**:461–475 (1978)

INDIVIDUAL DIFFERENCES BETWEEN ANIMALS AND THE NATURAL REGULATION OF THEIR NUMBERS

By ADAM ŁOMNICKI*

SUMMARY

(1) A model of population regulation by emigration is developed based on (i) spatial and temporal heterogeneity of natural environments; (ii) unequal resource partitioning among individuals of single species populations. This latter phenomenon has not previously been taken into account by theoretical ecologists, although it is of basic importance for regulation of population density.

(2) Unequal resource partitioning increases the stability of a confined population and allows emigration of some population members into suboptimal and hostile areas to evolve by individual selection.

(3) In a local habitat supplied by a constant inflow of food, animal density is lower and food density is higher when emigration occurs, than without emigration.

(4) Regulation of population density by emigration allows for the adjustment of this density to the amount of food, in such a way that mortality is not directly due to food shortage but to other agents (predators, adverse weather conditions, etc.).

(5) Animals with overlapping generations and cyclic changes in population density can emigrate at lower density level, than that determined by food supply at any given point in time. At high densities such populations can also exhibit reduced reproduction, which under certain circumstances can be considered as an adaptation.

(6) Population outbreaks are thought of as a failure of population regulation by emigration resulting from spatial homogeneity of the environment. Homogeneity diminishes both differences among individuals in local populations and differences between local habitats and their surroundings, these in turn can make regulation by emigration impossible

INTRODUCTION

According to the present state of ecological theory, emigration of animals into hostile or suboptimal areas is not considered as an important mechanism of natural regulation of animal numbers because such emigration, as pointed out by MacArthur (1972), cannot evolve by individual selection being of selective disadvantage for migrants. The evolution of emigration and other behavioural mechanisms of population regulation have been explained by group selection (Wynne-Edwards 1962; Van Valen 1970; Gilpin 1975). It seems that all the difficulties in accepting emigration as a mechanism of population regulation which has evolved by individual selection result from an assumption, widely accepted by theoretical ecologists, that animals within a single species population are like identical molecules which differ only in sex and age. This paper is an attempt to demonstrate, without invoking group selection, that animal density can be regulated by

* Present address: Institute of Environmental Biology, Jagiellonian University, ul. Krupnicza 50, 30-060 Krakow, Poland.

emigration if differences among animals, other than sex and age, and not necessarily genetical, are taken into account.

Unfortunately, the long-held theoretical assumption that animals are identical has had a profound effect on ecology, so that very little is known about resource division amongst individuals or about differences in their reproduction and survival. It seems that because of this assumption, field ecologists report much less about individual differences among animals in their learned journals and textbooks, than they actually know. The assumption mentioned above appears to be confirmed to a certain degree by ecological laboratory investigations. Laboratory populations are kept under standard conditions, which usually means that they are maintained in very homogeneous artificial habitats and the methods by which the animals are fed and counted homogenize these populations even more. In such habitats large differences between individuals can hardly arise, and if they do they are often lost. For these reasons the results obtained in laboratories can be highly biassed, showing smaller individual differences than those existing in the field, and conforming better to present ecological theory than field data.

The differences between individuals will be considered here in terms of food partitioning among population members. If all individuals take about the same share of food, then a decrease in the amount of food per individual below its maintenance cost brings about the death of the entire population. Populations behaving in this way were defined by Nicholson (1954) as showing 'scramble' competition, they are subject to violent fluctuations and they can easily become extinct through shortage of food. Such a situation can be easily found in the laboratory due to small differences among individuals.

Although no good data concerning resource partitioning within a single population can be found in the ecological literature, it seems that the scramble type of competition is a rare phenomenon, and that both plants and animals exhibit rather a 'contest' (Nicholson 1954) type; some individuals take as much or almost as much as they require, while others get nothing or almost nothing. The resources seem unequally partitioned and this inequality increases when resources become scarce. A mechanism of unequal resource partitioning can be clearly seen in competition for light among plants; an individual plant which by a random accident happens to become a little taller, subsequently takes more light and overgrows its competitors. There are no good reasons to reject the idea that very similar mechanisms of positive feed-back operate among animals; those which are slightly better fed or are in a somewhat better place can, therefore more easily obtain a higher proportion of food and a much better place to live than their competitors.

A formal description of food partitioning among animals can be given by assuming that the probability of an individual encountering a food particle is an increasing function of food density and of the amount of food taken previously by this individual. This assumption leads to food partitioning which approaches a lognormal distribution. If there is a constant maximum number of food particles which can be taken by each individual and if food is abundant, then all individuals take approximately the same number of food particles irrespective of how many they encounter. Therefore if food is abundant, individual differences in food intake are small and they increase if food is scarce.

As mentioned, although resource partitioning within a single species population is a much neglected area of ecological research, some evidence confirming unequal partitioning does exist. Among many vertebrates unequal resource division is brought about by dominance hierarchies which are especially well pronounced when resources are scarce. Inequality of resource partitioning is well expressed in differential growth of population members. For example salamander *Ambystoma maculatum* larvae exhibit a skewed distribution of body size, more skewed when the median of body size for the whole population decreases (Wilbur & Collins 1973). An increase in coefficient of variation with decreasing body size during bad years was found in a field population of wood pigeons (Murton, Isaacson & Westwood 1966). Differences in breeding success of *Drosophila* males (Bateman 1948) suggest that there are large individual differences among insects, which can also influence the competition for food.

The model of the regulation of animal population density presented here is based on two phenomena: (i) unequal resource partitioning described above and (ii) spatial and temporal heterogeneity of the natural environments. While the study of the former phenomenon is neglected, the importance of the latter is now becoming recognized (Wiens 1976; Levin 1976). A recent study by Taylor & Taylor (1977) has elucidated mechanisms by which population regulation due to movement in relation to spatial and temporal heterogeneity can be brought about.

THE BASIC MODEL

Consider an animal population of a single species with non-overlapping generations, in a unit of space where the intrinsic rate of natural increase r is positive. Changes in number of animals N and amount of food V within this space unit are given in discrete units of time t, equal to the generation time. There is a constant inflow C of food per unit of time, per unit of space.

A rank $x = 1, 2, \ldots, N$ is assigned to each individual and this rank determines the amount, y of food it takes, as described by

$$y(x) = a\left(1 - \frac{a}{V}\right)^x \qquad (1)$$

where a is a parameter determining the maximum amount of food an individual can take and it is assumed that the total food available $V > a$. The function $y(x)$ as defined by eqn (1) and shown in Fig. 1 is a simplified description of food partitioning and it is applied to the model presented here for its two properties: (i) it describes both unequal food partitioning and increase of this inequality when food is scarce, and (ii) its application to the model makes analytical solution possible, which is not the case when food partitioning is based on a lognormal or another more realistic distribution. In order to make the presentation of the model more general it is assumed that the number of animals N is large enough so that, when calculating food intake of N individuals, integration from $x = 0$ to $x = N$ instead of summation from rank $x = 1$ to rank $x = N$ can be applied.

Let R denote the amount of food required by one individual to attain reproductive age and to produce one offspring. From eqn (1) the number L of animals (Fig. 1) which receive enough food to produce at least one offspring is given by

$$L = \frac{\ln(R/a)}{\ln(1 - a/V)}. \qquad (2)$$

Note that L denotes both the number of successful animals per unit of space considered, and a rank such that if $x > L$, an individual of rank x dies without leaving progeny, due to the shortage of food.

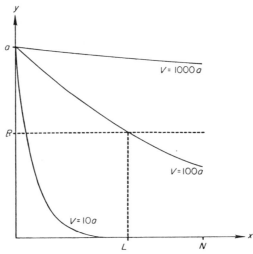

FIG. 1. Amount of food y taken by an individual as a function of its rank x, as defined by eqn (1), for three different food densities. For the density $V = 100a$, the graph shows also the potential number of animals L which could take at least R units of food. Note that L can be either smaller or larger than the number of animals present $N(N = 100)$, depending on food density V.

The amount of food V in the next unit of time $t+1$ is given by

$$V_{t+1} = V_t + C - \int_0^{N_t} y\,dx$$

which, after substituting (1) and integrating, yields

$$V_{t+1} = V_t + C - \frac{a(q_t^{N_t} - 1)}{\ln q_t} \tag{3}$$

where $q_t = 1 - a/V_t$.

Reproduction of animals is assumed to be linearly related to the amount of food taken and since an amount R is used to maintain the individual and to produce one offspring, each animal of rank $x \leq L$ produces $1 + (y - R)h$ offspring, where h is the efficiency of turning food eaten into progeny. L_t does not depend on the density of animals N_t but it is a function of V_t defined by eqn (2).

When $L_t \leq N_t$ the number of animals in the next generation is given by

$$N_{t+1} = L_t + h \int_0^{L_t} (y - R)\,dx$$

while $(N_t - L_t)$ animals die without leaving any progeny. After substituting eqn (1) and integrating, the above equation yields

$$N_{t+1} = \frac{1}{\ln q_t} [(1 - hR)\ln(R/a) - h(a - R)]. \tag{4}$$

Note that if $L_t \leq N_t$ then N_{t+1} does not depend on N_t but is determined by V_t.

When $L_t > N_t$ the number of animals in the next generation is given by

$$N_{t+1} = N_t + h \int_0^{N_t} (y - R)\,dx$$

while there is room for $(L_t - N_t)$ animals to immigrate and to reproduce.

After substituting eqn (1) and integrating, this equation yields

$$N_{t+1} = N_t(1-hR) + \frac{ah}{\ln q_t}(q_t^{N_t} - 1). \quad (5)$$

Note that if $L_t > N_t$ population increases $N_{t+1} > N_t$.

An example of changes in animal number N and the amount of food V, as determined by eqns (3)–(5) is shown in Fig. 2. From eqn (3) the isocline of the equation $\Delta V = 0$ ($\Delta V = V_{t+1} - V_t$) is given by

$$\hat{N}_{\Delta V=0} = \frac{1}{\ln q} \ln\left(1 + \frac{C}{a} \ln q\right).$$

When $L_t \leq N_t$ the isocline $\hat{N}_{\Delta N=0}$ of the equation $\Delta N = 0$ ($\Delta N = N_{t+1} - N_t$) is given by the right hand side of the eqn (4). When $L_t > N_t$ no isocline for the positive value of V and N exists, because $N_{t+1} > N_t$. The isoclines (Fig. 2), as well as numerical calculations, suggest that the system is stable and this was confirmed by local stability analysis. The stability is due not only to a constant inflow of food, which makes the isocline $\hat{N}_{\Delta V=0}$ a decreasing function of V, but also to the isocline $\hat{N}_{\Delta N=0}$ being an increasing function of V.

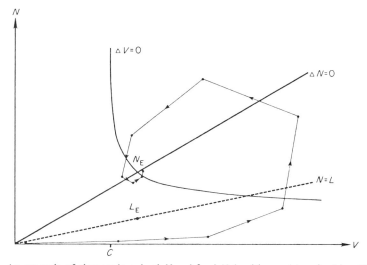

Fig. 2. An example of changes in animal N and food V densities as determined by difference equations (thin lines with arrows), presented on a phase plane with coordinates V and N. Above and on the right of the isocline $\Delta V = 0$ food density V decreases, while above and on the left of the isocline $\Delta N = 0$ animal density N decreases. The isocline $N = L$ (broken line) determines the density L of animals which reproduce at a given food density V. N_E shows the animal equilibrium density after reproduction, while L_E denotes the equilibrium density of the reproducing part of the population. The following values of parameters were used in this graph: $a = 10$, $R = 6$, $C = 1000$, $h = 1$. The initial values are $V_0 = 100$, $N_0 = 1$.

Consider a set of local habitats, i.e. parts of space where the intrinsic rate of natural increase r is positive, surrounded by a hostile environment such that the survival s of migrants moving from one local habitat to another is relatively low. With the constant inflow of food C per unit of space, each local habitat supports a stable local population, which after reproduction has an equilibrium density N_E (Fig. 2). Random causes acting with probability p per local habitat per unit of time reduce the population density from N_E

to density M, where $0 \leq M < L$. This implies that there is room for $(L-M)$ immigrants per unit of space, to enter and reproduce. Competition among immigrants can be neglected if migrant survival s is low. If a migrant which left a local habitat can, during its lifetime encounter only one other local habitat, then the probability Q_E that a migrant will succeed in leaving at least one offspring is

$$Q_E = sp. \tag{6}$$

The probability Q_S that an individual which does not emigrate will produce at least one offspring is determined by its rank: $Q_S = 1$ for ranks $x \leq L$ and $Q_S = 0$ for ranks $x > L$. Emigratory behaviour, as determined by an individual rank x in relation to the rank L is assumed to be hereditary, and therefore, in the global population containing individuals from all local populations, one can expect selection for emigratory behaviour such that animals would emigrate if their rank $x > L$ and would refrain from emigration if $x \leq L$, providing that $Q_E > 0$.

If individuals of rank $x > L$ emigrate outside the local habitats and do not take their share of food, eqn (3) has to be modified, so that

$$V'_{t+1} = V'_t + C - \int_0^{L_t} y\,dx.$$

After substituting eqns (1) and (2) and integrating, the above equation yields

$$V'_{t+1} = V'_t + C - \frac{R-a}{\ln q}. \tag{7}$$

If $L_t > N_t$ no emigration takes place and therefore the amount of food V in the next unit of time is given by eqn (3). It does not matter, for the dynamics of a single local population whether animals die or emigrate, therefore the dynamics of animal density are given by eqns (4) and (5).

An example of changes in animal N and food V densities, when emigration occurs, is given in Fig. 3. The isocline of equation $\Delta V' = 0$ is from eqn (7) given by

$$\hat{V}_{\Delta V' = 0} = \frac{a}{1 - e^{(R-a)/C}}.$$

and is perpendicular to the V-axis, making the system even more stable than this system without emigration (Fig. 2). Note that emigration considerably increases the food density V at equilibrium. It also increases the equilibrium animal density after reproduction N_E and the equilibrium density L_E of the reproducing part of the population (Fig. 3), because the reproducing population can use food which in the absence of emigration would have been used by individuals with rank $x > L$. On the other hand N_E without emigration is higher than L_E with emigration; therefore, if no emigration occurs, the local population density will be higher, with mortality due to food shortage.

APPLICATION AND EXTENSIONS OF THE MODEL

The unit of space within which the model is considered should be large enough to represent the entire population (including all ranks) and small enough so that there is competition among animals within the unit, more precisely so that all individuals of higher rank influence the food intake of lower ranks.

The concept of rank was introduced here to make the model manageable but it does not necessarily imply social ranks or a social hierarchy. Animals can be ranked according to their weights, the sizes of their territories or other indices of their individual performance,

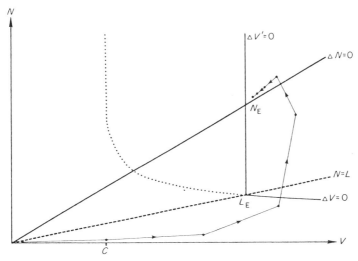

FIG. 3. The same as Fig. 2 but with emigration of all individuals of rank $x > L$. Note that the emigration changes isocline $\Delta V = 0$ into the isocline $\Delta V' = 0$, within the part of the phase plane where $N > L$.

which determine their individual food intake and the number of offspring they produce. For the evolution of emigration, well defined ranks are of importance for ranks close to L only.

It is assumed that the ranks are assigned at random in relation to the genetic basis of the emigratory behaviour. This does not exclude a possibility of a correlation between the rank of parents and the rank of their progeny. Such a correlation may be or may not be hereditary in character. If a strong selection against a genetic trait occurs in the population, one can expect the individuals with this trait to have lower ranks. In such a situation the migrating individuals can differ genetically from nonmigrating ones.

Equation (1) relating individual food intake y to rank x should not be considered as the best description of food partitioning, but as a first approximation which makes investigation into the consequences of inequality among members of a single population, possible. A more realistic description of food partitioning when an analytical solution is not required, can be based on the lognormal distribution. The particular form of the function $y(x)$ given by eqn (1) does not seem to be of great importance. An attempt was made to develop the model presented here using the function $y(x)$ given by equation $y(x) = aV/x^w$, where a and w are constants between zero and unity. This form of the function $y(x)$ assumes that: (i) there is no constant maximum food intake by an individual, (ii) individual food intake is linearly related to food density V and (iii) when food is scarce inequality among individuals does not increase. In spite of these differences, the model based on this equation exhibits the same essential properties which make the evolution of emigration possible, as the model developed from eqn (1). Equation (1) seems to conform better with empirical data and for this reason the model based on it is put forward here.

The model presented here can also be applied to populations with overlapping generations by redefining R as the maintenance cost per individual per unit of time. In such populations, ranking of individuals can be age-dependent, with young individuals having the lowest ranks and migrating as the first ones if there is no room for them. Other consequences of overlapping generations will be discussed later.

Local habitats can be of any size, as long as the size does affect the possibility of emigration outside, and the local habitats do not cover so large a proportion of the whole area, that the reduction of migrants outside the local habitats cannot occur. This is an important difference between this model and all models of group selection in which the size of the local habitats and their genetic isolation is important (Levins 1970; Gilpin 1975). No genetic isolation was assumed here and the evolution of emigration is due to individual selection only.

When studying a natural population it is often difficult to determine the boundaries between the local habitats and the hostile environment outside. In such a case the model explains the evolution of an interesting phenomenon, namely that animal population density is adjusted to its food supply but death due to hunger is rarely an agent which brings about this adjustment. If there is a reduction of food density V, this decreases L and increases the number of migrants $(N-L)$. These migrants die from adverse weather conditions, predator pressure or other nonspecific agents.

The reduction of the population density from equilibrium N_E to the density M ($0 \leq M < L$) can be the result of (i) an appearance of new unpopulated local habitat, (ii) an increase in the amount of food supply or a fast return to the normal level after a temporary decrease, or (iii) a reduction of animal density due to predators, diseases or adverse weather conditions. The cases mentioned above are very common in nature and they can also occur in very stable ecosystems.

The probabilities of survival and leaving at least one offspring for nonmigrating Q_S and migrating Q_E individuals, as used in the model, are very crude measures of individual fitness. If a habitat colonized by a migrant is empty, the migrant can leave many more than one offspring, therefore its fitness should be much higher than the probability Q_E as defined by eqn (6). A much more complicated model is required to estimate this fitness, but as long as it is higher than zero the evolution of emigration is possible.

Q_S is a function of individual rank x decreasing with ranks close to rank L. To decide whether to migrate or to stay and attempt to reproduce, an individual has to perceive its own rank x in relation to the rank L. This involves estimation of (i) its own physiological and ecological status, (ii) the density of other individuals, especially those which are better fed and have a better chance of survival, (iii) the density of available food, and (iv) the possibility of exchanging rank with another individual. The inaccuracy of these estimations can be considered in terms of the variation of the real values of L around its estimated value $\bar{L}$. Variation of L changes the Q_S of individuals with rank close to $\bar{L}$; therefore Q_S can be presented as a cumulative probability distribution of L along the x-axis. This is shown in Fig. 4, using the normal probability distribution, which demonstrate that inaccuracy of the estimation of L increases the rank number above which animals emigrate from $\bar{L}$ to L'. Using similar arguments it is possible to determine Q_S as a function of varying maintenance cost R or food density V, since L is determined by R and V by means of eqn (2).

The mechanism of the evolution towards emigration above rank L' can be visualized as follows. Assume that the amount of food y taken by an individual during its lifetime is positively correlated with the number of food particles v it encounters per hour at an early stage of its life. According to this correlation a group of animals from all local populations which encountered v_0 particles per hour would get on the average $\bar{y}_0$ units of food in their lifetimes ($\bar{y}_0 < R$). If every individual in the group gets the same amount of food equal to the average $\bar{y}_0$ then all of them should emigrate from their local populations, because no one has a chance to leave a single offspring. If on the other hand the mean $\bar{y}_0$ is based on

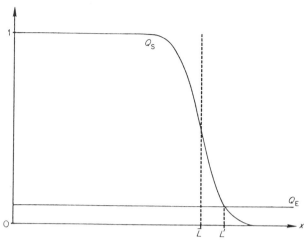

Fig. 4. Probabilities of producing at least one offspring for individuals which do not migrate Q_S and those which migrate Q_E as functions of individual rank x. If L is normally distributed random variable (standard deviation equal to 0·1 was applied in this graph), Q_S is given by cumulative distribution with mean $\bar{L}$. The graph shows that it is of selective advantage to emigrate above rank L' instead of $\bar{L}$.

different values of y, such that 10% of individuals from this group get $y > R$ units of food, and if $Q_E < 0·10$, then the probability of survival of nonmigrants from this group is higher than that of migrants. This implies that, under the circumstances described above, it is of selective advantage for the individual which encounters v_0 food particles per hour to refrain from emigration. Note that the error of the prediction of y from v can be due either to the environment or to the behavioural features of the animals.

If animals emigrate above rank L', instead of above $\bar{L}$, it moves the isocline $\Delta V' = 0$ (Fig. 3) to the left. In an unpredictable habitat this isocline can move leftward to the point of intersection of the isoclines $\Delta V = 0$ and $\Delta N = 0$ (Fig. 2), and then the density of animals is regulated by mortality due to a shortage of food instead of by emigration.

Emigration usually occurs at the early stages of an animal life; therefore the model assumes that migrants leave their local habitats before taking any food. If animals do take a certain amount of food before emigration, then either this food can be included in the reproductive cost of their parents or eqn (7) should be modified.

An important feature of the model is that at the equilibrium N_E there is high food density V when emigration occurs (Fig. 3) but low food density without the emigration (Fig. 2). This can help to answer an important ecological question discussed by Hairston, Smith & Slobodkin (1960), Wynne-Edwards (1962) and others, namely why do animals not deplete their food supply? When answering this question one has to take into account not only emigration but also other differences between field and laboratory populations. Animal populations in the laboratory are usually supplied with homogeneous food in high concentration, their density per space unit is high and directly regulated by food shortage, and there are many undernourished individuals which would eat any food particles left by other members of the population. Since no emigration is allowed, the best strategy for an individual is immediate survival, without taking into account whether future survival and reproduction are really possible. In the field, however, food of high quality is often dispersed and the animals are often time-limited. If an individual has a lower rank and it occupies a home range of inferior quality, if it were vacated, no individual of higher rank would occupy it. In addition, adverse weather conditions or predator pressure do not allow the survival of low rank individuals, which in the laboratory or under human protection would be able to survive and reproduce. If food quality and quantity is low, it is better to emigrate, but if no emigration is possible, maximum utilization of all available food is the best strategy.

If the model presented here has a counterpart in nature, one can expect that a steady state population density, regulated by emigration, could be maintained in the laboratory, by removing individuals which exhibit emigratory behaviour. If animals are not sufficiently different, either they should all emigrate or all of them should stay. The possibility of maintaining such 'free' populations was confirmed experimentally for *Hydra* (Lomnicki & Slobodkin 1966) and for *Tribolium* (Zyromska-Rudzka 1966). By removing floating individuals of *Hydra*, is was possible to maintain a stable population at a level several times lower than the level of a confined population with the same amount of food and space. Members of these free populations resembled individuals found in the field, in that they were well fed and budding, unlike those in the confined populations where they were starved. In the free populations after feeding, some food particles were left, while none remained in the confined ones. These results were obtained in spite of the homogeneous conditions in which the populations were kept, including random food distribution and the scrambling of the populations daily by changing the containers in which they were kept. To study other animal species in this way, it will probably be necessary to maintain more diverse and stable laboratory habitats and to use more sophisticated means of deciding which animals are migrants which should be removed.

SOME CONSEQUENCES OF OVERLAPPING GENERATIONS

Taking into account the unequal resource partitioning in populations with overlapping generations, two questions can be asked.

(1) How can overlapping generations influence the density level at which emigration from a local population occurs?

(2) Under what circumstances can one expect a reduction in births or complete absence of reproduction? Reduction in births is considered here as an absolute reduction in the number of progeny produced in a unit of time and surviving to reproductive age, not merely a reduction in the number of eggs laid.

I will consider these questions for a local population subjected to regular oscillations; they can either be brought about by innate cyclicity in the system of animals and their food supply or can be due simply to cyclic changes in food abundance. A simplified example of a local population with (i) unequal resource partitioning, (ii) overlapping generations, and (iii) cyclic oscillations in the amount of food, expressed as changes in L, is presented in Fig. 5. This example demonstrates that a reduction in number of progeny, as well as emigration of animals of rank $x < L$ are possible if two additional conditions are fulfilled: (iv) animals are able to recognize the stage of the cycle and (v) as far as reduction in births is concerned, such a reduction increases the possibility of longer survival and later reproduction. A more realistic model than that presented in Fig. 5 should also include a larger number of animals, mortality as a function of individual age, with the senile animals moved to lower ranks, number of progeny as a decreasing function of rank x, and increasing longevity with a decreasing natality. Such a realistic model would be difficult to present in a single figure, but it would not change the conclusions presented here.

To recognize the stage of the cycle, information about food density is not sufficient, and additional information about the density of other individuals is required (Fig. 5). This can explain why high population density brings about decreased reproduction in a situation where animals have enough food and there are no other physical reasons for reduced reproduction. Two hypotheses explaining the reduced reproduction at high densities can be given. The first hypothesis is that high density in the field is a signal for emigration, and if no emigration is allowed, emigratory behaviour degenerates into abnormal aggressive

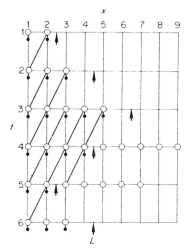
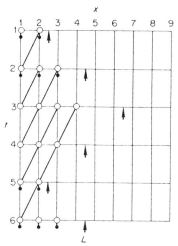

FIG. 5. Survival and reproduction of animals in a simplified local population assuming that the amount of food oscillate regularly, so that $L = 2, 4, 6, 4, 2, 4, \ldots$ Individuals (o) are ranked and maintain their position by changing rank x for a better rank $x-1$, in every unit of time. It is assumed for simplicity that in every unit of time one individual of the highest rank ($x = 1$) dies and all individuals of rank $x \leq L$ reproduce giving one offspring (•) each. In the next time unit newly born individuals take the lowest rank available. Individuals of rank $x > L$ do not reproduce and do not survive the next time unit. The diagram on the left considers the case of no reduction in births and no emigration. Note that the reproduction at time $t = 3$ and $t = 4$ is unnecessary because progeny produced then cannot survive in the local population. An individual which at time $t = 3$ has rank $x = 5$ can survive until time $t = 5$ only, but its progeny cannot survive. Due to low reproductive rate, rank 4 at time 2 and rank 6 at time 3 are not filled. It is of advantage for an immigrant to enter at time 2, but not at time 3.

The diagram on the right considers the same population with reduced reproduction and emigration below L level, to demonstrate that no loss of fitness within the local population occurs. For the sake of clarity possible advantages of individuals which reduce their reproduction and emigrate below the L level are not shown. Within the global population the selective advantage of an individual emigrating below the L level (the individual of rank 5 at time 3) is determined by probability Q_E of finding a better place to live and reproduce. Abstention from reproduction at time 3 and 4 depends on whether it is more advantageous to refrain from reproduction in order to reproduce later or to produce progeny which would emigrate with a possibility of finding less crowded local habitats.

behaviour which brings about decreased reproduction and increased mortality. For species without aggressive behaviour, this can be induced by abnormally high concentration of some chemical compounds, which in nature can serve to induce emigration, or by other agents which play a similar role. Here the reduced reproduction is a by-product of emigratory behaviour, resulting from an abnormal condition of the confined populations. Data on the behaviour of rodents exposed to high densities (Archer 1970) supports this first hypothesis.

The second hypothesis is that the reduced reproduction at high density is an adaptation, which can sometimes occur in the field, and which allows individuals to reproduce later when their population density is low and the conditions for leaving progeny are better. Such adaptation can evolve when migrant survival is low as compared with advantages of later reproduction which can allow the leaving of progeny in the same local habitat. Note that such reduced reproduction is of selective advantage when ranks are well defined, so that newly born individuals are in different positions than their parents (Fig. 5). If this hypothesis has a counterpart in nature one should expect increased reproduction after a population with well defined ranks has been scrambled. Petrusewicz (1957, 1963) has studied the occurrence of population growth in confined populations of laboratory mice both spontaneous growth and that induced by altering the population social structure. The social structure was altered by changing the cages in which the populations were kept (Petrusewicz 1957) and by temporary removals and introductions of several individuals (Petrusewicz 1963); both resulted in significant increases in the occurrence of the population growth.

This raises an interesting point: an increase in population density may not always be the result of better ecological conditions (more food, fewer predators, favourable weather) but can be due to any interference which scrambles the population or obstructs the flow of information among population members concerning their density and their ranks. These can be brought about by many different agents, which may not necessarily be related to animal mortality or natality. Theoretically it is possible that any human activity such as use of chemicals, ploughing or mowing could have this effect. Reduced reproduction is of selective advantage in an environment which is more or less predictable. When an animal encounters quite new conditions, which do not ensure later improvement, it is not advantageous to refrain from reproduction, but rather to reproduce at any cost. This can be one of the reasons for population stability in natural ecosystems, as compared with man-made ecosystems.

SPATIAL HETEROGENEITY AND ANIMAL POPULATION OUTBREAKS

The explanation of population outbreaks, which seem to be more common in homogeneous and cultivated areas than in natural and diverse ones, is an old and still important task of animal ecology. Arguments developed by ecologists to explain this phenomenon are based mainly on the concept of stability of the entire ecosystem (May 1973) and relations between species diversity and stability (Pielou 1975). Using the present model the population outbreaks can be explained by properties of the single species population and its environment, without invoking the dynamics of the entire ecosystem. This makes further theoretical investigations much simpler and generates hypotheses easier to test. A similar, more detailed model, explaining population regulation in terms of the differential movements of individuals in heterogeneous environments, but without taking into account unequal resource partitioning was developed by Taylor & Taylor (1977).

In the model presented here population outbreaks are thought not to be a result of low stability of the difference equations describing changes in density of animals N and their food V, but a failure of the population's regulation by emigration, so that the equilibrium is at low food density (Fig. 2) and the population is directly regulated by starvation.

In a heterogeneous environment one can expect large differences in animal reproduction and survival at different points of the space, and these differences may allow the existence of many species, resulting in a high species diversity. From the point of view of a single species, spatial heterogeneity can be thought of as macroheterogeneity, which divides the space into local habitats and the hostile environment outside, and microheterogeneity, which operates with a local habitat. Microheterogeneity makes the home range of an individual different from the home range of its neighbour and therefore it can promote and increase individual differences and inequality in resource partitioning. Large differences among individuals bring about well defined ranks, which in turn allow the regulation by emigration at the level L instead of higher level (Fig. 4). Macroheterogeneity makes possible the regulation of the global population by reducing the number of migrants before they reach other local habitats.

On the other hand, in a homogeneous environment ecological conditions are uniform, which is unfavourable for many species and favourable for only a few species. For these few species there is no hostile environment outside the local habitats, since the entire area forms one large local habitat. Therefore, migration does not reduce the population density, but simply moves animals to other similar places. Without the possibility of finding a less crowded place, animal try to survive and reproduce; this brings about depletion of the food supply, which eventually results in a population crash.

CONCLUSIONS

Models of population genetics are derived from the genetic properties of individuals. In a similar way, the ecological model of population exponential growth is derived from the biological properties of individuals. This seems to be a very good research strategy, since we know much more about individuals than about populations, and it should be more widely applied in ecology by deriving ecological models from physiological, behavioural and ecological properties of individuals. In contrast to this strategy, the logistic equation and models based on it are not derived from the properties of individuals but from the behaviour of the entire population; more precisely from a statistical relation between population density and per capita population growth. This relation is used in the logistic equation without taking into account its variance, so that nothing is known about the distribution of reproduction and death probabilities among population members at different densities. The assumption that all individuals are equally affected by increasing population density disagrees with empirical evidence, mentioned earlier in this paper, and leads theoretical population biology into a blind alley. Using this assumption, regulation of population density by means of emigration from optimal habitats cannot be explained by individual selection. In addition, population models with discrete regulation, which make use of the logistic assumption predict chaos for quite realistic sets of parameters (May 1976).

In order to get out from this blind alley one has to study resource partitioning among members of competing single species populations, as well as distribution of reproductive abilities and death probabilities within these populations. The present model is an attempt to show what are the consequences of unequal resource partitioning on ecological theory. Compared with the logistic equation the model is much more complicated; growth of a confined population is described by exponential increase when $N < L$, by reproduction of stronger population members and mortality of weaker ones when $N > L$, and by a possible adaptive reduction in natality as a reaction to high density. On the other hand, much field data, which describe adjustment of the population size to food supply by dispersion and other behaviourally induced phenomena (Wynne-Edwards 1962), can be explained by this model without invoking group selection. Also, data from the laboratory (Huffaker

1958) and from the field (Errington 1946) concerning predation suggest that a model of predator–prey system can be developed on the basis of the model presented here.

This model, like many other ecological models, is very general; however it generates predictions which can be easily tested in the field or in the laboratory. The model can be falsified by any of the following phenomena found in nature.

(i) Decreased variation with a decreasing mean of body size, weight, individual food intake or other indices of individual fitness, as a result of competition among population members.

(ii) Complete extinction, i.e. the death of all individuals, of a local population in the field due to depletion of resources by this population.

(iii) Higher fitness of emigrating individuals which can be determined by size, weight or other indices. This does not apply to long range seasonal migrations like those in birds but to the emigrations from the optimum habitats outside.

(iv) No increase in population density if part of an area, where animals are abundant, is enclosed for several generations, making emigration impossible.

In addition, the concept of population stability, in the sense of regulation below starvation level, and species diversity can be tested by comparing species diversity with the magnitude of spatial differences in reproduction and survival of animals of a given population.

ACKNOWLEDGMENTS

I am grateful to J. Kozlowski, M. Klinowska, B.C.R. Bertram, S.A. Corbet, R.A.J. Taylor, D.J. Rogers, L.B. Slobodkin, and R.A. Kempton for useful comments. This work was completed while the author was a visiting fellow at King's College, Cambridge.

REFERENCES

Archer, J. (1970). Effect of population density on behaviour in rodents. *Social Behaviour in Birds and Mammals* (Ed. by J. H. Crook) pp. 169–210. Academic Press, London.
Bateman, A. J. (1948). Intra-sexual selection in *Drosophila*. *Heredity*, **2**, 349–68.
Errington, P. L. (1946). Predation and vertebrate populations. *Quarterly Review of Biology*, **21**, 147–77, 221–45.
Gilpin, M. E. (1975). *Group Selection in Predator–Prey Communities*. Princeton University Press, Princeton.
Hairston, N., Smith, F. E. & Slobodkin, L. B. (1960). Community structure, population control, and competition. *American Naturalist*, **94**, 421–5.
Huffaker, C. B. (1958). Experimental studies on predation: dispersion factor and predator-prey oscillations. *Hilgardia*, **27**, 343–83.
Levin, S. A. (1976). Population dynamic models in heterogeneous environment. *Annual Review of Ecology and Systematics*, **7**, 287–310.
Levins, R. (1970). Extinction. *Some Mathematical Questions in Biology* (Ed. by M. Gertsenhaber) pp. 77–107. The American Mathematical Society, Providence.
Lomnicki, A. & Slobodkin, L. B. (1966). Floating in *Hydra littoralis*. *Ecology*, **47**, 881–9.
MacArthur, R. H. (1972). *Geographical Ecology*. Harper & Row, New York.
May, R. M. (1973). *Stability and Complexity in Model Ecosystems*. Princeton University Press, Princeton.
May, R. M. (1976). *Theoretical Ecology; Principles and Applications*. Blackwell Scientific Publishers, Oxford.
Murton, B. K., Isaacson, A. J. & Westwood, N. J. (1966). The relationship between wood-pigeons and their clover food supply and the mechanism of population control. *Journal of Applied Ecology*, **3**, 55–96.
Nicholson, A. J. (1954). An outline of the dynamics of animal populations. *Australian Journal of Zoology*, **2**, 551–98.
Petrusewicz, K. (1957). Investigation of experimentally induced population growth. *Ekologia Polska, seria A*, **5**, 281–309.
Petrusewicz, K. (1963). Population growth induced by disturbance in the ecological structure of the population. *Ekologia Polska, seria A*, **11**, 87–125.

Pielou, E. C. (1975). *Ecological Diversity*. John Wiley & Sons, New York.
Taylor, L. R. & Taylor, R. A. J. (1977). Aggregation, migration and population mechanics. *Nature, London*, 265, 415–21.
Van Valen, L. (1970). Group selection and the evolution of dispersal. *Evolution*, 25, 591–8.
Wiens, J. A. (1976). Population response to patchy environment. *Annual Review of Ecology and Systematics*, 7, 81–120.
Wilbur, H. M. & Collins, J. P. (1973). Ecological aspects of amphibian metamorphosis. *Science N.Y.*, 182, 1305–14.
Wynne-Edwards, V. C. (1962). *Animal Dispersion in Relation to Social Behaviour*. Oliver & Boyd, Edinburgh.
Zyromska-Rudzka, H. (1966). Abundance and emigration of *Tribolium* in a laboratory model. *Ekologia Polska, seria A*, 14, 491–518.

(*Received* 9 *May* 1977)

Aggregation, migration and population mechanics

L. R. Taylor
Rothamsted Experimental Station, Harpenden, Hertfordshire, UK

R. A. J. Taylor
Imperial College Field Station, Silwood Park, Ascot, Berkshire, UK

A concept is developed for the regulation of populations by density-dependent movement, rather than by overt competition alone. Fitness is seen as maximising the reproductive advantage of a balance between migratory and congregatory behaviours. Population density is shown to be spatially, as well as temporally dynamic and a mechanism is proposed that accounts for observed spatial behaviour.

THE Malthusian concept of population is contained in the basic proposition that, with full census information, population change can be assigned to either reproduction or mortality[1]. This proposition is consequently the basis of Darwinian fitness measured in terms of reproduction and survival[2], and of population models for competition[3,4] and predation[5,6] that are primarily concerned with equilibrium conditions[7,8]. Populations are seen as persisting through time owing to the check on multiplication imposed by such familiar Malthusian[9] restraints as disease, predation and resource limitation. Key factors[10,11] direct our attention to the causes and time of death, but not to the place of death nor to its distance and direction from the place of birth. Lacking this component of movement in real space, it seems to us that populations are static[12], not dynamic. They can neither function nor evolve unless they include the controlled mobility that enables organisms to select a place to live and so survive and reproduce. This mobility we shall equate largely, but not entirely, with internal or external migration using the word in a wider sense than that familiar in ornithology.

Migration and static population concepts

Until recently migration has often been regarded as an idiosyncrasy to be included with unaccountable mortality in life tables based on census information from a restricted area. Yet migration is not synonymous with mortality. Re-examination of old census records now reveals more extensive, successful, internal migration in man than was formerly appreciated[13,14]. Among insects, many species formerly regarded as sedentary have been found to be vagile[15]. For example, *Drosophila* seem non-migratory as compared with the monarch butterfly[16], and their flight behaviour is well adapted to minimising dissemination[17,18]. Even so, their migratoriness is now known to be considerably underestimated in the Dobzhansky–Wright models that so influenced concepts of population mobility in genetics[19,20]. New evidence from bird ringing shows long distance movements in territorial species formerly considered extremely sedentary, the European blackbird, for example[21], and, since young birds rarely occupy their parents' nests, they must migrate to establish a new territory. Dispersal in small mammals has been related to population cycles[22] and the zoo syndrome has been called "frustrated dispersal"[23] in species not commonly thought of as migratory.

Segregation of migration from dispersal obscures their common population function. In birds, fish and especially in insects, whose complex life-histories enable the various aspects of migration to be analysed in most detail[24,25], much is now known about ontogenetic[25], physiological[26] and behavioural[27] processes that initiate movement and about its evolutionary origins[28] and ecological function[29], as well as about the more striking, but in this context less relevant, orientation[30,31] and navigation[32] behaviour. Gene migration models[33–35], like population models[36], have not yet made due allowance for the sophisticated migration control mechanisms in which the experience of parents and grand-parents[37], as well as population density[38], affect the migrant status of the individual. These are most clearly evident in the labile[39] migratory polymorphisms, remote from genetic heterogeneity, in clonal populations of aphids[40]. As a result, little light has been shed on the role of migration in the dynamics, and hence in the fundamental concept, of population.

Concepts of population are not consistent[41]. A lack of rigour in distinguishing total population from population

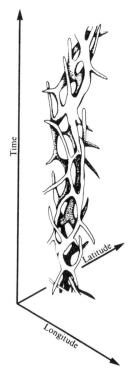

Fig. 1 Conceptual model for population anatomy based on Rostowzew's[75] drawing of the stelar structure of the adder's tongue fern. The stele represents the space-time reticulum created by the distribution of an arbitrary density level through time.

Fig. 2 Measured annual (1969–73) adult population density distribution of two moth species showing the changing patterns typical of the 623 species sampled by the Rothamsted Insect Survey[17]. Each map is based on 62–88 sample points; there were inadequately distributed in Scotland, hence the poor definition there in (*b*). Density layers 0, 1–2, 3–9, 10–31, 32–99, 100–315, 316–999, individuals per sample unit. *a*, *Agrochola lychnidis* (Denis and Schiff.) a marginal species near its northern limit. *b*, *Xanthorhoe fluctuata* (L.) has a very stable total population. The maps are visualised as serial sections through the conceptual model in Fig. 1. In both instances the appearance and disappearance of 'holes' in the population is clearly visible so that all parts of the population are essentially 'marginal'.

density can imply uniform distribution of density in space, whereas the actual spatial distribution of few organisms is known[12]. Again, key factors differ for the same species in different places and affect different stages in the life history. This makes models for the whole species population incomprehensible. If, however, a model is considered to represent a separate element of the population, equivalent to the mean density at a certain place, then growth rate can be equated with unity, as required by basic theory[9,13], only if there is no population flow between such places.

But we doubt that populations are often like that. All populations are spatially fluid in some measure because movement is a fundamental biological response to adversity. Nor is population density distributed uniformly in space. A valid concept of population must encompass both the more settled species, like the red grouse say[11], to which the current models appear to be most suited, and the highly volatile species, like aphids, for which local stability in time seems almost irrelevant because of the persistent change of place.

A spatial concept for population and some evidence

The classical model for the whole population of a species changing only through time was the phylogenetic tree. If we treat the anatomy of a real population as being in three dimensions, latitude × longitude through time, we visualise its internal structure to be reticulate, not unlike the fern stele in Fig. 1 in which density changes constantly through time and space. So, spatial sections through this anatomical model at successive time intervals differ slightly, or markedly, depending upon the time scale. (This has much in common with MacArthur's[15] r–K continuum, and with Southwood's[16] τ/H ratio.)

After monitoring the changing distribution of density of hundreds of species of Lepidoptera throughout Great Britain for almost a decade[17], we can show that growth cycles in different local population elements are normally asynchronous, as in Ford's marsh fritillaries[18], combining with movement to change the pattern of density in space like serial sections through the fern stele model, even when overall abundance remains relatively constant (Fig. 2). The moths we have sampled range from known migrants to species considered localised, but density distribution is always changing. The time-scale is, of course, short compared with vertebrates. Our conceptual model predicts that once a density element has been defined in space, its life-span is limited. Our field samples confirm this, and hence that the use of spatial averages, in fitness say[35], makes questionable assumptions about the nature of spatial distribution, for density can diminish locally to create what are usually regarded as marginal conditions[19,50] even in the heart of a population (Fig. 2).

The individual's dilemma, and the population consequence

In an environment changing through time and space, the most probable strategy for a new individual to adopt to survive and reproduce is not necessarily to stay to compete with its parents or congeners[51] but may be to go elsewhere, to find an empty environmental hole to inhabit. Its ability to do this depends on its malleable migrant status. In a doomed habitat, one offspring capable of migrating at the right time may well have more survival value than a litter of more sedentary morphs[40]. Thus the migrant status of offspring is an essential component of fitness controlling the production of grand-children, which we take to be the relevant criterion. This point was originally made by Grinnel[52] but he assumed population centres to remain static and, like Gadgil[53] and others[35], regarded migration as compulsive for a genetically fixed proportion of the population. MacArthur[54] argued, legitimately, that such migration would be from an equilibrium population to a less hospitable place and involve a loss of fitness. Much environmental hostility which results in death or dispersal is transient, however, especially that due to biological restraints such as predation and disease or the loss of nesting sites or food supply. This creates vacant holes in the environment which subsequently recover their hospitality and are re-occupied by searching migrants so that population centres are mobile when seen at the appropriate time scale. These more fluid populations make MacArthur's argument unconvincing and, from the individual viewpoint, offer alternative courses of action. It was to an apparently social option that Hamilton[55] addressed his attention which he so graphically illustrated by his selfish frogs. The environmental menace was a predator and the choice considered was where to go within the herd so as "not to be nearest the snake". There is an additional option which Hamilton precluded, to abandon the herd and find a pond without a snake. This individual choice creates recognisable spatial properties at population level which we suggest lead to population homeostasis without group selection[56,57].

A dynamic mechanism and a postulate

We would argue that all spatial dispositions can legitimately be regarded as resulting from the balance between two fundamental antithetical sets of behaviour always present between individuals. These are, repulsion behaviour, which results from the selection pressure for individuals to maximise their resources and hence to separate, and attraction behaviour, which results from the

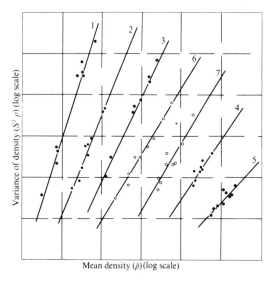

Fig. 3 The relation between variance of spatial disposition and annual mean density for five moth species from the same source as in Fig. 2, and two simulated quasi-stable populations. Each point is one years sample, equivalent to a map in Fig. 2. The fitted regression is $\log S_{p^2} = a - b \log \bar{p}$. The range of aggregatory response to overall population level differs widely between species but is highly consistent within species. 1, *Lycophotia varia* (Villers) ($a = 0.02$; $b = 3.02$). 2, *Xanthorhoe ferrugata* (Clerck) ($a = -0.29$; $b = 2.41$). 3, *Malacosoma neustria* (L.) ($a = 0.94$; $b = 1.98$). 4, *Xestia xanthographa* (Denis & Schiff) ($a = 1.16$; $b = 1.48$). 5, *Epirrhoe alternata* (Müller) ($a = 1.56$; $b = 1.04$). 6, From Fig. 5*a*; heterogeneous environment ($a = 1.38$; $b = 1.65$). 7, From Fig. 5*b*; homogeneous environment ($a = 1.35$; $b = 1.66$). Axes are staggered to save space. Grid is at ×10 intervals. Symbols are (●) real samples, (○) from simulation.

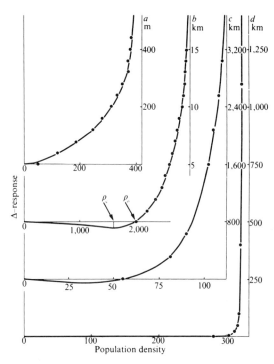

Fig. 4 Δ-functions reflecting different life styles with superimposed density × distribution data from *a*, sedentary insect, *Drosophila pseudoobscura*; fluorescent dust recoveries at attractant traps in California[20]. *b*, Social primate, man; census data from Sweden[13]. *c*, Migrant insect, monarch butterfly; wing-label recoveries in North America[16]. *d*, Territorial bird, blackbird; leg-ring recoveries in Great Britain[21].
Parameters

	ε	ρ_0	p	q
a	0.03	6	0.97	0.72
b	5.2	2000	7.3	6.7
c	11.0	54	7.2	1.1
d	7.3×10^{-22}	225	151	2.1

ρ_0 is the population density at which migratory and congregatory behaviour balance and no net movement ensues. At $\rho > \rho_0$ the population is over-dense and migratory behaviour predominates; at $\rho_c < \rho < \rho_0$ density is below par, congregatory behaviour predominates and leads to rapid recovery; at ρ_c, minimum Δ, congregatory behaviour is maximised; at $\rho < \rho_c$ spatial recovery is greatly impeded, T_R is prolonged and relies on high reproductive rate. Density units are arbitrary as in the original data. Densities below ρ_0 are difficult to measure and short-lived, hence the data points are inadequately represented.

selection pressure to make the maximum use of available resources and hence to congregate wherever these resources are currently most abundant. The balance between these two conflicting behavioural tendencies operating on each individual determines its movements and hence the resulting spatial pattern of the population at any instant in time. It is the response of this balance to changing internal and external environmental conditions that, we suggest, constitutes the dynamic element in populations.

From this basis we have sought to bridge the gap between models and real data with a functional mechanism for the distributive processes in a population that would lead to the diffusion rates and spatial dispositions observed in nature.

The only hard spatial evidence we know of, that relates equally to populations of all kinds of organisms, is the statistical measure of fine-scale spatial disposition of individuals[58]. Randomness is by definition atypical of the microdistribution of living organisms and one of us has previously shown empirically[59] that the spatial variance (S_ρ^2) characteristic of species, at a particular point in the environment, is proportional to a fractional power of mean population density ($\bar{\rho}$) at that place;

$$S_\rho^2 = a\bar{\rho}^b, \quad (1)$$

where b is a measure of the density dependence of aggregation. Our field monitoring now yields convincing evidence that macrodistribution, at least in Insecta, still has the same property of density-dependent aggregation when mean density relates to an area the size of Great Britain instead of a few square metres (Fig. 3). From this it is clear that, to return to the ferri stele model, successive stelar sections characterise the species spatial behaviour on a geographical as well as on a local scale.

The other relevant experimental evidence relates to rates of diffusion, or migration, of individuals from population concentrations, but for this there has yet been no generally accepted functional relation[60,61]. Biological diffusion is no more random than spatial disposition, for it is the outcome of equally sophisticated behaviour, and the problem has been to find a distance function common to such diverse organisms as, for example, the migrant monarch[16] and the sedentary *Drosphila*[20] the territorial blackbird[21] and social man[13].

The working model does not deal with the underlying complexities of social behaviour[62] and bizarre adornment associated with sex[63], which can be regarded largely as visible evidence of the congregatory behaviour, nor with the equally complex behaviour and locomotory mechanisms associated with migration[25], for they are too difficult to measure. It deals with the net displacement resulting from these. When the motion is centrifugal to the nearest population concentration it is here classified as 'migratory' to distinguish it from the random motion implied by 'dispersive'. Alternatively, when the motion is centripetal fulfilling the function of "gravitational attraction" in Skellam's[64] population concept, but missing from Lidicker's[65], it is here called 'congregatory' to distinguish its active motion from the resulting instantaneous static condition or aggregation. Migratory motion, like congregation, is overlaid by elaborate behaviour patterns and can be almost obscured by movements concerned with reproduction, fighting and feeding, especially daily and seasonal commuting in man and certain highly mobile and long-lived birds whose feeding and reproductive sites are far apart[60,66]. Nor does it necessarily imply displacement right outside occupied territory, for much migration is internal. The displacement crucial to fitness is that between the birthplace of one generation and that of the next and it is the distribution of sexually mature adults as in Fig. 2 that is, therefore, relevant.

We postulate that the net centrifugal displacement of an individual (Δ_m) generated by migratory behaviour as defined above, is proportional to a fractional power of the population density (ρ) it experiences; $\Delta_m = G\rho^p$. So also is net centripetal displacement (Δ_c) due to congregatory (i.e. anti-migratory)

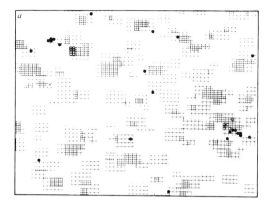

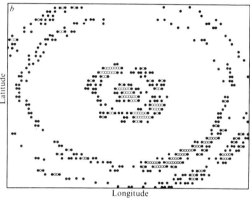

Fig. 5 *a*. Simulated population complex of a highly migrant species, something like an aphid, founded by three individuals and quasi-stable after five generations: Δ-response parameters, $\varepsilon = 0.333$; $\rho_0 = 40$; $p = 4$; $q = 2$, in an unchanging heterogeneous, mainly hostile, environment. Blank background is hostile, faint hatching tolerant but unproductive, denser hatching permits reproduction. At this stage two nuclei are reproducing and excess offspring migrate. Individuals on non-reproductive sites await an environmental change that will either release reproduction or kill them; their prospects depend on Southwood's[16] τ/H ratio, currently being held in suspense by the models fixed environment. *b*, When the environment is homogeneous and the universe unlimited, density annuli radiate outwards from a point source in successive synchronised generations; parameters similar to (*a*). In this form the model describes the annular pattern of movement familiar, on a short time scale, as the radar ring angels[76] produced by the commuting flight of starlings in response to a daily cycle in the changing balance between feeding and flocking behaviours. The figures are distorted to conserve space. Hollow circles are occupied by more than one individual.

behaviour; $\Delta_c = H\rho^q$. The exponents p and q are rate constants for density-dependent migration and congregation respectively which may be specific. G and H are constants of proportion for which we offer no biological interpretation.

If we adopt the sign convention that any tendency to increase the mean distance between individuals is positive while reduction in mean separation is negative, the intuitive sum of the two conflicting density-dependent behaviour sets is

$$\Delta = G\rho^p - H\rho^q \quad (2)$$

Equation (2) may be written in the parametric form

$$\Delta = \varepsilon\left[\left(\frac{\rho}{\rho_0}\right)^p - \left(\frac{\rho}{\rho_0}\right)^q\right] \quad (3)$$

where G and H are replaced by the biologically identifiable parameter ρ_o, the density at which migration exactly balances congregation[67], and

$$\varepsilon = (H^p G^{-q})^{1/(p-q)} \quad (4)$$

which is a scale factor containing the dimensional units. Congregation (negative displacement) now occurs when $\rho < \rho_o$, and migration when $\rho > \rho_o$ (see Fig. 4). The condition $p < q$ cannot evolve because growing populations would continue to congregate indefinitely and condense into a point in space.

Simulation and interpretation

For simplicity in the primitive simulations presented here (Fig. 5), reproduction is by binary fission at random in space among those individuals reproductively mature, and at regular intervals in time. In an environment consisting of a permanent two-dimensional random mosaic of areas that are lethal, or tolerable but prevent reproductive maturation, or hospitable, the individuals of a founder population divide and each daughter organism responds, by movement, to the population density at the time of birth. This Δ-response is to the density at a fixed population centre and is assumed to be radial. Mortality occurs only when movement ends in a hostile part of the environment. In practice the Δ-response may be to the individual's or to its parent's experience, to the distance from the nearest neighbour, or to number of social contacts, level of aggression or competition, or to other epideictic[56] or environmental cues relating to population density, and may not be radial. In a few generations population nuclei on hospitable areas reach a quasi-stable condition when migration balances natality (Fig. 5a). Satellite populations, repeating the process by responding to their own population centres, would ultimately occupy all the habitable areas.

Successive simulations with the environmental mosaic changing between generations (not shown here) generate additional movement and mortality. The resulting redistribution of population centres becomes a game of hide-and-seek, emulating the response of local population elements to predation, disease and other restraints, akin to Huffaker's survival by refuge[68]. The net area population density now depends on the balance between rates of environmental change, and the movement and reproductive rates of the organism, which fails to use some resources because they are not disovered in time. Hence the population is stabilised below carrying capacity. Environmental change releases the reproductive potential of surplus individuals that wait in survival areas to mature if conditions improve, or to die if they worsen. The balance is set by the parameters in the Δ-function with respect to the environment into which further non-random elements are introduced because the Malthusian restraints are themselves biological and hence non-random[36,69]. This now generates further local contraction[70] and small-scale geographical shifts of density like those clearly visible in successive years in Fig. 2.

The simplifying conditions adopted for simulation appear not to be restrictive. Age-dependent or facultative Δ-behaviour and sexual reproduction add complexity in the computing but without fundamental change in the outcome, so far as we can tell, for the resulting distance function appears to apply quite well to sexual animals with high levels of organisation (Fig. 4). Behaviour inherited or imprinted from the parents' or grandparents' experience or modified by experience throughout life, further complicates the computing but seems not to affect the outcome. Refinements of social and migratory behaviour, together with increase in size, minimise migratory risks by providing more accurate sensory information on which to make the choice. Again parental care and altruism add sophistication but seemingly make no fundamental change in the logic. Non-radial migration, by controlled orientation or navigation, or even environmentally assisted motion, affects individual migratory risks and hence proportionate mortality. It affects the ultimate spatial disposition only in large-scale geographical trends, because final settling place depends on the successful location of environmental holes, not just on navigation[71]. Death due solely to aging is a minor factor and all other causes are included in environmental adversity.

If each new recruit responds independently to density, a continuous generation model is formulated; synchronous response generates a discrete system as in the examples shown here.

The ecological results and the evolutionary consequences

Analysis of the spatial dispositions generated by simulation yields the power relation between variance and mean population density in equation (1) that is required by the field evidence (see Fig. 3). While the power relation could arise from mathematical models based on other functional mechanisms, this result suggests that our conceptual model for a balance between specific distributive behaviours is reasonable.

The Δ-function for individuals results in a density distributing process which generates a hollow frequency distribution of numbers against displacement in a population which matches the experimental evidence on dispersal for organisms as disparate in behaviour as *Drosophila* and man (see Fig. 4). Again, other functional mechanisms could generate appropriate distributions but no previous model has comprehended spatial behaviour over so wide a range of distances or organisms. We suggest that, if migration and congregation are seen in the collective sense used here, the dispersive behaviour of other animals and plants[72,73] will be found to fit the pattern. The non-migratory condition when $\rho < \rho_o$ is perhaps most readily recognised in island life where ρ_o is high due, for example in insects, to flight-inhibiting behaviour and even morphological adaptation such as brachyptery.

The only biological premises involved are that an organism can move at some stage in its life history before reproduction and can detect directly or indirectly, by any means, including those listed by Wynne-Edwards[56] and Wilson[62], the presence of another of its own kind and hence can respond to it. Reproductive success then often lies in the avoidance of overt competition by movement and goes to the individual with the best Δ-adaptation, rather than to the more superficially successful competitor, and so provides an essential[70,74] counterbalance to excessive competitiveness.

As with other population models, migration and mortality can in some respects be equated. The emphasis is here shifted from the key factors that cause death to the place where they operate, because this reflects the real space hostility pattern of the total environment, including other organisms, and to the subsequent replacement of individuals by movement as well as by birth. Thus the return time for populations displaced from equilibrium, T_R (ref. 8), can be short-circuited by migration. A demon with perfectly developed Δ-response and ability to locate environmental holes could survive almost any catastrophe. If survival is thus seen as a spatial exercise in eluding adversity, evolution depends on learning the geometry, not just of Hamilton's selfish herd[55], but of the changing Darwinian landscape of which the herd is a part.

Received June 15; accepted December 6, 1976.

1 Varley, G. C., Gradwell, G. R. & Hassell, M. P. *Insect Population Ecology* (Blackwell, Oxford, 1973).
2 Lewontin, R. C. *A. Rev. Ecol. System.* **1**, 1–16 (1970).
3 Hassell, M. P. *J. Anim. Ecol.* **44**, 283–295 (1975).
4 Hassell, M. P. & Comins, H. N. *Theor. Popul. Biol.* **9**, 202–222 (1976).
5 Beddington, J. R., Free, C. A. & Lawton, J. H. *J. Anim. Ecol.* **45**, 791–816 (1976).
6 Hassell, M. P. & May, R. M. *J. Anim. Ecol.* **42**, 693–726 (1973).
7 May, R. M. *Nature*, **261**, 459–467 (1976).
8 May, R. M., Conway, G. R., Hassell, M. P. & Southwood, T. R. E. *J. Anim. Ecol.* **43**, 747–770 (1974).
9 Fisher, R. A. *The Genetical Theory of Natural Selection* (Clarendon, Oxford, 1930).
10 Morris, R. F. *Ecology* **40**, 580–588 (1959).
11 Varley, G. C. & Gradwell, G. R. *J. Anim. Ecol.* **29**, 399–401 (1960).
12 Holling, C. S. *A. Rev. Ecol. System.* **4**, 1–23 (1973).
13 Hägerstrand, T. *Les Deplacements Humains* (ed. Sutter, J), 61–83 (Union Europeéne D'Editions, Monaco, 1962).
14 Küchemann, C. F., Boyce, A. J. & Harrison, G. A. *Human Biol.* **39**, 251–276 (1967).

15 Most of the papers in *Insect Abundance* (ed. Southwood, T. R. E.) (Blackwell, Oxford, 1968).
16 Urquhart, F. A. *The Monarch Butterfly* (University Press, Toronto, 1960).
17 Lewis, T. & Taylor, L. R. *Trans. R. ent. Soc. Lond.* **116**, 393–469 (1965).
18 Taylor, L. R. *J. Anim. Ecol.* **43**, 225–238 (1974).
19 Crumpacker, D. W. & Williams, J. S. *Ecological Monographs* **43**, 499–538 (1973).
20 Dobzhansky, T. & Powell, J. R. *Proc. R. Soc. B* **187**, 281–298 (1974).
21 Greenwood, P. J. & Harvey, P. H. *J. Anim. Ecol.* **45**, 887–898 (1976).
22 Krebs, C. J. & Myers, J. H. *Adv. Ecol. Res.* **8**, 267–399 (1974).
23 Lidicker, W. Z., Jr. *Small Mammals: their Productivity and Population Dynamics* (eds Golley, F. B., Petrusewicz, K. & Ryszkowski, L.), 103–128 (University Press, Cambridge, 1975).
24 Williams, C. B. *Migration of Butterflies* (Oliver and Boyd, Edinburgh, 1930).
25 Johnson, C. G. *Migration and Dispersal of Insects by Flight* (Methuen & Co., London, 1969).
26 Johnson, C. G. *Physiology of Insecta* (ed. Rockstein, M.), **2**, 187–226 (Academic, New York, 1965).
27 Kennedy, J. S. *Nature, Lond.* **189**, 785–791 (1961).
28 Southwood, T. R. E. *Biol. Rev.*, **37**, 171–214 (1962).
29 Dingle, H. *Science*, **175**, 1327–1335 (1972).
30 Jones, F. R. H. *Fish Migration* (Edward Arnold, London, 1968).
31 Williams, C. B. *Insect Migration* (Collins, London, 1958).
32 Matthews, G. V. T. *Bird Navigation*, 2nd ed. (University Press, Cambridge, 1968).
33 Wright, S. *Genetics* **28**, 114–138 (1943).
34 Kimura, M. & Weiss G. H. *Genetics* **49**, 561–576 (1964).
35 Karlin, S. in *Population Genetics and Ecology* (eds Karlin, S. & Nevo, E.), 617–657 (Academic Press, New York, 1976).
36 Hassell, M. P. & Rogers, D. J. *J. Anim. Ecol.* **41**, 661–676 (1972).
37 Lees, A. D. *Insect Polymorphism* (ed. by Kennedy, J. S.), 68–79 (Blackwell, Oxford, 1961).
38 Bonnemaison, L. *Ann. Epiphyt.* **2**, 1–308 (1951).
39 Johnson, C. G. in *Insect Flight* (ed. by Rainey, R. C.), 217–234 (Blackwell, Oxford, 1976).
40 Taylor, L. R. *J. Anim. Ecol.* **44**, 135–163 (1975).
41 Thoday, J. M. *Heredity, Lond.* **32**, 406–409 (1974).
42 Preston, F. W. in *Diversity and Stability in Ecological Systems* (eds Woodwell, G. M. & Smith, H. H.) 1–12 (Brookhaven National Laboratory, New York, 1969).
43 Haldane, J. B. S. *New Biology* **15**, 9–24 (1953).
44 Watson, A. & Moss, R. *Proc. XV Int. Ornith. Congr.*, 134–149 (1972).
45 MacArthur, R. H. *Am. Natur.* **94**, 25–36 (1960).
46 Southwood, T. R. E. *Trop. Sci.* **13**, 275–278 (1971).
47 Taylor, L. R. *Rep. Rothamsted expt. Stn. for* **1973**, *Part* **2** 202–239 (1974).
48 Ford, E. B. *Ecological Genetics* 4th Edn. (Chapman and Hall, London, 1975).
49 Levins, R. in *Some Mathematical Questions in Biology* **2**, 76–107 (American Mathematical Society, Providence, 1970).
50 Boorman, S. A. & Levitt, P. R. *Proc. natn. Acad. Sci. U.S.A.* **69**, 2711–2713 (1972).
51 Kennedy, J. S. *J. Aust. ent. Soc.* **11**, 168–176 (1972).
52 Grinnell, J. *Auk*, **21**, 364–382 (1904).
53 Gadgil, M. *Ecology*, **52**, 253–261 (1971).
54 MacArthur, R. H. *Geographical Ecology* (Harper and Row, New York, 1972).
55 Hamilton, W. D. *J. theor. Biol.* **31**, 295–311 (1971).
56 Wynne-Edwards, V. C. *Animal Dispersion in Relation to Social Behaviour* (Oliver and Boyd, Edinburgh, 1962).
57 *Group Selection* (ed. Williams, G. C.) (Aldine: Atherton, Chicago, 1971).
58 Taylor, L. R. in *Statistical Ecology* (eds Patil, G. P., Pielou, E. C. & Waters, E.) **1**, 357–377 (Pennsylvania University Press, University Park, 1971).
59 Taylor, L. R. *Nature* **189**, 732–735 (1961).
60 Cavalli-Sforza, L. in *Les Deplacements Humains* (ed. Sutter, J.) 139–158 (Union Européene D'Editions, Monaco, 1962).
61 Wolfenbarger, D. O., Cornell, J. A. & Wolfenbarger, D. A. *Res. Popul. Ecol.* **16**, 43–51 (1974).
62 Wilson, E. O. *Sociobiology: the New Synthesis* (Belknap, Cambridge, Massachusetts, 1975).
63 Williams, C. G. *Sex and Evolution* (University Press, Princeton, 1975).
64 Skellam, J. G. *Biometrika* **38**, 196–218 (1951).
65 Lidicker, W. Z., Jr. *Am. Natur.* **96**, 29–33 (1962).
66 Hamilton, W. J., III & Watt, K. E. F. *A. Rev. Ecol. System.* **1**, 263–286 (1970).
67 Kennedy, J. S. & Crawley, L. *J. Anim. Ecol.* **36**, 147–170 (1967).
68 Huffaker, C. B. *Hilgardia* **27**, 343–383 (1958).
69 Varley, G. C. *Parasitology* **33**, 47–66 (1941).
70 Cole, L. C. *Science* **132**, 348–349 (1960).
71 Taylor, L. R., French, R. A. & Macaulay, E. D. M. *J. Anim. Ecol.* **42**, 751–760 (1973).
72 Harper, J. L., Lovell, P. H. & Moore, K. G. *A. Rev. Ecol. System.* **1**, 327–356 (1970).
73 Scossiroli, R. E., Cavicchi, S. & Puppi, G. *Atti Acad. Sci. Ist. Bologna Memorie* **3**, 109–132 (1973).
74 Richards, O. W. & Southwood, T. R. E. *Insect Abundance* (ed. Southwood, T. R. E.), 1–7 (Blackwell, Oxford, 1968).
75 Rostowzew, S. *Overs. K. danske Vidensk. Selsk. Forh.* 54–83 (1891).
76 Eastwood, E., Isted, G. & Rider, G., *Proc. R. Soc. B.* **156**, 242–267 (1962).

ERRATA

Figure 3, the ordinate should read: "Variance of density (S_ρ^2) (log scale)."

Figure 3, line 5 of legend should read: "fitted regression is log $S_\rho^2 = a + b \dots$"

Figure 3, line 9 of legend should read: "(Clerck) ($a = -0.29$; $b = 2.41$). 3, *Malacosoma neustria* (L.). . . ."

Page 243, left-hand column, line 29 should read: "(Fig. 3). From this it is clear that, to return to the fern stele . . . ."

25

Copyright ©1964 by Macmillan Journals Ltd.
Reprinted from *Nature* **201**:728-729 (1964)

A FIELD'S CAPACITY TO SUPPORT A BUTTERFLY POPULATION

V. G. Dethier
Robert H. MacArthur

Department of Biology, University of Pennsylvania, Philadelphia 4.

The controversy among ecologists on 'density dependence' swelled out of all proportion due to differences among authors in the meaning and use of such words as 'carrying capacity', 'competition' and 'density'. Beneath these unimportant disagreements, however, there lies at least one fundamental question which has seldom been investigated: Does an area have for a species of organism a carrying capacity with the property that the population decreases when it exceeds the carrying capacity and increases when it is less ? Few doubt that each area has such a carrying capacity in theory, but there is little evidence that the carrying capacity is ever reached or has much bearing on the actual size of population. Rather, as Andrewartha and Birch[1] have claimed, it is often more useful to ask how recently the population met a catastrophe, to what level it was reduced, and at what rate it is able to recover toward the theoretical carrying capacity until it is overtaken by the next catastrophe. (Smith[2], on the other hand, was able to detect evidence of a carrying capacity in Andrewartha's data, but even so it was rather poorly correlated with the actual population.) When extra vertebrates are stocked into an area their population quickly falls to the pre-stocked level[3]. Thus ecologists have long known that areas reached their carrying capacity in so far as vertebrate populations are concerned; the real question is whether the same is true with respect to invertebrates. It is the purpose of this communication to describe some experiments which not only demonstrate the existence of a carrying capacity for the checkerspot, *Melitaea harrisii*, but also to show, in general, what determines this carrying capacity.

Melitaea harrisii is a nymphalid butterfly the larvæ of which, in the New England area where this work was conducted, feed exclusively on *Aster umbellatus* Mill. Eggs are laid in July, the larvæ feed through the third instar, pass the winter in that stage, and complete their feeding and development the following June. The adults emerge in July. This butterfly lives associated with asters in old fields in appropriate stages of succession. The fields, scattered

Table 1. COUNTS OF INDIVIDUALS IN DIFFERENT DEVELOPMENTAL STAGES IN AND NEAR THE CENSUS AREA

Year	Spring larvæ	Adults	Autumn larvæ	No. of autumn larvæ per larva following spring	Egg masses per adult in field	Egg masses in adjacent roadside
1956	—	24	11,400	—	2·3	0
1957	—	—	14,200	418	—	0
1958	34	—	—	—	—	0
1959	34	19	8,600	261	2·3	0
1960	29	—	8,000	364	—	0
1961	22	6	800	257	0·66	0
(Stocked with 19,800 autumn larvæ)			20,600 (total)			
1962	80	22	400	—	0·09	> 10
1963	—	6	1,800	—	3·0	0

throughout a heavily forested area, represent a patchy environment; consequently, the butterfly occurs in fairly isolated populations. Because of the partial isolation of populations, the restricted larval feeding habits, the habit of the females of laying eggs in batches of about 200, and the gregarious web-spinning behaviour of the young larvæ it has been possible to maintain a fairly accurate census of the local populations in the town of East Bluehill, Maine. Such a census was conducted during the years 1956–63 inclusive[4].

The results of the observations are summarized in Table 1. The general decline in population from 1956 to 1961 was associated with the trend in succession of the old field which was becoming progressively less suitable. On the basis of this trend it was estimated in 1961 that the butterfly population would drop to zero by 1963.

In autumn of 1961, the field was stocked with an additional 19,800 larvæ representing a 25-fold increase over the 800 larvæ already present. Remarkably enough, their survival over the winter, as indicated in the fifth column, was not less than in previous years; in fact, it was somewhat greater. Thus 80 larvæ were present in the spring following the stocking, in contrast to the fewer than 20 which would have been expected in the absence of stocking; however, the adults in 1962, at least five-fold more common than they might have been in the absence of stocking, laid only 0·09 egg masses per adult in the field instead of a typical figure of about 0·66–3·0. Why did these adults lay so few eggs in the field? A clue to the answer is seen in the last column, which indicates that in this year alone a substantial number of eggs were laid on the asters along the roadside leading out of the forested area in which the field was located. In any event, the number of autumn larvæ in 1962, one year following the stocking, fell to 400, which is about what it would have been had there been no new larvæ introduced.

Although this completes the account of the field for which the census data are tabulated, two more general points can be made. First, all the egg masses for stocking of the census area in 1961 were taken from three fields that are 1–2 miles distant in the unforested environs of the village. Yet in 1962 these three fields had reconstituted approximately their normal numbers of Melitaea. In fact, counts of egg masses in 1958 in these three fields were 22, 1, 1, respectively, while the same three fields in 1962 had 18, 8, 9, respectively. (These fields are kept in fairly uniform habitat by yearly mowing or grazing.) This result, too, is consistent with the hypothesis that the main density dependent regulation of the population within a field takes place by dispersion of adults to and from the field.

It is an unfortunate fact that the weather in the summer of 1962 was the least suitable for butterflies of any of the years in which counts were made. There were only three days of sun during July and two sunny days during the first three weeks of August. Thus, within the town, fields which had 21,000 autumn larvæ in 1958 had only 10,200 in 1962; however, in these fields the number of eggs was depressed to 50 per cent of normal, while in the overstocked field the eggs were reduced to about 5 per cent of normal. Hence, the bad weather can account for only a small part of the reduction.

Thus Melitaea harrisii populations within a field seem to be governed by a carrying capacity of the field and quickly return to that level when additional individuals are stocked. Emigration occurs before larval food or oviposition sites become inadequate. The main density dependent factor operating within a separate field is not local mortality but emigration.

[1] Andrewartha, H. G., and Birch, L. C., The Distribution and Abundance of Animals (University of Chicago, 1954).
[2] Smith, F. E., Ecology, 42, 403 (1961).
[3] Allen, D. H., Our Wildlife Legacy (Funk and Wagnalls, New York, 1962).
[4] Dethier, V. G., Canad. Entomol., 91, 581 (1959).

Immigration and emigration as contributory regulators of populations through social disruption

E. D. BAILEY

Department of Zoology, University of Guelph, Guelph, Ontario

Received March 31, 1969

Previously isolated male laboratory mice were placed into groups of 16 mice per cage. Each day for 10 days, one animal was exchanged for another between two of the groups in a simulated balanced emigration–immigration. The animals in the third groups served as controls. Stress indicator organs were weighed to determine the effect of the balanced movement and the ensuing social disruption. Splenic and adrenal hypertrophy was greater in the mice exposed to movement than in the controls. Variance of hypertrophy followed the same pattern. Testicular development and size of seminal vesicles was greatly inhibited in the mice exposed to movement. Reproductive potential may be influenced by immigration and emigration through social disruption alone, with no change in actual available space, food, and other factors on a per animal basis.

Introduction

Individual populations of animals are constantly altered by forces of mortality, natality, and movement or migrality (combined immigration and emigration) (Davis and Golley 1963). If these forces are exactly balanced, the net change in population number will be zero. But there will be continuously changing composition in age and sex ratios as individuals are removed by death or emigration and replaced by birth or immigration. The replacement of removed individuals with new or strange individuals disrupts social organization (Calhoun 1962, 1963). Effects on stress indicator organs of mice increase as numbers of possible social interactions increase. These effects are similar to those caused by crowding even though space per animal is constant (Bailey 1966). The implications of the effects of stress on population growth have been well documented (Christian 1963; Christian and Davis 1964). However, most of the studies on the relationships between population levels and endocrine function as well as general physiology have dealt with crowding. Immigration of animals into a population, even with a compensating emigration, acting through disruption of the social organization, might result in a stress effect that is similar to that brought about by crowding or social interaction. Migrality then could act in a contributory fashion, as a population control through factors of stress. The purpose of this study was to determine the effects of immigration with a compensating emigration on stress indicator organs and the possible ultimate effect on population numbers.

Methods

White male laboratory mice were isolated in cages for 3 weeks after they were weaned. There was no visual or physical contact between animals during the isolation period. Auditory and olfactory contact were not controlled. After the isolation period, the mice were segregated into three groups. Each group contained six populations (cages) of 16 mice each. The groups were designated as A and B, the experimental groups, and C, the control group. Before the isolation period each animal was assigned a number and marked by toe clipping for later individual recognition.

Each day for 10 days one randomly chosen animal from A was moved to B and one from B moved to A. As a control on the effects of handling, one randomly chosen group C animal was moved each day, held in a spare cage for 15 minutes, and returned to the C group cage. A tuft of hair was clipped from the back of each animal as it was moved or handled to avoid repeated moving or handling of any individual.

The mice were weighed, measured, and autopsied at the end of the 10-day period. The spleen, adrenals, left testis, and seminal vesicles from each animal were preserved for weighing at a later time. The organs were dried in a circulating oven at 40 C until their weights were constant. Weighings were made on a direct reading milligram balance. An average of three final weighings was used. The A and B groups, both alone and combined, were compared to the C group.

Food and water were supplied ad libitum; space per group and conditions of light, temperature, and other environmental factors were presumed to be equal for the three groups. Only the variable of the effect of balanced migrality between populations was measured. No effort was made to separate dominant and subordinate animals. All individuals in each group were included in mean values and presented as the total effect.

248

Results

Splenic Hypertrophy

The mean weight of the spleens in respect to total body weight was greater in the experimental groups A and B than in the control group C. The increase in splenic hypertrophy was not significant ($p > .05$) (Table I). A t-test analysis showed the A and B groups were more similar to each other than either was to group C. The variance was greater in both A ($p < .05$) and B ($p > .05$) than in C (Table II).

Adrenal Responses

The mean weight of the adrenal glands expressed on a per body weight basis was lower in the C than in the A and B groups. Adrenals of A animals were significantly larger ($p < .05$) than in C and B. The adrenals of groups B were not different from the C animals. The mean adrenal weight of A and B combined was greater than C with a p value approaching .10, but not significant at the .05 level (Table I). Variance of A and B was greater, but not significantly, than C (Table II).

Testes Development

The mean weight of the testes of C group, corrected for body weight, was greater than mean corrected weight of A group ($p < .01$) and B group ($p < .01$). Testes weights of A and B were not different. The mean testes weight of A and B groups combined, when compared to C, was also highly significant ($p < .01$) (Table I). The variance of B group was larger than C ($p > .05$) and, the C group was higher than A ($p < .05$) (Table II).

Seminal Vesicle Development

The mean weight of the seminal vesicles, adjusted for body weight, was greater in C than in A ($p < .05$) and greater than in B ($p < .10$). The A group was not different from B group. The overall mean of the A and B groups combined was significantly lower than in the C group (Table I). The variances of both the A group and B group were identical and smaller than the C group ($p < .05$) (Table II).

The variability within groups for all organs measured was not significant. The moved animals of A and B groups did not differ significantly from the unmoved animals. Similarly, the handled animals in the C groups were not different from non-handled animals. The organs of animals that were moved on any specific day within all groups did not differ significantly from those moved on any other day. No within-group variability was significant, though the adrenal weights of B group approached significance (Table II).

Discussion

The degree of hypertrophy and (or) hyperplasia, or the degree of involution of the stress indicator organs is to a large extent dependent upon time elapsed between application of a

TABLE I

Mean organ size (as percentage of body weight) with statistical comparisons between experimental and control groups

Groups	Spleens	Adrenals	Testes	Seminal vesicles
A ($N=96$)	0.0034 (NS)	0.0000629 ($< .05$)	0.000411 ($< .01$)	0.0009 ($< .05$)
B ($N=96$)	0.0036 (NS)	0.0000585 (NS)	0.000403 ($< .01$)	0.0010 ($< .10$)
C ($N=96$)	0.0032	0.0000584	0.000428	0.0011
A+B	0.0035 (NS)	0.0000606 (NS)	0.000407 ($< .01$)	0.0010 ($< .05$)

TABLE II

Variance of mean corrected organ weights

Groups	Spleens	Adrenals	Testes	Seminal vesicles
A	1.77×10^{-6}	1.51×10^{-10}	0.0030×10^{-9}	0.7×10^{-8}
B	1.35×10^{-6}	1.82×10^{-10}	0.0055×10^{-9}	0.7×10^{-8}
C	1.07×10^{-6}	1.57×10^{-10}	0.0046×10^{-9}	1.1×10^{-8}
A+B	1.63×10^{-6}	1.77×10^{-10}	0.0052×10^{-9}	0.8×10^{-8}

stressor and measurement of organs (Selye 1951, 1956). Hyperplasia would continue for a longer period of time and repressed development would continue for longer if a stressor was applied repeatedly than if applied only one time. Introduction of a strange animal into a population acts as a stress factor on the stranger and on the resident until the introduced animal either is removed or is integrated (Calhoun 1962).

Mice moved on day 1 of a 10-day experiment would have a greater possibility of becoming integrated into the population than mice introduced on days 10, 9, or 8. Presumably, the adrenals and spleens of animals moved early in the experiment could return to normal and animals moved later could be in various stages of adaptation. Also, the animals moved on days 10 or 9 might not yet show the responses. But, if stress did occur, the variance would be expected to be greater in the experimentals than in the controls though means of organ weights were not significantly different. The comparison of variances of spleens supports this hypothesis. However, similar analysis of adrenal weights does not support it, though differences do approach significance. Presumably, the A and B groups should show similar means and variances. However, the B group had a mean adrenal weight almost identical with the C group though variance was much wider. No reason can be given with certainty at present for this disparity.

A complicating factor and possible explanation is that the moved animals were also exposed to newly introduced animals. Those animals moved on day 1 were subjected to stress from animals immigrating on subsequent days. Similarly, animals moved into a population on day 2, day 3, and so on, were subjected to later immigrants. The removal or emigration of residents also acted to disrupt the social structure by creating openings in the hierarchy. So, social disruption occurred repeatedly and probably in varying degrees and the adaptation picture would not be expected to resemble a once-and-done stress.

The more important indicators of potential reproductive implications in maturing male mice are the responses of testes and seminal vesicles. The depressed development of both these organs in the experimentals as compared to the controls indicates that the disruption of social structure through immigration and emigration does have a suppressive effect on reproductive potential, at least to the point of retarding development of reproductive organs while the movement is occurring.

The movement of an animal into a population of established social structuring acts as a stressor on the new animal even without change in actual number of animals per area in a laboratory situation. The stress is probably greater on the immigrant than on the residents, but does occur in the residents also. Any movement of animals from one local population to another in the wild could act to regulate reproductive potential of both populations for some period though actual numerical change may be negligible. All the stress factors acting in nature simultaneously with movement would tend to mask the effects of immigration and emigration. But movement with the resulting social disruption is a contributory factor.

Acknowledgments

Support for this work was granted by the National Research Council of Canada, grant number A-2354.

BAILEY, E. D. 1966. Social interaction as a population control mechanism in mice. Can. J. Zool. **44**: 1007–1012.
CALHOUN, J. B. 1962. The ecology and sociology of the Norway rat. U.S. Public Health Serv. Publ. **1008**.
―――― 1963. The social use of space. In Physiological mammalogy. Vol. 1. Edited by M. V. Meyer and R. G. Van Gelder. Academic Press, New York. pp. 1–187.
CHRISTIAN, J. J. 1963. Endocrine adaptive mechanisms and physiologic regulation of population growth. In Physiological mammalogy. Vol. 1. Edited by M. V. Meyer and R. G. Van Gelder. Physiological mammalogy. Vol. 1. Academic Press, New York. pp. 189–353.
CHRISTIAN, J. J. and DAVIS, D. E. 1964. Endocrines, behaviour and population. Science, **146**: 1550–1560.
DAVIS, D. E. and GOLLEY, F. B. 1963. Principles in mammalogy. Reinhold Publ. Corp., New York.
SELYE, H. 1951. Annual report on stress. Acta, Inc. Medical Publishers, Montreal, Canada.
―――― 1956. The stress of life. McGraw-Hill Book Co. Inc., New York.

Part V

THE EVOLUTION OF DISPERSAL AND MIGRATION

Editors' Comments
on Papers 27 Through 32

27 DINGLE
Excerpt from *Ecology and Evolution of Migration*

28 GILL
Effective Population Size and Interdemic Migration Rates in a Metapopulation of the Red-spotted Newt, Notophthalmus viridescens *(Rafinesque)*

29 HAMILTON and MAY
Dispersal in Stable Habitats

30 LIDICKER
Emigration as a Possible Mechanism Permitting the Regulation of Population Density below Carrying Capacity

31 VAN VALEN
Group Selection and the Evolution of Dispersal

32 HANSEN
Larval Dispersal and Species Longevity in Lower Tertiary Gastropods

Part V concerns perhaps the most important and at the same time the most difficult set of questions concerning movements of organisms. It is an area of extremely active research at the present time. To what extent, under what conditions, and by what mechanisms do dispersal and migratory behaviors evolve? In Part I we argued that not only do such behaviors generally have a genetic basis, but that genetic variation for these traits frequently occurs as well. Hence the substrate for evolution is available. In Parts II, III, and IV, we explored the possible advantages and risks to both the moving individuals themselves and to the populations they have left as emigrants and have joined as immigrants. The evolutionary process juggles this multiplicity of variables, and if successful produces a population of individuals that possesses an adaptively appropriate dispersal "strategy." It is important to understand which variables contribute significantly to the evolution of movement behavior, how these variables interact, and how this evolutionarily significant matrix varies over space and time. Of course, to construct models with predictive power, it is generally important that all these variables and interactions be quantified.

Although considerable progress has been made, especially in recent years, regarding our understanding of the evolution of dispersal and migration, there remains much work ahead. The largest current need is for good empirical studies designed to test particular theoretical models. Dingle (1980) has provided us with a comprehensive and extremely useful review paper on dispersal and migration. We recommend the entire paper, but are reprinting here only his discussion section (and appropriate references) as Paper 27. The reader should be aware that Dingle uses the term "migration" in the general sense of "specialized behavior especially evolved for the displacement of the individual in space."

Recent research concerning the evolution of dispersal and migration can be grouped for convenience around the following general questions:

1. In what ways and to what extent does environmental heterogeneity in space and time contribute to the evolution of dispersal?
2. Are there particular life history modalities associated with dispersal or migration?
3. What is the role of interspecific biotic interactions in influencing dispersal and migration?
4. Are there circumstances in which mixed strategies (polymorphisms) are appropriate, and do they occur?
5. Can dispersal evolve by individual-level selection, by various group selection mechanisms, by neither, or by both?
6. What are the consequences of dispersal for local demic differentiation, and conversely, for species cohesion?
7. Do species with good dispersal capabilities have a different (a) probability of long-term evolutionary success, and (b) speciation potential as compared to more sedentary forms?

Environmental heterogeneity, a nearly ubiquitous attribute of species' ranges, is clearly of fundamental importance in the evolution of dispersal. Numerous authors have addressed this issue mathematically (Lewontin, 1965; Levins, 1968; Gadgil, 1971; Roff, 1974, 1975; Gillespie, Paper 21, 1976, 1981; Łomnicki, Paper 23, 1980; Comins et al., 1980; Andersson, 1980). Good empirical examples are provided by Brown (1951), Southwood (1962), Newsome and Corbett (1975), Johnston and Heed (1976), Hansson (1977), and Gill (Paper 28). Paper

28 is a particularly good study addressing this issue. Useful reviews can be found in Gaines and McClenaghan (1980) and Stenseth (1982). Somewhat in contrast to this large body of theory and data supporting the importance of environmental heterogeneity is the interesting paper by Hamilton and May (Paper 29) who argue that dispersal can evolve even in stable habitats. Their argument stems from potential competition between parents and their offspring and leads to the conclusion that it is advantageous to have offspring that attempt to establish away from their birthplace even if considerable mortality is suffered in the process.

Life history strategies clearly interlock with matters of environmental variability. Nevertheless, organisms have unique properties that give them life history features that may transcend patterns of environmental variability. It is therefore useful to consider whether such factors as body size, fecundity, survival rates, sexual dimorphism, sex ratios, vagility, and others are compatible with dispersal behavior. A related issue concerns what life history modalities lead to dispersal of different age, sex, or condition components of a population. Good examples of attempts to treat life history attributes and dispersal comparatively are those of Brussard and Ehrlich (1970), Dingle (Paper 22), Caldwell (1974), Greenwood and Harvey (1976, 1977), and Greenwood (1980).

The possibility that various interspecific interactions can influence dispersal and migration was alluded to in Part III in our discussion of how the success of movements may be affected by competition and predation. One of the earlier generalizations relating dispersal ability to interspecific competition was that of the "fugitive species." This term was first coined by Hutchinson (1951) and refers to a species with high reproductive potential and dispersal capabilities but with weak powers of interspecific competition. Such weedy species survive by rapidly colonizing patches of relatively empty habitat, but later succumb to superior competitors. The role of competition in influencing the success of dispersers and hence the evolution of dispersal behavior is also discussed by Harper (1965) and Sakai (1965). The increased vulnerability of individuals to predation when outside their home range or when attempting to colonize a new area can certainly act to discourage the evolution of dispersal (see discussion and references in Part III and for colonizing plants, see Harper, 1965). On the other hand, where predators tend to direct their attention on foci of high prey density, dispersing may be of distinct selective advantage. This situation illustrates a "diplomatic" advantage for dispersers as described by Lidicker (Paper 30). On a zoogeographic scale, the success of colonizers reaching islands or new continents is spectacularly affected by the presence or absence of appropriate predators and parasites. There are examples from all over the world, but Australia and New Zealand have provided particularly rich arenas for this phenomenon (Birch, 1965; Wodzicki, 1965). Moreover, the literature on biological control provides many examples of how the success of introduced species depends on the presence or absence of particular predators and/or parasites.

The possibility that polymorphisms for dispersal tendency may evolve was one of the first dispersal questions to receive attention (see discussion of this issue in Part I). Insects provide the richest source of examples of dispersal polymorphisms (Southwood, 1962; Vepsäläinen, 1978; Harrison, 1980). A few authors, however, have attempted to argue against the plausibility of this strategy evolving in both vertebrates and in insects (for example, Murray, 1967, and Davis, 1980, respectively). The search for and analysis of such polymorphisms, especially for noninsects, should carry rich rewards, not only for dispersal theory, but also for our understanding of the evolution and maintenance of polymorphisms in populations. Lavie and Ritte (1978) provide an instructive example from their work on reproductive fitness in high and low dispersal lines of the flour beetle *Tribolium castaneum*. Beetles of the disperser morph possess shorter developmental times and higher fecundity than do non-dispersers.

A major philosophical issue in evolutionary biology today concerns the level of complexity at which selection can operate, that is, what are the operating units of selection. The study of dispersal has the potential for contributing to this important discussion. With the growing awareness that dispersal could be adaptive, more specific models of evolution were sought that went beyond the vague notion that dispersal was somehow good for the species because it allowed it to expand its range and occasionally to occupy empty habitat. Such nonrigorous statements were common in the dispersal literature until recent decades and still occur. In 1962, Lidicker suggested how dispersal could evolve through individual-level selection (Paper 30). In this paper, advantages to the individual disperser were grouped into three categories: quantitative (increased opportunities for mating); qualitative (possibility of being involved in new and advantageous genetic recombinations); and diplomatic (leaving sites likely to suffer economic, predatory, parasitic, or socially mediated disasters) (see also Table 1 in Part II). To these, Bengtsson (1978) has added the additional qualitative advantage of avoiding inbreeding. In an important contribution (Paper 31), Van Valen shows how group selection can lead to the evolution of dispersal. Some authors have opposed the operation of group selection in this context (for example, Murray,

1967), and others have pointed out how individually selected dispersal can lead to reduced population-level fitness (Gadgil, 1971; Paper 29).

The relation between demic differentiation and dispersal (gene flow, in this case) is a major area of inquiry in population genetics and hence is beyond the scope of this book. Clearly, quantitative information on dispersal and the nature of dispersers will contribute importantly to this larger area of population biology.

The final question is one that bridges evolutionary biology, ecology, and paleontology. It concerns the long-term potential for survival and speciation of lineages given various levels of dispersal ability. In a sense it is the ultimate question of the adaptive significance of dispersal. Wilson (1965) discusses this ultimate adaptive value of dispersal tendencies and concludes that: "For most evolving species there is a point short of the maximum attainable dispersal power at which fitness is greatest" (p. 15). In a pioneering effort to relate dispersal to speciation rates, Jackson (1974) proposed a model in which species of marine bivalves with hardy planktonic larvae are long-lived as species but have low speciation rates. Hansen further explores the relationship between evolutionary rates and larval dispersal in the volutid gastropods of the North American Gulf Coast region (Lower Tertiary); his excellent study constitutes Paper 32. He concludes that species with highly dispersive larvae (planktonic) persisted twice as long in the geologic record as those with nondispersing larvae (nonplanktonic). Moreover, geographic range was substantially larger in the former group, and they had broader environmental tolerances. On the other hand, rapid environmental changes in the Upper Middle Eocene resulted in the extinction of the species with planktonic larvae, which Hansen speculates was due to their being out-competed by the more specialized nonplanktonic forms and/or by their greater genetic inertia and hence longer evolutionary response times. A somewhat contrasting but equally intriguing possibility is that, as already pointed out, dispersal may make local adaptation more difficult leading to a reduced overall population or species-level fitness, but at the same time it may force continued genetic experimentation that will improve the long-term chances of lineage success. Thus, the challenge for survival is clearly not purely a question of to disperse or not to disperse, but to find the appropriate level of dispersal for a given circumstance. Perhaps a successful strategy would be to alternate episodes of panmixia with those of local adaptation. Aside from hedging one's bets on the pluses and minuses of dispersal, such a strategy permits the synergistic benefits of individual and group selection.

REFERENCES

Andersson, M., 1980, Nomadism and Site Tenacity as Alternative Reproductive Tactics in Birds, *J. Anim. Ecol.* **49:**175-184.

Bengtsson, B. O., 1978, Avoiding Inbreeding: At What Cost?, *J. Theor. Biol.* **73:**439-444.

Birch, L. C., 1965, Evolutionary Opportunity for Insects and Mammals in Australia, in *The Genetics of Colonizing Species*, H. G. Baker and G. L. Stebbins, eds., Academic, New York, pp. 197-213.

Brown, E. S., 1951, The Relation Between Migration-Rate and Type of Habitat in Aquatic Insects, with Special Reference to Certain Species of Corixidae, *Zool. Soc. London Proc.* **121:**539-545.

Brussard, P. F., and P. R. Ehrlich, 1970, Contrasting Population Biology of Two Species of Butterfly, *Nature* **227:**91-92.

Caldwell, R. L., 1974, A Comparison of the Migratory Strategies of Two Milkweed Bugs, *Oncopeltus fasciatus* and *Lygaeus kalmii*, in *Experimental Analysis of Insect Behaviour*, L. B. Browne, ed., Springer-Verlag, Heidelberg, pp. 304-316.

Comins, H. N., W. D. Hamilton, and R. M. May, 1980, Evolutionarily Stable Dispersal Strategies, *J. Theor. Biol.* **82:**205-230.

Davis, M. A., 1980, Why Are Most Insects Short Fliers? *Evol. Theor.* **5:**103-111.

Gadgil, M., 1971, Dispersal: Population Consequences and Evolution, *Ecology* **52:**253-261.

Gaines, M. S., and L. R. McClenaghan, Jr., 1980, Dispersal in Small Mammals, *Ann. Rev. Ecol. Syst.* **11:**163-196.

Gillespie, J. H., 1976, The Role of Migration in the Genetic Structure of Populations in Temporally and Spatially Varying Environments. II. Island Models, *Theor. Pop. Biol.* **10:**227-238.

Gillespie, J. H., 1981, The Role of Migration in the Genetic Structure of Populations in Temporally and Spatially Varying Environments. III. Migration Modification, *Am. Nat.* **117:**223-233.

Greenwood, P. J., 1980, Mating Systems, Philopatry and Dispersal in Birds and Mammals, *Anim. Behav.* **28:**1140-1162.

Greenwood, P. J., and P. H. Harvey, 1976, The Adaptive Significance of Variation in Breeding Area Fidelity of the Blackbird (*Turdus merula* L.), *J. Anim. Ecol.* **45:**887-898.

Greenwood, P. J., and P. H. Harvey, 1977, Feeding Strategies and Dispersal of Territorial Passerines: A Comparative Study of the Blackbird, *Turdus merula*, and the Greenfinch, *Carduelis chloris*, *Ibis* **119:**528-531.

Hansson, L., 1977, Spatial Dynamics of Field Voles, *Microtus agrestis*, in Heterogeneous Landscapes, *Oikos* **29:**539-544.

Harper, J. L., 1965, Establishment, Aggression, and Cohabitation in Weedy Species, in *The Genetics of Colonizing Species*, H. G. Baker and G. L. Stebbins, eds., Academic, New York, pp. 243-268.

Harrison, R. G., 1980, Dispersal Polymorphisms in Insects, *Ann. Rev. Ecol. Syst.* **11:**95-118.

Hutchinson, G. E., 1951, Copepodology for the Ornithologist, *Ecology* **32:**571-577.

Jackson, J. B. C., 1974, Biogeographic Consequences of Eurotopy and Stenotopy among Marine Bivalves and Their Evolutionary Significance, *Am. Nat.* **108:**541-560.

Johnston, J. S., and W. B. Heed, 1976, Dispersal of Desert-adapted *Drosophila;* Saguaro-breeding *D. nigrospiracula, Am. Nat.* **110:**629-651.

Lavie, B., and U. Ritte, 1978, The Relation Between Dispersal Behavior and Reproductive Fitness in the Flour Beetle *Tribolium castaneum, Can. J. Genet. Cytol.* **20:**589-595.

Levins, R., 1968, *Evolution in Changing Environments,* Princeton University Press, Princeton N. J., 120p.

Lewontin, R. C., 1965, Selection for Colonizing Ability, in *The Genetics of Colonizing Species,* H. G. Baker and G. L. Stebbins, eds., Academic, New York, pp. 77-94.

Łomnicki, A., 1980, Regulation of Population Density due to Individual Differences and Patchy Environments, *Oikos* **35:**185-193.

Murray, B. G., Jr., 1967, Dispersal in Vertebrates, *Ecology* **48:**975-978.

Newsome, A. E., and L. K. Corbett, 1975, Outbreaks of Rodents in Semi-arid and Arid Australia: Causes, Preventions, and Evolutionary Considerations, in *Rodents in Desert Environments,* I. Prakash and P. K. Ghosh, eds., Junk Pub., The Hague, pp. 117-153.

Roff, D. A., 1974, The Analysis of a Population Model Demonstrating the Importance of Dispersal in a Heterogeneous Environment, *Oecologia* **15:**259-275.

Roff, D. A., 1975, Population Stability and the Evolution of Dispersal in a Heterogeneous Environment, *Oecologia* **19:**217-237.

Sakai, K., 1965, Contributions to the Problem of Species Colonization from the Viewpoint of Competition and Migration, in *The Genetics of Colonizing Species,* H. G. Baker and G. L. Stebbins, eds., Academic, New York, pp. 215-241.

Southwood, T. R. E., 1962, Migration of Terrestrial Arthropods in Relation to Habitat, *Biol. Rev.* **37:**171-214.

Stenseth, N. C., 1982, Causes and Consequence of Dispersal in Small Mammals, in *The Ecology of Animal Movements,* I. Swingland and P. J. Greenwoods, eds., Oxford University Press, Oxford, pp. 63-101.

Vepsäläinen, K., 1978, Wing Dimorphism and Diapause in *Gerris:* Determination and Adaptive Significance, in *Evolution of Insect Migration and Diapause,* H. Dingle, ed., Springer-Verlag, New York, pp. 218-253.

Wilson, E. O., 1965, The Challenge from Related Species, in *The Genetics of Colonizing Species,* H. G. Baker and G. L. Stebbins, eds., Academic, New York, pp. 7-27.

Wodzicki, K., 1965, The Status of Some Exotic Vertebrates in the Ecology of New Zealand, in *The Genetics of Colonizing Species,* H. G. Baker and G. L. Stebbins, eds., Academic, New York, pp. 425-460.

27

Copyright © 1980 by Academic Press, Inc.
Reprinted from pages 1-2 and 78-83 of *Animal Migration, Orientation, and Navigation,*
S. A. Gauthreaux, Jr., ed., Academic, New York, 1980, 387p.

Ecology and Evolution of Migration

HUGH DINGLE

I.	The Phenomenon of Migration and Some Problems of Definition	2
II.	The Vertical Migration of Plankton	5
	A. The Phenomenon of Vertical Migration	6
	B. The Adaptive Significance of Vertical Migration	8
III.	The Dispersal of Planktonic Larvae	15
IV.	The Migrations of Demersal and Terrestrial Crustaceans	18
V.	Migration of Insects and Other Terrestrial Arthropods	19
	A. The Behavior of Migrants	19
	B. Variation within and among Species	21
	C. Migration and Life Histories	33
	D. Conclusions: The Adaptive Significance of Migration in Terrestrial Arthropods	37
VI.	Migration in Fishes	38
	A. Oceanodromous and Potomadromous Migrations	38
	B. Catadromous Migration	41
	C. Anadromous Migration	41
	D. Migration and Life Histories	48
	E. Conclusions: The Evolution of Fish Migration	50
VII.	Migration in Amphibians and Reptiles	51
	A. Amphibians	51
	B. Reptiles	52
VIII.	Migrations of Birds	54
	A. The Diversity of Bird Migration	54
	B. The Evolution of Bird Migration	56
	C. Conclusions: Evolution and Coevolution of Migration Strategies	62
IX.	Migrations of Mammals	64
	A. Marine Mammals	64
	B. Large Terrestrial Mammals	67
	C. Bats	69
	D. Small Terrestrial Mammals	71
X.	Discussion: The Evolution of Migration	78
	A. Theoretical Models	78

	B. Some Empirical Comparisons	80
	C. Prospects	82
XI.	Addendum	83
	A. Vertical Migration	83
	B. Demersal Crustaceans	84
	C. Insects	85
	D. Fish	87
	E. Amphibians	87
	F. Birds	88
	G. Mammals	89
	H. General Dynamics	90
	References	91

[Editors' Note: Material has been omitted at this point.]

X. Discussion: The Evolution of Migration

In attempting to arrive at general principles concerning the evolution of migration, one at first is likely to feel overwhelmed at the bewildering diversity of patterns and processess. Is there indeed any common ground for understanding the movements of a vertically migrating copepod, a Pacific salmon, a juvenile vole, or a trans-Saharan flying bird? Can we put order into our thinking about the selective mechanisms involved? Or is the migration of each species unique requiring a unique explanation? Such questions have been examined both theoretically and with models derived from empirical comparisons.

A. Theoretical Models

Ultimately selection acting on migration is a function of the relative survival and subsequent reproductive success of migrant and nonmigrant individuals. In the simplest sense this can be expressed as the relative replacement rates (R_0 of the population ecologist) of the two types. Replacement rate is a function of survivorship, l_x, and the birth rate, m_x, and is expressed as $R_0 = \Sigma l_x m_x$ over the reproductive life of the individual. Clearly if R_0 (migrant) $> R_0$ (nonmigrant) within a population, migrants will be favored, and the key to understanding the adaptive value of migration is understanding the selective forces acting on l_x and m_x. As I have repeatedly emphasized, this makes migration an integral part of life history strategies. Migration will be favored if survival and reproduction are greater in a new habitat in spite of the risks of migrating; this frequently occurs because the risks of remaining sedentary increase as a consequence of habitat changes (Southwood, 1977; Baker, 1978). The relative stability and suitability of the habitats are functions of generation time (H/τ of Southwood, 1977). What then constitutes an optimal migration strategy?

The problem has been considered with respect to optimization theory by Cohen (1967) and Parker and Stuart (1976). Cohen examines migration as a problem of optimal choice between alternatives with randomly varying outcomes. Migration can occur as a "pure strategy," with all individuals migrating, or as a mixed strategy with only a portion doing so. Conversely remaining sedentary is also a possible pure strategy. A migratory pure strategy is favored when the variance in the viability coefficient of nonmigrants increases, i.e., risks of remaining sedentary become greater, and vice versa. If viability coefficients for migrants and nonmigrants are independent, then a mixed strategy is more likely, for example, where the variability in winter survival between years is high. When some environmental cue allows prediction of unfavorability, the population will be expected to respond to that cue by emigrating. Seasonal migration in response to photoperiod is an example. As a final conclusion, Cohen's model indicates that migration can only be optimal when survival and reproduction at any one place do not remain constant; if they do organisms gain no advantage by changing habitats. However, while agreeing that migration is clearly advantageous where habitats are transient and patchy, Hamilton and May (1977) demonstrate that constancy is not necessarily a constraint and that dispersal in stable habitats can also be selectively advantageous.

Parker and Stuart (1976) devise models that consider emigration thresholds from resource patches encountered by given search strategies. They use the concept of evolutionarily stable strategies (ESS's), introduced by Maynard Smith (1974), and consider optimal investment durations, gain accumulated, and search costs between patches. Various pure and mixed strategies are possible depending on factors such as competition or resource sharing and ability to assess resources (as in Cohen's models). As an example, for patches with decreasing returns and increasing investment, the ESS is usually for all to stay until a critical threshold is reached, whereupon individuals should leave at a rate which balances the value of staying and of leaving. The model is applied to male dung flies, where the resource is availability of females, but has yet to be considered for more explicitly migratory situations (although cf. Baker, 1978). The colonization dispersals of rodents, for example, may be a possible case for wider applicability.

In a comprehensive attempt to analyze dispersal, Roff (1975) used computer simulation to determine the influence of heterogeneous environments on populations in which dispersal is genetically determined. In both simple and quantitative genetic models changes in the probability of dispersal produced changes in population size, spatial distribution of genotype frequencies, and proportion of dispersers per habitat. Roff found, however, that increasing environmental stability did not always decrease dispersal rate, nor was increased dispersal necessarily the result of an increase in dispersal genotypes. The influence of environmental stability was dependent on the nature of the genetic mechanism influencing dispersal (polygenic or "simple"), the type of dispersal strategy (density dependent or independent), and the form of the environmental changes occurring. The models also demonstrated a major role for sexual reproduction in determining genotype frequencies. In heterogeneous environments dispersers are generally at an advantage when environmental variance is high (cf. also Cohen, 1967), but with long-term stability there is a continuous loss of dispersal genotypes which are not replaced. The presence of dispersers permits colonization of new areas and persistence of the population as a whole (cf. also Taylor and Taylor, 1977). Somewhat similar conclusions with respect to populations were reached by Gadgil (1971). The various dispersal strategies of the gerrids studied by Vepsäläinen (1978) are good examples of outcomes predicted by Roff's models.

B. Some Empirical Comparisons

The most obvious conclusion from comparing diverse organisms is that migration strategies are extraordinarily flexible. Movement is a fundamental biological response to adversity (Taylor and Taylor, 1977), but the timing, duration, and other variables are subject to a variety of intrinsic and extrinsic constraints. To take an obvious but important example, long distance annual movements are most likely in large, long-lived, mobile animals such as whales and birds. Similar patterns also evolve in similar habitats; desert locusts, desert birds, and desert mammals all display migrations to ensure breeding in areas of rainfall. Other common patterns are the north-south migrations of birds, butterflies, bats, and cervids; annual altitudinal migrations of birds and mammals; and the inshore-offshore breeding migrations of whales, fish, and crustaceans. Nevertheless selective regimes do differ, and it is hardly surprising to observe differences in pattern between ocean and desert. Proximate and ultimate selective forces must also be kept distinct. Thus in Cox's (1968) competition model for the evolution of bird migration is it the case that competition leads to migration or to the occupation of marginal habitats by some species which then adopt a migration strategy? There is evidence that many migratory birds may be fugitive species in this sense (Karr, 1976), and some migratory insects may be also (Southwood, 1962). No single factor theory of migration is thus ever likely to be adequate.

Some generalizations nevertheless seem possible. Cohen (1967) and Ziegler (1976) have suggested that if cues for habitat variation are absent or unreliable, genetic determinism should be favored. Greater environmental determinism should evolve in the presence of reliable cues such as photoperiod which, if they occur at the appropriate time in the life cycle, can lead to behaviorally or morphologically specialized morphs (e.g., brachyptery–macroptery in insects). When cues can occur at any time during the life cycle, a more generalized emigration response may evolve with movement a function of current conditions as in the *Tribolium* studied by Ziegler (1976) or in various temperate zone birds in winter (Lack, 1968). With respect to optimality theory, the cost/benefit ratio for migration should vary predictably from season to season, but relatively unpredictably from year to year. The former selects for physiological responses, e.g., to photoperiod, and the latter for behavioral flexibility as indeed has been observed (Lack, 1968).

In an extension of some ideas of Southwood (1977) and Solbreck (1978), I have attempted in Fig. 6 to summarize the role of migration in life histories. Included are diapause and hibernation as additional responses to environmental adversity. In general organisms can choose to breed "now" or "later" either "here" or in "near" and /or "far" environments. Immediate breeding in the current habitat requires no migration and often leads to strategies such as winglessness, asexual reproduction, and territorial behavior (used here broadly to include agonistic dominance). Variations in space and time parameters lead to other available strategies. Note that these are not mutually exclusive; for example, delayed maturity (and large body size) and round trip migration can occur in the same species (e.g., whales, sea birds). Note further that the same species, populations, or even individuals can adopt different tactics in different places at different times. These can be sequential (e.g., round trip bird migrants which are also territorial; insects which lose their wings following migratory flight) or more or less haphazard depending on circumstances. The model is clearly not all encompassing, but may serve as a useful generalization. The reader will have further conceptual modifications.

Various migration strategies may also serve as examples for the model of evolution described by Slobodkin and Rappaport (1974). They conceive of evolution as an "existential game" in which the strategy is to stay in the game (since there can be no external "payoffs") by minimizing the "stakes" of an environmental perturbation. Migration would seem to be the quintessential response in such a game, since successful evolution requires maintaining flexibility of response in the most parsimonius way. Thus organisms faced with short-term fluctuations should develop rapid response systems dependent primarily on immediate behavioral adjustments. Longer-term predictable fluctuations such as seasons will result in the evolution of physiologic responses to cues such as photoperiod and a deeper "commitment" on the part of the organism. Finally, essentially permanent alterations will result in genetic changes leading up to mor-

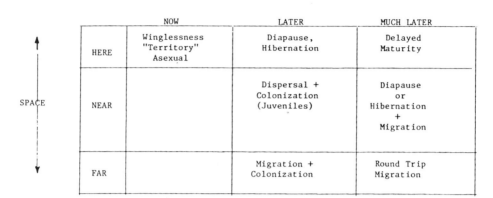

Fig. 6. Generalized "reproductive matrix" for incorporation of migration and other "escape" behaviors (hibernation, diapause) into life histories. Organisms can breed "here and now" or can postpone to avoid deteriorating habitats in space, time, or both. Categories are not mutually exclusive for species as different populations at different times and places adopt various strategies, e.g., winter diapause winged and summer wingless gerrids. Also the same populations may adopt different strategies at different times, e.g., migrant birds territorial in the breeding season. (See text for further discussion.)

phological variation such as wing polymorphism or brachyptery. Suitable adjustments in environmental periodicity should produce genetic variation influencing adaptive responses ranging through behavioral, physiological, and finally morphological mechanisms. Clearly there are numerous examples of migratory responses which have evolved at all these levels (Table II).

C. Prospects

I have tried throughout this chapter to point out areas where I felt additional research effort should be concentrated to obtain the data to fill lacunae in our knowledge. Three areas which seem to be of special importance are listed below.

1. We need more tests of theory and especially ways to test theory experimentally. For the latter, there must be suitable indices of migratory capability. For insects tethered flight is such an index if not always a satisfactory one. In spite of its limitations, however, its use has resulted in significant advances in our studies of insect migration. Suitable indices ought to be possible for mammals and fish, and a start has indeed been made (Myers and Krebs, 1971; Rasmuson et al., 1977). It is also possible to fly birds (Tucker, 1972) and bats (Pennycuick, 1971) in the laboratory which

suggests indices for these animals may be possible if tedious to obtain. Investigators have already taken advantage of *Zugunruhe* in birds for studies of physiology and orientation, and it may also be possible to do so in consideration of ecological and evolutionary constraints.

2. We need to examine the interaction between genetic and environmental influences in more detail. Since migration is likely to be polygenic in most cases, we need a metric character to index it, another reason for appropriate indices. Both empirical data (Dingle *et al.*, 1977; Istock, 1978) and theoretical models (Roff, 1975) indicate the importance of understanding the underlying genetic architecture of migration. That architecture can be understood only through cognizance of the appropriate environmental context for the expression of genetic variance.

3. We need to know more about the role of migration in life history strategies. The concept of migration as an integral part of life histories is relatively new, but its importance cannot be overemphasized. New impetus is given to its development by the recent elaboration of the role of migration in population dynamics in both space and time (Krebs *et al.*, 1973; Southwood, 1977; Taylor and Taylor, 1977). Here especially we seem on the threshold of exciting discoveries and conceptual advances, and the need for data and theory is clear. Our understanding will be much more profound if we can also include the role of genetic influences.

It would be foolish to underestimate the difficulties in undertaking such studies as I have adumbrated, and I do not do so. I make no apologies, however, for the complexities of nature, since with those complexities come exciting challenges to the abilities, energies, and intuition of ecologists and evolutionary biologists.

[*Editors' Note:* Material has been omitted at this point.]

REFERENCES

[*Editors' Note:* Only the references cited in the preceding excerpts are reprinted here.]

Baker, R. R. (1978). "The Evolutionary Ecology of Animal Migration." Holmes & Meier, New York.
Cohen, D. (1967). *Am. Nat.* **101,** 5-17.
Cox, G. W. (1968). *Evolution* **22,** 180-192.
Dingle, H., Brown, C. K., and Hegmann, J. P. (1977). *Am. Nat.* **111,** 1047-1059.
Gadgil, M. (1971). *Ecology* **52,** 253-261.
Hamilton, W. D., and May, R. M. (1977). *Nature (London)* **269,** 578-581.
Istock, C. A. (1978). *In* "Evolution of Insect Migration and Diapause" (H. Dingle, ed.), pp. 171-190. Springer-Verlag, Berlin and New York.
Karr, J. R. (1976). *Wilson Bull.* **88,** 433-458.
Krebs, C. J., Gaines, M. S., Keller, B. L., Myers, J. H., and Tamarin, R. H. (1973). *Science* **179,** 35-41.
Lack, D. (1968b). "Ecological Adaptations for Breeding in Birds." Chapman & Hall, London.
Maynard Smith, J. (1974). *J. Theor. Biol.* **47,** 209-222.
Myers, J. H., and Krebs, C. J. (1971). *Ecol. Monogr.* **41,** 53-78.
Parker, G. A., and Stuart, R. A. (1976). *Am. Nat.* **110,** 1055-1076.
Pennycuick, C. J. (1971). *J. Exp. Biol.* **55,** 833-845.
Rasmuson, B., Rasmuson, M., and Nygren, J. (1977). *Hereditas* **87,** 33-42.
Roff, D. A. (1975). *Oecologia* **19,** 217-237.
Slobodkin, L. B., and Rappaport, A. (1974). *Q. Rev. Biol.* **49,** 181-200.
Solbreck, C. (1978). *In* "Evolution of Insect Migration and Diapause" (H. Dingle, ed.), pp. 195-217. Springer-Verlag, Berlin and New York.
Southwood, T. R. E. (1962). *Biol. Rev. Cambridge Philos. Soc.* **37,** 171-214.
Southwood, T. R. E. (1977). *J. Anim. Ecol.* **46,** 337-365.
Taylor, L. R., and Taylor, R. A. J. (1977). *Nature (London)* **265,** 415-421.
Tucker, V. A. (1972). *Am. J. Physiol.* **222,** 237-245.
Vepsäläinen, K. (1978). *In* "Evolution of Insect Migration and Diapause" (H. Dingle, ed.), pp. 218-253. Springer-Verlag, Berlin and New York.
Ziegler, J. R. (1976). *Evolution* **30,** 579-592.

EFFECTIVE POPULATION SIZE AND INTERDEMIC MIGRATION RATES IN A METAPOPULATION OF THE RED-SPOTTED NEWT, *NOTOPHTHALMUS VIRIDESCENS* (RAFINESQUE)[1]

Douglas E. Gill

Department of Zoology, University of Maryland, College Park, Maryland 20742

In a species that is subdivided into discrete breeding demes, three unique dynamic forces may exert strong influence on its evolution: genetic drift, demic extinction, and interdemal migration. Despite the extensive development in evolutionary and population genetics theory (e.g., Wright, 1931; Levins, 1970; Boorman and Levitt, 1973), the importance of these forces relative to those of demographic structure and natural selection that normally operate within large panmictic populations remains controversial. The controversy stems largely from the lack of quantitative data from natural populations. For example, we are generally ignorant of the actual effective sizes of populations and the magnitude of interpopulational migration rates in nature because of the overwhelming difficulties facing the field biologist in their measurement. Indeed, how does one distinguish a panmictic population from one composed of "isolated or semi-isolated" subpopulations in the first place? The determination of population size usually depends on some mark-recapture technique because direct censuses are typically impossible. The measurement of migration rates requires the indelible marking of all residents in a way that distinguishes them from all migrants. Moreover, the correct interpretation of the magnitude of interpopulational migration rates depends upon the relative size and extent of self-regeneration of each subpopulation. Empirical studies from nature that meet all these requirements are understandably rare. The illuminating study of the checkerspot butterfly, *Euphydryas editha,* on Jaspar Ridge reviewed by Ehrlich et al. (1975) provides one of the exemplary exceptions.

A natural metapopulation of the red-spotted newt, *Notophthalmus viridescens* (Rafinesque), in a series of mountain ponds of Virginia, satisfies the requirements for measurement of genetic drift, demic extinction, and interdemal migration. A study of the dynamics of this metapopulation was begun in 1974 and continues to the present. The purpose of the study was quantification of the dynamics of a species structured as a metapopulation. One emphasis of the study was the measurement of effective population sizes and magnitude of interpopulational migration rates. The study took advantage of the fact that newts are restricted to small ponds as breeding adults and that the location of every pond in the region was known. The total numbers of breeding adults and progeny produced at each pond were easily determined by a combination of capture techniques. Each individual was uniquely identifiable, so that migrants were readily distinguishable from residents. Details of the natural history of the red-spotted newt in this region are reported elsewhere (Gill, 1978).

Study Site and Methods

Mountain ponds at an elevation of about 1,000 m in the George Washington National Forest, Rockingham County, Virginia were chosen because of their sim-

---

[1] This paper is one of a series of contributions from the Laboratory of Amphibian Biology at the University of Maryland.

ilar small size and natural appearance. The locations of all ponds which harbored adult populations of newts in the region were known. A co-adjacent set of ponds, Pond Ridge Pond (PR), Cline's Hacking Pond (CH), Dictum Ridge Pond (DR), Second Mountain Pond (SM), and White Oak Flat Pond (WOF), was selected for study without prior knowledge of the actual sizes of the breeding adult populations or the reproductive capacities in each pond. For these reasons, the results from those five ponds reported here are believed to be representative of any group of ponds in the region. Detailed description of the study site is found in Gill (1978).

A drift fence, with pitfall traps regularly spaced alongside, completely surrounded each pond. All amphibian movement in and out of each pond was monitored continuously from February through November each year. Adult newts were intercepted and identified in heaviest concentration during their spring entrance (March–April) and during their fall departure (August–September). Overwintering in terrestrial hibernacula, adult newts and efts were very rarely seen or captured in the ponds during the winter. The year-round censuses by the drift fence not only provided accurate determinations of breeding population size but also screened for migrant individuals from other ponds. In like manner the post-metamorphic juveniles that left the ponds every fall were identified and released.

Each adult newt and juvenile eft was uniquely identified by the number and pattern of dorsal crimson spots. In addition, a toe-clipping technique marked adults according to pond and year of first capture. Newts rapidly regenerate lost toes, so that this technique was effective on an individual basis for over one year. However, regenerated toes were reclipped whenever an individual was recaptured. Specific details of these techniques and others employed during the course of this population study are described elsewhere (Gill, 1978). This paper reports the results of the 1974, 1975, and 1976 field seasons.

Results

Interpopulational Migration Rates

The rate of exchange of adult newts among breeding ponds was zero. Of the 2,974 adult newts recorded in the five ponds during the three years, not a single one migrated from one pond to another. Year after year adult red-spotted newts faithfully returned to the pond in which they were first captured as adults. The annual departure from the ponds was apparently limited to hibernacula located relatively close by the breeding ponds.

Adult newts not only displayed philopatry (pond fidelity) but they were also capable of accurate homing behavior. In an experiment performed in 1974 elsewhere in the Shenandoah Mountains, the entire breeding population from one pond was experimentally transplanted to another pond about 0.3 km away. After one year had elapsed, most of the transplanted individuals had successfully homed back to the pond of original capture. In contrast, none of the residents of the recipient pond migrated to the experimental source pond. The striking contrast in behavior between the transplanted and the resident groups of newts could not be explained by the negative density effects demonstrated in the experimental transplantation (Gill, in press). Together, the philopatry data and homing experiment support the conclusion that adult migration rates between breeding subpopulations are in general about zero.

All the ponds showed a consistent pattern of high breeding population turnover rate (about 50% per year), resulting in a short adult life expectancy (Gill, 1978). Density dependence was demonstrable in adult female survival, so that life expectancy of mature females was only 1.7 breeding seasons compared to 2.1 in males. The paradox of persistent reproductive failure at most ponds and high recruitment rates at most ponds lead Gill (1978) to conclude that the recruits were colonists from other areas. As a consequence the recruitment of new adults at ponds provided an estimate of interdemic

TABLE 1. *Breeding population sizes at five mountain ponds censused year-round by drift fences. Estimations are based on total number of adults known present in the breeding season. Effective sizes N_o and N_e are computed from Wright's (1938) formulae, where $N_o = 4N_mN_f/(N_m + N_f)$, and $N_e = t/\sum_{t=1}^{t}(1/N_t)$ (see Li, 1975).*

	Numbers observed			Effective size N_o	Effective size N_e
	♂♂	♀♀	Total		
Pond Ridge Pond					
1974	122	53	175	148	
1975	153	57	210	166	149
1976	109	50	159	137	
Clines Hacking Pond					
1974	56	35	91	86	
1975	68	33	101	89	97
1976	67	58	125	124	
Dictum Ridge Pond					
1974	20	7	27	21	
1975	22	13	35	33	26
1976	17	11	28	27	
Second Mountain Pond					
1974	3	3	6	6	
1975	18	7	25	20	12
1976	30	14	44	38	
White Oak Flat Pond					
1975	674	344	1,018	911	
1976	868	586	1,454	1,399	1,103
Subtotals					
1974 (w/o WOF)	201	98	299	261	
1975 (w/o WOF)	261	110	371	308	
(w WOF)	935	454	1,389	1,219	
1976 (w/o WOF)	223	133	356	326	
(w WOF)	1,091	719	1,810	1,725	

migration rates via juvenile dispersal. This conclusion was strengthened by the inpouring of new adults into newly constructed ponds. In this study area the existence of breeding populations of newts on most ponds was utterly dependent upon immigration (Gill, 1978).

Effective Population Size

The size of each breeding newt deme remained stable over the three years of the study, but the variation among ponds ranged from six to over 1,000 in any year. If first the unequal sex ratio (N_o) and then the fluctuation in size over years (N_e) is taken into account, the initial estimates of effective population size in each pond were only modestly less than the direct population censuses (Table 1). These estimates were still generous because they did not take into account actual reproductive variation among individuals within each pond.

As a consequence of the demic breeding structure, it was expected that some level of inbreeding depression should occur. However, the impact of inbreeding is opposed by the extent of interpopulational exchange of juveniles. The exact magnitude of the inbreeding coefficient F is given as

$$F = \frac{(1 - m)^2}{2N - (2N - 1)(1 - m)^2}$$

where N is the population size by census, and m is migration rate (Li, 1975). In this case the effective size N_o from Table 1 was used for the formula N, and the an-

TABLE 2. *Inbreeding coefficients F for five newt demes and the metapopulation as a whole.* m = *annual immigration rate for population size* N, *for which the values* N_0 *from Table 1 were used.*

	1975		1976	
	m	F	m	F
Pond Ridge Pond	0.600	0.0014	0.415	0.0019
Cline's Hacking Pond	0.703	0.0005	0.568	0.0009
Dictum Ridge Pond	0.543	0.0040	0.536	0.0051
Second Mountain Pond	0.840	0.0007	0.682	0.0015
White Oak Flat Pond	—	—	0.512	0.0001
Total	0.637	0.0002	0.512	0.0001

nual rate of recruitment of new adults was used for the value m. It is clear (Table 2) that the enormous annual immigration rates essentially eliminated any inbreeding effect due to the subdivision of the newt metapopulation into discrete breeding demes. As a consequence, depression of effective population because of inbreeding ($N_{ei} = N/(1 + F)$, Li, 1975, p. 562) was negligibly small.

In order to calculate the variance effective population size, N_{ev} (Crow and Kimura, 1970; Crow and Morton, 1955), it was necessary to obtain an estimate of the average gametic contribution per parent, $\bar{k}$, and the variance in per capita reproduction, V_k. In the usual theoretical derivations (Crow and Kimura, 1970; Li, 1975) these terms explicitly refer to the mean and variance in reproduction of individuals *within* a single, often idealized population. In my case, the data provided only the total production of juveniles from each pond for each year of the study. Thus, I have estimates of the annual, average, per capita production from each pond. I do not have an estimate of within-pond reproductive variance, but I do have data demonstrating striking among-pond variation in reproductive output.

In any one year the among-pond variation in reproductive output was as great as the variation in population size (Table 3). The correlation between reproductive output and breeding adult population size was negative but not significant (Spearman rank $r_s = -0.26$, $P > 0.05$). In the three years of the study, ponds PR, CH, and DR produced relatively few, if any, juveniles. In contrast, SM, with the smallest number of breeding adults, consistently produced relatively large numbers of juveniles. The variation in reproductive output among ponds was more strongly emphasized when the data were expressed as the number of juveniles produced per breeding female (column J/f in Table 3). Per capita juvenile production in PR, CH, and DR fell far short of simple adult replacement; in three years 367 adults in PR

TABLE 3. *Number (#) of post-metamorphic juvenile newts produced at five mountain ponds in the years 1974–1976. Production of juvenile newts per breeding female (J/F) is also presented. WOF was not fenced or censused in 1974.*

	1974		1975		1976		Total	
	#	J/F	#	J/F	#	J/F	#	J/F
Pond Ridge Pond (PR)	1	0.02	39	0.68	20	0.40	60	0.38
Cline's Hacking Pond (CH)	54	1.54	18	0.55	0	0.00	62	0.49
Dictum Ridge Pond (DR)	9	1.29	0	0.00	0	0.00	9	0.29
Second Mountain Pond (SM)	113	37.67	197	28.14	216	15.43	526	21.92
White Oak Flat Pond (WOF)	—	—	191	0.56	2,490	4.25	2,681	2.88
Regional average	44.25	1.81	89.00	0.98	545.20	3.79	665.6	2.63
Regional variance	2,645	41.53	9,382	11.59	1.2×10^7	5.00	1.3×10^7	8.30

TABLE 4. *Variance effective population size of newt metapopulation in each of three years 1974–1976.* k = *number of gametes* (= *twice number of young*) *produced per effective adult,* $\bar{k}$ = *regional average number of gametes produced per effective adult,* V_k = *variance in number of gametes contributed to region, weighted by effective population size in each pond, where* $V_k = [\Sigma N_o(k - \bar{k})^2]/(\Sigma N_o - 1)$, N_{ev} = *variance effective population size where* $N_{ev} = (N_o - 1)\bar{k}/(1 + V_k/\bar{k})$. *Formulae from Crow and Kimura (1970).*

	1974	1975	1976
Pond Ridge Pond (k)	0.014	0.470	0.292
Cline's Hacking Pond (k)	1.256	0.404	0
Dictum Ridge Pond (k)	0.858	0	0
Second Mountain Pond (k)	37.666	19.700	11.368
White Oak Flat Pond (k)	—	0.420	3.556
$\bar{k}$ (w/o WOF)	1.356	1.309	1.448
(w WOF)	—	0.685	3.161
V_k (w/o WOF)	31.475	18.470	13.044
(w WOF)	—	5.673	3.143
N_{ev} (w/o WOF)	14.568	33.541	47.013
(w WOF)	—	95.821	2,732.139
N_E (w/o WOF, 1974–1976)		25.057	
(w WOF, 1975–76)		185.149	

produced only 50 young, 233 adults in CH produced only 62 young, and 61 adults in DR produced only nine young in total. Correction for adult longevity and other aspects of age-specific demography (Gill, 1978) confirmed the conclusion that breeding populations PR, CH, and DR were not replacing themselves autogenically.

With the data of total annual juvenile output from each pond, the critical parameters for the calculation of variance effective population sizes (N_{ev}) were available. In Table 4, the $\bar{k}$ value is the average gametic production per effective adult (N_o from Table 1) per pond, and V_k is variance in per capita reproduction attributable to among-pond variation. Assuming diploidy, I computed gametic production as twice the juvenile counts. Because WOF was not censused in 1974, the values of $\bar{k}$, V_k, and N_{ev} for the years 1975 and 1976 were calculated with and without the data from WOF. This allowed direct comparison of PR, CH, DR, and SM for the years 1974 through 1976.

The value of N_{ev} expresses reproduction in terms of the number of adults of equivalent effectiveness in contribution to the next generation. There was an enormous depression in regional effective population size from the total census count due to the large among-pond variation in reproductive output (Table 4). For example, in 1974 the contrast between the 299 censused adults and the N_{ev} of 14.6 was the result of most of that year's reproduction being concentrated in SM. The steady increase in N_{ev} from 1974–1976 (when WOF is excluded) was attributable to the growing population and consistently high per capita reproduction in SM.

Comparing 1976 with 1975, a staggering tenfold increase in reproductive output occurred in WOF. Nevertheless, on a per capita basis, the average female in WOF in 1976 produced just enough juveniles for adult replacement barring any significant future mortality during the eft stage. Despite the fact that WOF produced over 91% of all juveniles that year, the reproductive success of the average female in WOF was still one-fourth that of her counterpart in SM. In other words, the individuals in SM consistently had the highest reproductive fitness all three years.

The enormous reproductive output from WOF in 1976 led to an anomalous calculation of N_{ev} when all five ponds are considered. The variance in gametic contribution among all the adults in the region was exceedingly small because the impact of WOF swamped out the contributions from the other ponds. Not only did WOF produce virtually all the young in 1976, but also it contained the vast majority of breeding adults. Without an estimate of within-pond reproduction variance, my estimate of total variance in 1976 suggested a per capita reproductive rate that was more uniform than random (V_k less than $\bar{k}$). This situation is similar to the textbook example of effective population size being greater than the census value (Li, 1975, p. 562; Crow and Kimura, 1970, p. 358). A biological interpretation of this anomalous result is needed.

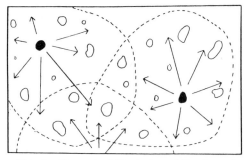

FIG. 1. Conceptual illustration of the dynamics of newt metapopulations. All demes contain active adult populations, but only one or two are highly reproductive. The juveniles disperse (arrows) from the reproductive centers and colonize other demes. The location of the reproductive center shifts with time.

DISCUSSION

The Metapopulation Model

The metapopulation of the red-spotted newt in my study region is structured around the pattern of ponds in a way that subdivided the species into numerous, small, breeding populations (the concept of Wright, 1952). The faithful philopatry and homing behavior of adults appears to render the breeding structure very restrictive and not panmictic. However, the consistent lack of juvenile production from most ponds in most years suggests that such demes are not generally self-perpetuating; individuals simply pour in, attempt to breed, and die there as reproductive failures. The evolutionary importance of such breeding populations is nil—the adults within are not contributing substantially to future generations.

The metapopulation of the red-spotted newt is pictured as a constellation of subpopulations, most of which function as reproductive sinks, knit together by the juvenile dispersal emanating from a central, highly reproductive center (Fig. 1). As a consequence of a moderately rapid turnover of ponds themselves, the center of a metapopulation presumably shifts spatially through time as an old center declines in productivity and a new pond gains reproductive dominance (Gill, 1978).

It follows from the model of the newt metapopulation that the meaningful interpretation of the phrase "migration rates" in a population genetic sense is the rate of exchange of juveniles between two metapopulation centers. The regular recruitment rates at ponds is primarily the regional flux of juvenile dispersal within the metapopulation. However, every pond presumably lies in the intersection of two or more metapopulations, and the new recruits arriving at the pond would derive from many sources. The genetic contributions from the various sources would presumably be inversely weighted according to the distance from the sources. Thus, the estimates of migration rates obtained from most ponds within a metapopulation are necessarily greater than the true genetic migration rates between two reproductive centers. In this light it is critical to discover the proportion of juveniles that derive from the same pond of origin as well as the extent of long-distance immigration from other ponds. These data will be available as soon as the marked 1974, 1975 and 1976 cohorts of juveniles from all my ponds return as breeding adults.

My model of the dynamics of newt metapopulations contrasts with models from classical population genetics (Wright, 1931, 1938, 1940, 1943, 1952) and the more recent ones of Levins (1970) and Boorman and Levitt (1973). All the previous models assumed a large degree of demic autogenicity and did not consider the possibility of a metapopulation consisting of a large number of reproductive sinks. Whereas Wright's models allowed varying degrees of migration among subpopulations, the models of Levins (1970) and Boorman and Levitt (1973) restricted attention to particular cases where migration rates are substantially smaller than intrademic dynamics. Levins (1970) considered the dynamic situation of a species colonizing a few islands in a large archipelago and most colonies going extinct sometime later. Boorman and Levitt (1973) discussed the genetic consequences of high extinction and turnover rates in small fragmentations at the periphery of a large, panmictic population. In both models demic extinction rate played the controlling dynamic role.

There is an important distinction between outright extinction of local populations and the persistence of reproductive sinks in a large metapopulation. Most discussions concerned with the evolutionary importance of interdemic selection conceive of it in terms of wholesale extinctions of those demes with unstable properties (Cook, 1961; Boorman and Levitt, 1973; Levins and MacArthur, 1966; MacArthur, 1965; Lewontin, 1962; Lewontin and Dunn, 1960; Wright, 1948). Extinction implies the final stage of the population cycle, which consists of birth by a founder propagule, logistic growth, senescence, and death. By definition, demic turnover would occur at the rate population cycles run their course and begin anew. In contrast, demes that persist because of high immigration may abruptly change from states of no reproduction to reproductive centers. In this way individuals are not totally eliminated from reproductive potential as they would be when a deme goes extinct. This point is illustrated by the tenfold change in reproductive output that occurred in WOF between 1975 and 1976. The fact that nearly 50% (58.02% of the males and 34.77% of the females) of the adults in 1976 were the same as 1975 is potentially significant in light of the observation that WOF produced over 91% of all progeny from the study ponds in 1976 but only 43% in 1975. Thus, emphasis on differential extinction of demes has treated only the survival component of interdemal selection. Differential output of young by persistent subpopulations adds the reproductive component of interdemal selection.

Effective Population Size

Given the model of metapopulational structure and its dynamics, the genetic structure of newt metapopulations should depend upon the effective breeding population size of a metapopulation. On the basis of the data presented, the effective breeding population size for this metapopulation was tiny and well within the range in which genetic drift could be important. For example, there is no way the 12 alleles per locus (assuming diploidy) in the six breeding adults in SM in 1974 could have fairly represented the true genetic variation in the region as a whole. Although the sizes of the breeding populations in most ponds were irrelevant because those demes were reproductive failures, the size of SM was particularly important because it was the putative reproductive center of the group of ponds that year.

The astonishingly low estimates of annual and overall effective population sizes are in fact overestimates of the actual effective sizes of the newt metapopulation for at least two reasons, and may be underestimates for one reason at least. The first reason is that the calculations of reproductive variance do not take into account any variance in reproductive success among individuals within the metapopulation center. Several features of the breeding populations, including age structure (Gill, 1978; Wright, 1938) territoriality (Bennett, 1970; Bellis, 1968), variability in goodness of oviposition sites, etc., could realistically engender high variance within a deme. Merrell (1968) has shown through field studies that these kinds of factors modestly reduced the effective population size of breeding leopard frogs, *Rana pipiens,* from census estimates. The second reason is that a certain level of inbreeding may occur as a consequence of siblings colonizing the same future metapopulation center. Inbreeding would depress inbreeding effective population size (N_i, Crow and Kimura, 1970) from census values, and would in turn reduce the estimates of variance effective size.

My calculated value of effective population size may be an underestimate because the formulae used assume discrete, non-overlapping generations. Sibling cohorts of efts may reach maturity at several ages, between 3–8 in western Massachusetts (Healy, 1973, 1974). Despite the fact that adults have short mean life expectancies, some individuals may breed several times in their lives. However, on the basis of their preliminary theoretical calculations, Hill (1972), Felsenstein (1969, 1971), and Sewall Wright (pers. comm.) suggest that the error in the estimates using discrete generations is probably small. In the absence of specific data, the reasons for overestimating effective size appear to outweigh the reason for underestimating it.

The frequency distribution of individual newt reproductive rates was not Gaussian (with a mean 1.0 for a stationary dynamics), nor was it even Poisson distributed as assumed by Crow (1964), Crow and Mortin (1955), and Boorman and Levitt (1973). When all the breeding females in the region were counted, the shape resembled a steep negative exponential (Fig. 2). For most years the among-pond differentials in reproductive success resulted in the variance in progeny per female being substantially greater than the mean.

One would expect low levels of heterozygosity in a species with the metapopulation structure described in this newt model. Allelic variation would be eroded at a high rate each generation because of large scale reproductive failure across ponds. Moreover, with effective population sizes as small (or variable) as I calculate, the rate of loss of heterozygotes by random walk processes would also be very

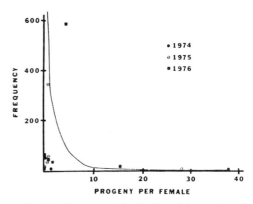

FIG. 2. Frequency distribution of the number of progeny per breeding female per year. A negative exponential frequency distribution ($\theta = 2.63$) based on the data for all three years is portrayed.

high (Wright, 1931; Li, 1975; Crow and Kimura, 1970). On the other hand, loss of heterozygosity would be counterbalanced by several factors, including migration between metapopulation centers, and differential selection during development. Maintenance of high allelic variation from migration could be important only to the degree that two or more metapopulational centers had differentiated genetically by the action of either drift or selection. Genetic polymorphisms would also be promoted by alternative selective pressures that probably occur during the three different life stages; certainly the environments in which the larval, eft, and adult stages occur are strikingly different.

A second counterbalance to loss of genetic variability available to the red-spotted newt is contained in the mating system of adults. Female newts mate as many as 20 to 30 times, laying only 6–10 eggs per oviposition bout, in a breeding season (Bishop, 1941; Gage, 1891). These matings appeared to be random with respect to the male population in a pond despite the reported male territoriality (personal observation). Contrasting sharply with the single fertilization mating systems of other pond-breeding amphibians, this mating behavior would promote high levels of genetic diversity among progeny because of recombination among the available adult genotypes.

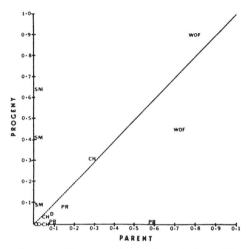

FIG. 3. The proportion of the region's total juveniles produced each year per pond plotted against the proportion of the region's total adults. The 45° line = per capita replacement. Points above the line = increase in genetic representation in a future generation; points below the line = quasi-selective disadvantage.

Genetic Drift vs. Interdemal Selection

The evolutionary impact of variation in reproductive success among ponds can be viewed in several contrasting ways. Superficially, extreme among-pond variation suggests that the genetic contributions of individuals to future generations is deme dependent. Indeed, year-to-year consistency of the variation tends to validate the use of the phrase "interdemal selection" in this case (Fig. 3). This characterization would be true either for total reproduction or for per capita reproduction, whether or not both values coincide in the same pond (which they did in 1974 and 1975, but not in 1976).

Although those adults in the right pond in the right year are indeed the progenitors of virtually all the next generation, this position seems to be the consequence of stochastic processes. It is inconceivable that supergenotypes have the ability to evaluate the reproductive potential of a set of ponds and selectively choose the best pond. The controlling factor of reproductive success appeared to be totally unrelated to adult breeding competence but rather was attributable to differential survival of larvae (Gill, 1978). Since newts do not brood their young or in any other way exhibit post-embryonic parental care (in fact they cannibalize larval newts), adult newts have no way of assessing their own reproductive success. From an adult newt's point of view, all ponds are more or less equivalent so long as ample water, food, and mates are available. Thus, it would be improper and misleading to label as "interdemal selection" the persistent reproductive differentials among ponds, because they appear to be stochastically independent of genetic constitution. Abrupt fluctuations in genetic structure of metapopulations are likely to result from large interdemal variation in reproductive output (Hanson, 1966).

In spite of the potential importance of drift, particularly as a result of the very small effective sizes measured in this system, Gill (1978) offered solutions to two problems using arguments of natural selection. The first problem was why do adult newts exhibit homing behavior at consistently poor reproductive ponds. The second problem was why do adults produce dispersing juveniles (efts) at high reproductive ponds. To answer the first, given continuously high reproductive potential at the central pond for a few years, and given heritable variation in behavioral tendencies to wander, natural selection will favor those adults at the center pond that faithfully return to the same pond year after year because they will have capital reproductive success. Those adults that tend to wander or disperse to other ponds between breeding seasons are likely to arrive at a pond with low reproductive potential. This argument is similar to the mechanism of selection for migration modification discussed by Balkau and Feldman (1973) and Van Valen (1971). Thus, selection favors individual pond recognition and preferential return by adults each year, in effect homing behavior. The homing behavior of the adults at the poor reproductive ponds is manifest because they are the progeny of the metapopulational center deme. Having experienced at least one bad reproductive season at a bad pond, such adults can never match the future reproduction of the adults which initially colonized a good pond.

For the second problem, Gill (1978) suggested that selection would favor a dispersal phase in the life cycle if the pond habitats themselves had finite duration. If former metapopulation centers sharply declined in quality or disappeared altogether, as beaver ponds do, there would be an absolute selective premium on juvenile dispersal to other ponds, especially the next center pond. Gill (1978) has shown that probability of survival and per capita reproduction was highest in new, low density ponds, and argued that these features added additional selective advantages to colonizing new, young ponds. These hypotheses are currently being tested by field transplantation experiments.

Summary

An empirical study of the dynamics of a metapopulation of the red-spotted newt, *Notophthalmus viridescens,* was undertaken in order to evaluate actual magnitudes of effective population size and interpopulational migration rates. The newt in mountain ponds of Virginia was chosen because of the precision possible in determining the number of breeding adults and their progeny, in monitoring the activities of known individuals, and in measuring the extent of migration between breeding ponds. The data reported here cover the years 1974–1976.

It was found that adult newts faithfully returned to the same pond every year and never migrated to other ponds. Censuses of adults revealed great variation in breeding population size among ponds, but little variation between years. Preliminary calculations of effective size were modestly reduced from census values because of the 2:1 (male:female) sex ratios and annual fluctuations in breeding numbers. Most ponds proved to be non-reproductive most years, so that among-pond variation in per capita reproduction was very large. Regional effective population size was only about 25 individuals for three years, a tiny number relative to the hundreds of breeding adults present in all ponds.

A model of the dynamics of the newt metapopulation is presented. The evolutionary consequences of the model, especially the relative importance of genetic drift versus interdemal selection and intrademal natural selection are discussed.

Acknowledgments

The research reported here was funded by NSF Grants #BMS 74-19664 and #DEB 76-20326. Numerous individuals assisted or contributed to the work and I am grateful to all, in particular, Mr. Keith Berven. Mr. Christopher Molineaux and Mr. Douglas Darling faithfully assisted in both the field and laboratory aspects of the newt processing. Thanks are due to the staff of the National Forest Service for their permission, cooperation, and encouragement to pursue this project in the George Washington National Forest.

The sharpening of the ideas about newt dynamics benefited greatly from discussions with Keith Berven, Douglas Fraser, Nelson Hairston, William Healy, Robert Jaeger, Robert Merritt, Steve Tilley, and Henry Wilbur. Constructively critical feedback came from the evolutionary ecology group at the University of Maryland, especially from David Allan, Richard Highton, Douglass Morse, and Geerat Vermeij.

Literature Cited

Balkau, B. J., and M. W. Feldman. 1973. Selection for migration modification. Genetics 74:171–174.

Bellis, E. D. 1968. Summer movement of red-spotted newts in a small pond. Jour. Herp. 1:86–91.

Bennett, S. 1970. Homing, density and population dynamics in the adult newt, *Notophthalmus viridescens.* Ph. D. Dissertation, Dartmouth College, University Microfilms, Ann Arbor, Michigan.

Bishop, S. C. 1941. The salamanders of New York. New York State Museum Bulletin 324:54–81.

Boorman, S. A., and P. R. Levitt. 1973. Group selection on the boundary of a stable population. Theor. Pop. Biol. 4:85–128.

COOK, L. M. 1961. The edge effect in population genetics. Amer. Natur. 95:295–308.
CROW, J. F. 1964. Breeding structure of populations. II. Effective Population Number, p. 543–556. *In* Kempthorne, Bancroft, Gowen, and Lush (eds.), Statistics and Mathematics in Biology. Hafner Publ. Co., New York.
CROW, J. F., AND M. KIMURA. 1970. An Introduction to Population Genetics Theory. Harper and Row, Publishers, New York.
CROW, J. F., AND N. E. MORTON. 1955. Measurement of gene frequency drift in small populations. Evolution 9:202–214.
EHRLICH, P. R., R. R. WHITE, M. C. SINGER, S. W. MCKECHNIE, AND L. E. GILBERT. 1975. Checkerspot butterflies: A historical perspective. Science 188:221–228.
FELENSTEIN, J. 1969. The effective size of a population with overlapping generations. Genetics 61:s19.
———. 1971. Inbreeding and variance effective numbers in populations with overlapping generations. Genetics 68:581–597.
GAGE, S. H. 1891. Life history of the vermilion-spotted newt (*Diemyctylus viridescens* Raf.) Amer. Natur. 25:1084–1110.
GILL, D. E. 1978. Dynamics of a metapopulation of the red-spotted newt. Ecol. Monogr. *In Press*.
HANSON, W. D. 1966. Effects of partial isolation by distance, migration, and different fitness requirements among environmental pockets upon steady state gene frequency. Biometrics. 22:453–468.
HEALY, W. R. 1973. Life history variation and the growth of juvenile *Notophthalmus viridescens* from Massachusetts. Copeia 1973(4):641–647.
———. 1974. Population consequences of alternative life histories in *Notophthalmus viridescens*. Copeia 1974(1):221–229.
HILL, W. G. 1972. Effective size of populations with overlapping generations. Theor. Pop. Biol. 3:278–289.
LEVINS, R. 1970. Extinction. *In* Some Mathematical Questions in Biology, Lecture on Mathematics in the Life Sciences, Vol. 2. American Mathematical Society, Providence, R.I.
LEVINS, R., AND R. H. MACARTHUR. 1966. The maintenance of genetic polymorphism in a spatially heterogeneous environment: Variations on a Theme by Howard Levene. Amer. Natur. 100:585–589.
LEWONTIN, R. C. 1962. Interdeme selection controlling a polymorphism in the house mouse. Amer. Natur. 96:65–78.
LEWONTIN, R. C., AND L. C. DUNN. 1960. The evolutionary dynamics of a polymorphism in the house mouse. Genetics 45:705–722.
LI, C. E. 1975. First Course in Population Genetics. The Boxwood Press, Pacific Grove, CA.
MACARTHUR, R. H. 1965. Ecological consequences of natural selection, p. 388–397. *In* T. H. Waterman and H. J. Morowitz (eds.), Theoretical and Mathematical Biology. Blaisdell Publishing Co., New York.
MERRELL, D. J. 1968. A comparison of the estimated size and the "effective size" of breeding populations of the leopard frog, *Rana pipiens*. Evolution 22:274–283.
VAN VALEN, L. 1971. Group selection and the evolution of dispersal. Evolution 25:591–598.
WRIGHT, S. 1931. Evolution in Mendelian populations. Genetics 16:97–159.
———. 1938. Size of population and breeding structure in relation to evolution. Science 87:430–431.
———. 1940. Breeding structure of populations in relation to speciation. Amer. Natur. 74:232–248.
———. 1943. Isolation by distance. Genetics 28:114–138.
———. 1948. On the roles of directed and random changes in gene frequency in the genetics of populations. Evolution 2:279–294.
———. 1952. The theoretical variance within and among subdivisions of a population that is in a steady state. Genetics 37:313–321.

29

Copyright ©1977 by Macmillan Journals Ltd.

Reprinted from *Nature* **269**:578-581 (1977)

Dispersal in stable habitats

W. D. Hamilton
Imperial College Field Station, Silwood Park, Ascot, Berkshire, UK

Robert M. May
Department of Biology, Princeton University, Princeton, New Jersey 08540

Simple mathematical models show that adaptations for achieving dispersal retain great importance even in uniform and predictable environments. A parent organism is expected to try to enter a high fraction of its propagules into competition for sites away from its own immediate locality even when mortality to such dispersing propagules is extremely high. The models incidentally provide a case where the evolutionarily stable dispersal strategy for individuals is suboptimal for the population as a whole.

IN nature, adaptations for dispersal are ubiquitous and are often applied in almost suicidal ventures or in ventures which seem too feeble to be worthwhile (such as dehiscent seeds that fall only a few feet from the parent plant). This behaviour is clearly advantageous if the habitat is unstable or offers many empty, if transient, patches. We discuss here some simple models that help to explain why such dispersal is also advantageous even in stable and saturated habitats.

To begin, we consider a wholly parthenogenetic species in an environment that provides a fixed number of sites, at each of which just one adult can live. In a fixed season at the end of its life each adult produces a certain constant number of offspring, m. A fraction v of these are programmed to be migrants (for example, insects provided with wings and appropriate instincts, seeds with a pappus for wind dispersal). The remaining fraction $(1-v)$ are destined to be sedentary competitors for the home site. The mother's genotype, by means of some maternal influence on each ovum (or testa or fruit), determines the fraction v. When ready, at about the time of the mother's death, the migrant offspring take off and after mixing with all other migrants, and suffering a mortality such that only a fraction p survive, they are distributed equally to all the sites. There they compete on equal terms with the resident young which, it is supposed, up to this stage suffer no mortality (alternatively, if they do, p is the relative survival of the migrants). In effect, one offspring is chosen at random from among the young present on a site to become the adult at that site in the new generation.

We acknowledge that this simple model probably has few close parallels in the real world. Nevertheless it may usefully force a re-examination of some widely held ideas about migration.

For example, it has been claimed as "intuitively obvious" that in a saturated and time-invariant environment "organisms can never gain any advantage by changing their locations"[1]. The schematic illustration presents a counter-example based on the above model

		X			X			
	o	o	o	o	o	o	o	o
Offspring	o	o	o	o	o	o	o	o
(after dispersion)	o	o	o	o	o	o	o	o
	o	o	o	o	o	o	o	o
	o	o	o	X	o	o	o	o
Adults	O	O	O	X	O	O	O	O
Site labels	a	b	c	d	e	f	g	h

The environment (represented by the eight sites labelled a to h) here is saturated and time-invariant. In the absence of any migrating mutant like X, and with full survival of stay-at-home propagules, the majority genotype O which keeps all propagules at its own site can perfectly maintain the population. But the 'O' strategy is not evolutionarily stable. The genotype O will be replaced by the mutant X, which keeps only one propagule at home, even though it loses two of its remaining four propagules due to mortality in migration (assumed to be 50%; $p = 0.5$): obviously X has a chance of 1/6 of winning each of sites c and f; meanwhile, at least against so ill-advised a genotype as O, it certainly retains its base at d. Hence X is certain to become the established type. Of course the particular migration probability of X ($v = 4/5$) illustrated here will not itself prove to be the evolutionary stable strategy, or "ess"[2], except for some special value of the survival factor p. Normally other mutations would supervene, after the spread of X, until finally a migration probability that was evolutionarily stable was approximated.

In general, the one or more ess migration probabilities which the model might have can be determined as follows. The population is imagined to contain two types, using migration fractions v and v'. An expression is written for the fitness, w', of one adult of type v'. This fitness, or expected number of sites to be gained by offspring, will consist of the chance of retaining the home site plus the expectation of sites to be gained elsewhere by migrant offspring. From this we find the value of v which has the property that $w' \geq 1$ for all $v' \neq v$ (mean fitness in the model is unity, so that $w = 1$). The value of v so found, symbolised v^*, is unbeatable[3] in the sense that any genotype with strategy $v' \neq v$ will have a diminishing frequency in any mixture.

For the simple limiting case where the number of sites and the number of propagules per parent are large, we find by this method that the ess or unbeatable migration probability is

$$v^* = 1/(2-p) \qquad (1)$$

Because this formula shows no dependence on the composition of the mixture, stable mixtures are not possible, and the ess can be considered safe against both rare mutations and any massive invasion by a different genotype.

A striking conclusion to be drawn from equation (1) is that even when migrant mortality is extremely high (small p), and the environment offers no vacant sites for colonisation, it is still advantageous to commit slightly more than half of the offspring to migration.

Pre-eminent among the artificialities of this model are: (1) insistence on death and replacement of every parent in each generation; (2) absence of vacant sites (stemming from the deterministic description of the propagation processes); (3) pure parthenogenesis. We now indicate how the model may be extended to encompass such effects, paying particular attention to (2).

Death and replacement of parents

There is one simple and often realistic assumption that allows for perennation of the parents, while preserving the result (1). The assumption is that all parents, irrespective of age, have

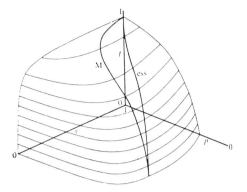

Fig. 1 The surface depicted here gives f (the equilibrium fraction of sites occupied) as a function of v (the fraction of offspring that migrate) and p (the relative survival probability for migrants), under the assumptions delineated in the text, for a mean 'brood size' $m = 3$. The contour lines are for constant values of f, spaced at intervals 0.1 apart (that is $\Delta f = 0.1$). On this surface, the trajectory labelled ess shows the dispersal strategy that is evolutionarily stable, $v^*(p)$, as a function of p. The ess trajectory is to be contrasted with that labelled M, which shows the strategy which maximises site occupancy for given p, $\hat{v}(p)$.

some constant probability q to survive and retain their sites into the next season. This corresponds to the so-called 'Type II survivorship curve', which is known to be not far from true for many organisms. For such organisms, both stay-at-home and migrant offspring have their expectations of inheriting a vacant site devalued by exactly the same factor, namely $1-q$, which consequently cancels itself out of the analysis. (On the other hand, for organisms whose chance of surviving from one year to the next diminishes with age, the ess will be age-dependent, if this is biologically feasible. A likely outcome is a 'bang-bang' strategy[1], with all propagules dispersed until their parent reaches a certain age, whereupon all propagules stay at home.)

Vacant sites

In our simple model, the evolutionary pressure that commits at least half the offspring to migration, no matter how risky migration is, results from an advantage in arranging competitive interactions as far as possible with unlike genotypes. It can be imagined that in nature such a pressure to take risks could cause a wastage that was damaging to the population's chance of survival. This cannot happen in the simple model because its deterministic assumptions keep all sites occupied; to investigate wastage we need some version that relaxes this assumption (2). Such a version may have some bearing on our confidence in models of theoretical ecology in general, because these often take it for granted that the species discussed already have, or will evolve towards, maximal efficiency of resource utilisation.

To this end, we stochasticise the earlier model by assuming the number of propagules produced at any given site is given by a Poisson distribution, with mean equal to the previously constant value m. It follows that the numbers for emigrant and for stay-at-home offspring, and for immigrants, are all generated by Poisson distributions for which the same symbols and expressions as occurred in the simple model are now to be taken as the means. Note that m is the mean brood size and also the variance in brood size, so that the relative magnitude of statistical fluctuations goes as $m^{-\frac{1}{2}}$. The parameter m thus comes to have an important influence, especially when relatively small; when m is large, the present stochastic model tends to revert to the earlier deterministic one.

Now sites can become vacant. Vacancies begin or continue whenever a site happens to have no surviving stay-at-home offspring and receives no immigrants. Obviously, some migration is essential if extinction is to be avoided, and from the population point of view there will be some level of migration that keeps extinction at furthest reach[5].

For any specified combination of m, v and p, the system settles to some stable level of site occupancy f (that is $f =$ fraction of sites occupied), with a zero level signifying that extinction is inevitable. This stable level f is given (for $f > 0$) by the implicit relation

$$f/(1-f) = e^{m(1-v)}\,[e^{mvpf}-1] \qquad (2)$$

For given m, the site occupancy function f defined by equation (2) may be mapped over the unit square of admissable values of v and p: $f(v,p)$ is found to have the form of a smooth promontory (Fig. 1 for $m = 3$), or, for larger m, a squarish headland (Fig. 2 for $m = 10$). In any such figure, for fixed v the occupancy $f(p)$ falls convexly with decreasing p, and clearly must always hit extinction somewhere short of $p = 0$. The smallest p that still allows the population to exist may be shown to occur when $v = 1/m$, that is when just one migrant is dispatched from each site. The concomitant least value of p is given by $p_{min} = \exp(1-m)$, which is very small for moderate values of m; slightly larger values of p carry f up to a 'plateau' at a level of nearly complete occupancy. Conversely, for fixed p, the occupancy f as a function of v at first rises (albeit very slightly for large m) to a maximum as v decreases; then, as v decreases further, this is followed by a steepening fall in f, which goes to zero short of $v = 0$.

Thus we may trace out the locus of the strategy, v, that maximises the site occupancy f for given p; this locus moves up a 'cliff', and across a 'plateau' to $p = v = 1$ (Figs 1 and 2). This strategy $\hat{v}$ is that which is 'best for the population'.

This locus of the strategy for maximal occupancy, $\hat{v}$, is to be contrasted with the locus of the ess v^*, which is determined by the method described above. The ess migration probability, $v = v^*$, is given by

$$\exp[m(u+vpf)] = \frac{v^2p - vpf\,(u+vpf)}{v^2p + u\,(u+vpf)(1-mv)} \qquad (3)$$

Here $u = 1-v$, and f has the value implicity fixed by equation (2). The derivation of the results (2) and (3) will be set out

Fig. 2 As for Fig. 1, except that now the mean brood size is $m = 10$. Again the solid contour lines depict constant values of f, from $f = 0$ to 0.9 at intervals of 0.1; the increasingly broken contour lines on the 'plateau' are for $f = 0.99$, 0.999 and 0.9999, respectively. For a more full discussion of Figs 1 and 2, see text.

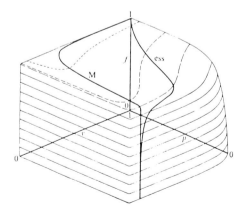

elsewhere (our work with H. N. Comins); unlike the earlier simple result (1), these equations are made more complicated if age-independent or 'Type II' perennation of the parents is introduced.

The locus of the ess migration fraction v^*, so determined, is shown in Figs 1 and 2. We note that ess migration is always higher than maximum-occupancy migration, although they converge at the two extremes (at perfect and at least-possible survival, $p = 1$ and $p = p_{min}$). This is true for all brood sizes m. As m is increased towards more usual biological values (from $m = 3$ in Fig. 1 to $m = 10$ in Fig. 2), the separation between the ess and maximum-occupancy loci with respect to v also increases strikingly. This separation results mainly from the way the changing shape of the site-occupancy surface $f(v,p)$ controls the course of $\hat{v}$; that is $\hat{v}(p)$ is much more affected by changing m than is $v^*(p)$. The separation between the two loci with respect to f, however, remains very small, even for populations that are on the 'cliff face' of the f surface and hence in some danger of extinction.

In other words, the ess v^* can demand far more migration than is 'best for the population' ($\hat{v}$), but such excess does not, in this model at least, substantially affect population safety nor create many extra vacant sites that would give other species opportunity to evolve as competitors for the niche. Notwithstanding the small difference in occupancy levels, there is usually strong selection to establish v^* rather than $\hat{v}$. One incidental consequence of the relatively abundant supply of migrants is that some must overflow from the system and occasionally have the luck to found new populations elsewhere. It therefore seems that our models show, after all, a case where a trait is positively adaptive at more than one level: the ess is best for the genotype, suboptimal at the level of an isolated population, and advantageous in carrying the population to new areas.

Sexual organisms

If the organisms are taken to be sexual rather than purely parthenogenetic, with migration probability depending on the genotype of the offspring, lower values of v^* are definitely expected.

This follows from the observation that our models directly treat the genetic success of a mother. A parthenogenetic offspring has an identical genotype, so that whatever maximises inclusive fitness[6] for the mother maximises it for the offspring. In contrast, the inclusive fitness of a sexual offspring normally differs from that of its mother, and, in some direct relation to the hazards of migration, will find its optimum at a lower migration rate. Hence there is a conflict of interest between parent and offspring[7,8]. Its outcome will depend on the biological circumstances: as regards applying a pappus to a seed to make it blow away it is clear that the parent has a strong position; as regards making wing buds become wings in an insect it is much more likely that the offspring can have its preference.

In nature, less migration generally means more inbreeding; such inbreeding reduces the genetic contrast between parent and offspring, and thus tends to diminish the difference between parthenogenetic models and sexual models with offspring-determined migration. Various situations can arise. If males mate before migration in a one-per-site situation, inbreeding is total and the model is again effectively as under parthenogenesis. Unfortunately, the small arthropods which almost meet this extreme of inbreeding and which have a dispersal polymorphism (for example, some mites of the genus *Pygmephorus* (ref. 9 and unpublished results of W.D.H.) are far from meeting the assumption of fixed unitary sites, being rather exploiters of patchy, ephemeral resources. If inbreeding occurs because males sometimes survive along with a sister, then our one-per-site assumption does not hold.

But if male offspring are assumed always to migrate, then outbred sex can be brought in and the one-per-site feature retained; moreover, this version may approach realism for some of the many insect species that have free-flying males while females are either wholly flightless or polymorphic for flight. For the simple, non-stochastic version of this model with allele semi-dominance, the ess migration probability for the female offspring is indeed lower than for the parthenogenetic model: in place of equation (1), we now have

$$v^* = 0 \qquad \text{for } \tfrac{1}{2} > p > 0$$
$$v^* = (2p-1)/(4p-1-2p^2) \quad \text{for } 1 > p > \tfrac{1}{2} \qquad (4)$$

The fact that $v^* = 0$ for $p < \tfrac{1}{2}$ suggests that when this model is made stochastic (along the lines followed previously) the ess may imply situations where the population and the species is dangerously liable to extinction. We are pursuing this open question.

(Roff[10] has used computer simulations to study migration in models which not only have sexual reproduction and multiple adults per site with a density-dependent migration structure, but also have temporal variability in site suitability. It is obvious from the facts of nature[11], and is confirmed in both Roff's model and several others[1,12–14], that erratic habitat suitability (including local extinction) is an extremely important factor in favouring adaptions for dispersal and migration. More relevant to the present discussion, however, is Roff's demonstration that his ess migration probabilities are considerably lower than those which would keep populations highest and most secure from extinction. Although partly attributable to the workings of sexuality and Mendelian inheritance in Roff's models, this effect arises primarily because there are many adults per site, which changes the calculation of inclusive fitness in such a way as to favour the short term, selfish option of not dispersing.)

It would be satisfying to conclude by listing a few biological illustrations of principles suggested by the models. Unfortunately most of the examples which have come to mind are more in the region of jokes than of reality: they range from *Caulobacter crescentus*[15], whose asexual fission always produces one stalked sessile cell and one motile one that swims away, through the inevitable lemmings and on to the non-first sons in Victorian families who joined the army or went to Australia. Both our simplest model, and the refined versions (1) and (2), support what these examples might illustrate, namely a minimum rule that 'at least one must migrate whatever the odds'. All three examples, however, fail to have fixed sites and, except for *Caulobacter*, fail also in regard to asexuality. There are flight dimorphisms in non-sexual insects (for example, aphids in summer: the tiny sub-cortical beetle *Ptinella errabunda*[16]) and probably parallels in some weeds (for example, Compositae)[17,18], but in being colonist species these fit badly with the assumptions of fixed sites and of no extinction or recreation of habitats on a large scale. But the fact that colonist plants in arid environments, where perennation of the whole plant (as opposed to survival of its seed) may be hardest to achieve, seem especially inclined to produce actually dimorphic or differently shed propagules with one class buried or otherwise retained beside or under the parent[19] (*Sieglingia decumbens* is the best example in Britain) does at least suggest that a plant's perennating body (or its tubers, corms or the like) should be considered its bid to retain the home site. We expect the investment in such aids to perennation, as opposed to investment in flowers and dispersing seed, to show positive correlation with the chance that seed if produced would end up destroyed or in sites that never could support an adult. For example, the more successful a specialised epiphyte or parasite is at getting its propagules to suitable sites, even when these are objects of intense competition, the more it should produce them. A mistletoe should flower more than an epiphytic orchid; and we should not be surprised to find that weedy perennial species put more net assimilation into seed production than do climax perennial species[20]. This may be so but is not particularly exciting; other explanations are available. It is not easy to test the quantitative conclusion that, because $v^* > \frac{1}{2}$ for most values of p, we expect to find at least one-half of net assimilation going into some form of growth or propagule production that extends competition beyond the space that the organism already occupies.

The value of our models lies primarily in their approach and method, in their well-determined conclusions, and in the novelty of having made plain a new reason for the ubiquity of dispersal. We have shown that the habitat does not have to be patchy and of erratic suitability; substantial dispersal is to be expected even when the habitat is uniform, constant, and occupied completely.

We thank H. N. Comins, H. S. Horn, T. R. E. Southwood and D. M. Waller for helpful comments. This work was supported in part by the NSF.

Received 9 June; accepted 5 July 1977.

[1] Cohen, D. *Am. Nat.* **101**, 5 (1967).
[2] Maynard Smith, J. & Price, G. R. *Nature* **246**, 15 (1973).
[3] Hamilton, W. D. *Science* **156**, 477 (1967).
[4] Macevicz, S. & Oster, G. F. *Behav. Ecol. Sociobiol.* **1**, 265 (1976).
[5] May, R. M. *Stability and Complexity in Model Ecosystems*, Second ed. (Princeton University, Princeton, 1974).
[6] Hamilton, W. D. *J. theor. Biol.* **7**, 1 (1964).
[7] Trivers, R. L. *Am. Zool.* **14**, 249 (1974).
[8] Hamilton, W. D. *J. theor. Biol.* **7**, 17 (1964).
[9] Moser, J. C. & Cross, E. A. *Ann. ent. Soc. Am.* **68**, 820 (1975).
[10] Roff, D. A. *Oecologia, Berl.* **19**, 217 (1975).
[11] Southwood, T. R. E. *Biol. Rev.* **37**, 171 (1962).
[12] Gadgil, M. *Ecology* **52**, 253 (1971).
[13] Van Valen, L. *Evolution* **25**, 5 (1971).
[14] Levins, R. in *Some Mathematical Questions in Biology* 2 (American Mathematical Society, Providence, Rhode Island, 1970).
[15] Kurn, N. & Shapiro, L. *Proc. natn. Acad. Sci. U.S.A.* **73**, 3303 (1976).
[16] Taylor, V. A. thesis, Univ. London, (1975).
[17] Harper, J. L., Lovell, P. H. & Moore, K. G. *A. Rev. Ecol. Syst.* **1**, 327 (1970).
[18] Stebbins, G. L. *A. Rev. Ecol. Syst.* **2**, 237 (1971).
[19] Zohary, M. *Plant Life of Palestine, Israel and Jordan* (Ronald, New York, 1962).
[20] Harper, J. L. & White, J. A. *Rev. Ecol. Syst.* **5**, 419 (1974).

EMIGRATION AS A POSSIBLE MECHANISM PERMITTING THE REGULATION OF POPULATION DENSITY BELOW CARRYING CAPACITY

WILLIAM Z. LIDICKER, JR.

The possibility that populations might at times regulate their densities below a level which would be dictated by the carrying capacity of their environments is a particularly controversial yet theoretically important question for the ecologist and population biologist. It is the purpose of this note to explore the possibility that emigration may under certain circumstances represent such a regulatory mechanism.

Ordinarily emigration is found to play an important secondary role in population regulation by acting as a safety valve which helps to adjust population density to some limiting resource or condition. In this capacity emigration acts as one of the regulatory processes, but is in a sense only a by-product of the regulatory mechanism, not a primary cause. In general, dispersal movements of this sort appear to be short-range (see discussions of Howard, 1960, and Johnston, 1961), but in any case this question is only of peripheral concern to the present thesis. Of more interest here is the possibility that a tendency to emigrate may be selected for in a given species and eventually come to play a primary role, even if only an uncommon or occasional one, in population regulation.

There seem to be at least three reasons why a tendency to emigrate might have adaptive advantages beyond the obvious benefits which may be bestowed on a species by dispersal alone. First of all, and probably of least importance, individuals who tend to wander between population subunits (demes) in a typical non-uniform environment would be more likely to come into contact with more individuals, and possibly therefore to mate more often. In this way the wanderer could have a quantitative advantage in that he would not only spread his genetic material more widely, but there would be more opportunities for advantageous recombinations of genetic material. The possibility of this occurring successfully would, of course, depend on the immediate circumstances as well as the particular species involved. Secondly, interdeme matings would tend to increase the probabilities of crosses between individuals which were more different genetically, and thus not only increase the chances that the resulting offspring will exhibit greater over-all fitness because of their relatively greater heterozygosity (Lerner, 1954), but also compound the likelihood of new and desirable genetic recombinations being developed (a qualitative advantage). And finally, individuals who managed to avoid getting involved in devastating population crashes (whatever their cause) by moving out of potentially congested places might have a better chance of survival than others who stayed put (a diplomatic advantage). That this last is a definite possibility has been supported by the suggestion of F. C. Evans (1942) that long-term survival in a population of the vole *Clethrionomys glareolus* may depend on the utilization of marginal habitats by a few individuals who can thus survive population crashes.

It seems apparent, then, that there are several potential advantages to vagrancy, and furthermore I think it need hardly be added that selection for emigratory behavior can proceed for reason of any one of these adaptive advantages. Other things being equal then, we may tentatively conclude that a vagrant individual may be more likely to pass on its genetic material to future generations than would a sedentary individual. Obviously, these advantages must be balanced in every case against the possible decreased chances of survival of the emigrant. This argument requires that emigratory tendencies can indeed be controlled genetically. That this is at least possible has been shown by Narise, Sakai, and Iyama (1960) for Drosophila. Also, some evidence for the genetic control of dispersal in vertebrates is reviewed by Howard (1960).

A brief word should perhaps be added concerning the effects of exchanges of individuals between demes and the possible influence of interdeme genetic homeostasis on this. Experimental evidence of genetic homeostasis in populations such as that presented by Lerner (1954) and Buzzati-Traverso (1955) clearly leads to the conclusion that some demes may exhibit a certain degree of homeostasis tending to inhibit interdeme gene flow. However, the extent and circumstances of this phenomenon under natural conditions have not been established, and furthermore the type of deme dealt with here is probably of a lower level of differentiation such that this would not ordinarily be of concern in the problem under discussion. However, even exchanges between demes possessing a certain amount of homeostasis might be advantageous in the long-run by helping to inhibit inbreeding degeneration (Lerner, 1954), as well as over-specialization of either deme. Experiments on the guppy (Lebistes) by Haskins and Haskins (1954) have demonstrated that it is possible in natural populations for invading genotypes of an unrepresented type to integrate into small populations and produce permanent changes in their genetic constitution. If there is an exchange of individuals between demes instead of emigration in only one direction, it would seem that even greater advantages are actually possible. This is because it would probably be easier for an emigrant to integrate into an adjacent deme if some members of that deme had recently left. Thus, exchanges may ease slightly the often characteristic deliberateness of gene flow, by improving its chances of fulfillment.

Even if we can accept at this point that a tendency to emigrate might confer a long-term selective advantage on individuals possessing it, at least two unanswered questions remain: namely, are emigratory movements which occur in different phases of population growth equally likely to be successful, and what is the relation between emigration and possible density regulation below the carrying capacity of the environment? It seems clear that emigratory tendencies will be strongest under two general types of conditions. Under pressure of a declining carrying capacity (carrying capacity is used here in a general sense so as to include short-term fluctuations), a certain surplus of individuals will result which will be faced with the alternatives of dying or emigrating. Emigration will also occur under conditions of increasing density if the emigratory drive increases with density, that is, it is positively responsive to density. Moreover, it seems probable that the waves of emigrants produced under these two types of conditions will be qualitatively different. In the former case, the individuals would, I think, tend to be in poorer condition and on the average to represent the otherwise most poorly adapted individuals. This is because the most poorly adapted individuals would seem to be the ones most likely to feel first the pinch of an increasingly unfavorable environment. On the other hand, individuals moving out under the conditions of increasing population pressure would tend to be in better condition, as the environment would still be favorable, and would not necessarily represent the most poorly adapted component of the population. Rather they would represent that group of individuals which was most sensitive to increasing density (often young). It is therefore my contention that the advantages accruing to the interdeme emigrant would tend to favor the evolution of the responsive type of emigration tendency. This type of emigration has perhaps been demonstrated experimentally by Strecker (1954) for house mice (*Mus musculus*)[1], and Lidicker and Anderson (1962) have evidence that this may occur in natural populations of the California vole (*Microtus californicus*). Furthermore, it has been shown by Sakai, Narise, Hiraizumi, and Iyama (1958) that genetically controlled thresholds to density-stimulated emigration do exist, at least in *Drosophila melanogaster*.

If we can then assume that there are species in which density-responsive emigratory tendencies are adaptively advantageous and hence selected for, we could reasonably expect that in such species a genetic balance (possibly a polymorphism) would develop between the perfection of responsive emigration tendencies, with its attendant selective advantages, and the tendency to be sedentary and to stay safely at home, with all of its attendant selective advantages. A logical corollary of this would seem to be that under some conditions this balance would be at a point where the population was actually regulated by density-responsive emigratory tendencies below a level which would be dictated by the food supply or other potential limiting requisite. Obviously this would happen most frequently in localities, or at times, when the environment was very favorable and the carrying capacity therefore great. In other words, certain populations of some species may be prevented from taking full advantage of a temporarily very favorable carrying capacity because of a built in sensitivity to increasing density, this sensitivity having evolved because of its long-term selective advantages.

In situations where emigration is prevented, or is rendered inadequate when compared to the emigratory drive, population growth may actually be

[1] Emigration occurred before the food resource was completely utilized, and also at low densities.

controlled by those built in forces which have been responsible for producing this emigration drive but which would now be hyperdeveloped. For example, intraspecific aggressiveness may have an important function in stimulating emigration. If, however, emigration is frustrated, this same aggressiveness may cause mortality through fighting or deleterious increases in energy output through non-productive pathways. Circumstances such as these may help to explain the behavior of some captive growing populations which have been shown to regulate their densities in spite of an overabundant food supply. One could speculate further that this kind of density sensitivity may be most widespread among species which are otherwise not wide-ranging. This is because relatively sedentary species are the ones most likely to break up into semi-isolated demes and presumably would thus profit most from interdeme contacts. A strongly mosaic habitat would provide the same sort of partial fractionation.

This analysis carries the important implication that all organisms have not succeeded in making an optimal adaptation to their energy resources, and so do not utilize completely the food in all portions of their range. To me this is not surprising for the following three reasons: (1) Many organisms can exist under somewhat varied conditions over a large geographical range, and yet must have a certain amount of genetic homogeneity throughout. This homogeneity can often be accompanied by a genetic homeostasis which may resist selection in any one specific direction. (2) Habitats (at least terrestial ones) are often subject to considerable variations, not only short-term changes due to seasonal and climatic fluctuations but also long-term ones due to geologic changes. (3) Survival of all organisms depends on a successful adaptation to an ordinarily complex environment with diverse selective pressures, and it seems not unlikely that some important adaptations may at times conflict with a completely optimal adaptation to the food resource. This merely suggests that adaptational comprises are sometimes necessary and that a density-responsive emigratory tendency may be sufficiently important to a species that under certain conditions (especially conditions of abundant food or in areas of unusually great need for mobility) this adaptation could interfere with the species' adjustments to its resources so as to cause "premature" density regulation. Obviously considerably more sophisticated field investigations are required before the extent to which this mechanism actually operates under natural conditions can be determined.

SUMMARY

A tendency to emigrate can be shown to have selective advantages under certain conditions. These advantages appear to be best fulfilled by a density-responsive type of emigratory drive, that is, one in which a sensitivity to increasing density levels may release emigratory behavior before the carrying capacity of the population's environment is reached or exceeded. Where this is true, a genetic balance is postulated between a tendency to emigrate and a tendency to be sedentary. Such a balance could then result in "premature" density regulation by emigration under circumstances or at times in which the carrying capacity was unusually great or in situations where high intensity of emigration was particularly important. Moreover, it seems likely that if the density-responsive emigratory impulse happened to be frustrated by barriers or lack of a density gradient, density regulation may occur through the hyperdevelopment of those very factors which ordinarily bring about emigration. Some possible implications of this proposed mechanism to theories of population regulation are discussed.

LITERATURE CITED

Buzzati-Traverso, A. A., 1955, Evolutionary changes in components of fitness and other polygenic traits in *Drosophila melanogaster* populations. Heredity 9: 153-186.

Evans, F. C., 1942, Studies of a small mammal population in Bagley Wood, Berkshire. J. Animal Ecol. 11: 182-197.

Haskins, C. P., and E. F. Haskins, 1954, Note on a "permanent" experimental alteration of genetic constitution in a natural population. Proc. Natl. Acad. Sci. U.S. 40: 627-635.

Howard, W. E., 1960, Innate and environmental dispersal of individual vertebrates. Am. Midland Naturalist 63: 152-161.

Johnston, R. F., 1961, Population movements of birds. Condor 63: 386-389.

Lerner, I. M., 1954, Genetic homeostasis. Oliver & Boyd, London.

Lidicker, W. Z., Jr., and P. K. Anderson, 1962, Colonization of an island by *Microtus californicus*, analyzed on the basis of runway transects. (in press).

Narise, T., K. Sakai and S. Iyama, 1960, Mass migrating activity in inbred lines derived from four wild populations of *Drosophila melanogaster*. Natl. Inst. Genet. (Japan), Ann. Rep. No. 10, 1959: 28-29.

Sakai, K., T. Narise, Y. Hiraizumi and S. Iyama, 1958, Studies on competition in plants and animals. IX. Experimental studies on migration in *Drosophila melanogaster*. Evolution 12: 93-101.

Strecker, R. L., 1954, Regulatory mechanisms in house-mouse populations: the effect of limited food supply on an unconfined population. Ecology 35: 249-253.

ERRATUM

Page 32, line 29 should read: "source. This merely suggests that adaptational compromises are sometimes. . . ."

GROUP SELECTION AND THE EVOLUTION OF DISPERSAL

Leigh Van Valen

Committee on Evolutionary Biology, University of Chicago, Chicago, Illinois 60637

Received March 15, 1971

Different animals and plants have widely differing dispersal abilities. This is a matter of major ecological importance, for it determines the gradational difference between colonizing and equilibrium species. The former sacrifice competitive ability and resistance to predation for the sake of dispersal, while for equilibrium species dispersal is secondary. There is also a third kind of species in this classification represented by such organisms as halophilic bacteria and rock-encrusting lichens, which emphasize survival in habitats that few species can tolerate. These may be called resistant species. Resistant and colonizing species form alternative beginnings to ecological successions.

There are several ways to look at the evolution of dispersal ability. One of these simply takes a species with its present characteristics and asks why the observed dispersal is adequate but not too large for the observed spatiotemporal distribution of density of individuals. A more sophisticated variant of this approach, as in the preceding paragraph, uses a balance between dispersal and other desirable features that would be reduced by an optimization of dispersal. Another approach adjusts dispersal so as to equalize the probability of survival for individuals that do and do not disperse.

The Theory

I will examine a rather different approach. We can first consider a population with individuals some of which have more offspring (e.g., seeds) that disperse outside the population than do others. However, the total number of offspring per individual is the same for each type, as is their fitness in comparable environments. If immigration is negligible, there will obviously be selection within the population against dispersal, the optimal condition being the absence of dispersal. Dispersal within the population may be advantageous, and to the extent that this is correlated with dispersal outside the population selection against the latter may be reduced. This may raise the optimum dispersal above zero. It cannot, however, eliminate selection against dispersal out from the population because all the latter dispersers are effectively dead. Furthermore, greater dispersal beyond the population is likely to be more costly (to the parent per offspring or to the offspring) than is less dispersal; this also reduces the optimal dispersal.

However, it may be that populations of this species sometimes become extinct. If there is more or less a long-term steady state over the entire species, extinctions must on the average be balanced by the founding of new populations or the expansion of old ones. Any new population must, by definition, be founded by dispersers outside an old population. This is then a selective advantage for dispersal. It is, moreover, a kind of group selection: group selection among populations favors populations with more dispersal, while individual selection within populations favors parents whose offspring do not disperse.

The conclusion that individual selection uniformly opposes dispersal beyond the population can be challenged in three ways. One is the correlation with within-population dispersal mentioned above. Another is that, from the viewpoint of an individual or a gene, it doesn't matter whether its offspring are in the same population or a new one as long as they survive. Biotic adaptation in the sense of Williams (1966) is not involved. This is a question of terminology which does not affect the existence of an opposition between two selective forces. The units of selection are populations; lineages of dispersers are not selectively preserved within populations. The third counter is that there is sometimes successful immigration to an established population. This means that within this established population some individuals will be present because they are dispersers. If their genotypes are equivalent, with respect to dispersal, to the genotypes of dispersers previously in the population, this situation is equivalent to the presence of some direct selection within the population for dispersal. We can look at this last case slightly differently. It is equivalent to the reflection back into the population of some dispersers. However, all nondispersers, not just some, stay in the population. Therefore, in order for within-population selection to do anything other than minimize dispersal, the probability of survival of an individual disperser, wherever it goes, must be as large as that of a nondisperser. This seems unlikely except for locally high densities and very ephemeral habitats, in both of which cases the probability of survival in the original population is low, but these are important cases.

The Model

I will present an oversimplified model from the above viewpoint and then see what bearing it may have on reality.

Consider an asexual species in which there is no successful dispersal from one population to another. The populations are numerous, large, equal in size, and discrete and consist of three genotypes: one in which all the offspring disperse, one in which few or none do, and one intermediate. (This is a kind of progeny selection, a mode familiar to breeders but not to my knowledge previously applied to natural populations, although it is widely applicable there.) These genotypes initially have equal frequencies in each population. Each genotype has the same number of offspring; non-dispersing offspring of each genotype are equivalent until dispersal in the next generation. Generations are discrete and each population has a constant probability of extinction each generation. Note that the fitness of a genotype within a population is the complement of its probability of dispersal, and that the relative frequency of new populations being formed at any generation equals the probability of a population becoming extinct.

For an analytic solution we must make some further restrictions. The probability of dispersal for genotypes (alleles) A, B, and C are 1, 0.5, and 0 respectively. Each population has the same probability of founding a new population, weighted only by its gene frequency (i.e., not by its recency of establishment or relative size). The expected composition of a new population is in the ratio of twice the frequency of C in the previous generation to one times the frequency of B. Then, if we let g denote the probability of extinction per population per generation, the gene frequencies for the first four generations are those in Table 1. These expressions can

TABLE 1. *Frequencies of alleles in the entire species in first four generations. Dispersal probabilities per generation: allele $A = 1$, $B = 0.5$, $C = 0$.*

Generaton	A	B	C
1	$\dfrac{1}{3}$	$\dfrac{1}{3}$	$\dfrac{1}{3}$
2	$\dfrac{2(1-g)}{3}$	$\dfrac{1}{3}$	$\dfrac{2g}{3}$
3	$\dfrac{4(1-g)^2}{5}$	$\dfrac{1 + 12g + 8g^2 - 16g^3}{5(1+4g)}$	$\dfrac{4g^2}{1+4g}$
4	$\dfrac{8(1-g)^3}{9}$	$\dfrac{1 + 36g + 312g^2 + 496g^3 - 1440g^4 + 768g^5 - 128g^6}{9(1 + 12g + 48g^2 - 16g^3)}$	$\dfrac{40g^3}{1 + 12g + 48g^2 - 16g^3}$

be used for initial results in other cases than those to be simulated, but it is clear that simulation will be necessary for much progress.

For the simulation, the populations can be divided into three kinds: I, populations present in generation 1; II, populations newly formed in the present generation; and III, populations formed in intermediate generations, the latter being heterogeneous. For each kind the contribution to total gene frequency at any generation is the product of the expected frequency of an allele within that kind and the relative frequency of that kind. The treatment is deterministic. The symbols are as follows:

$q_{n,m}(i) \equiv$ relative frequency of allele i at generation n in populations started at generation m.

$w_i \equiv$ proportion of offspring with allele i that do not disperse.

$e_{n,m} \equiv$ relative frequency at generation n of populations started at generation m.

$g \equiv$ probability of extinction per population per generation.

$v_m \equiv$ relative effectiveness with which a genotype contributes to successful dispersal if it is in a population started at generation m.

v_m measures the advantage of being in an older population; $\Sigma v_m = 1$. In one run it was equal for all populations ($v_m = n^{-1}$); in the other it was directly proportional to the age of the population. There are n kinds (ages) of populations in generation n.

The difference equations are as follows:

Population-type I

$$q_{n+1,1}(i) = \frac{w_i q_{n,1}(i)}{\sum_i w_i q_{n,1}(i)}$$

$$e_{n,1} = (1-g)^{n-1}$$

Population-type II

$$q_{n+1,n+1}(i) = \frac{\sum_{m=1}^{n} v_m (1-w_i) q_{n,m}(i) e_{n,m}}{\sum_{i=1}^{3} \sum_{m=1}^{n} v_m (1-w_i) q_{n,m}(i) e_{n,m}}$$

$$e_{n,n} = g$$

Population-type III

$$q_{n+1,m}(i) = \frac{w_i q_{n,m}(i)}{\sum_i w_i q_{n,m}(i)}$$

$$e_{n,m} = g(1-g)^{n-m}.$$

Results

The simulation was done with $w_3 = 0$, w_1 ranging from 1 to 0.8, w_2 ranging from 0.999 to 0.01, and g ranging from 10^{-6} to 0.9, for all combinations of these variables and the two states of v_m. The ranges encompass quite different kinds of species. The runs continued for 50 or, for a few combinations, 100 generations. The two

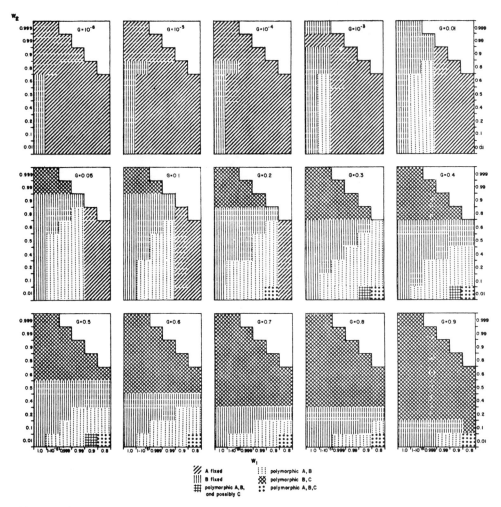

FIG. 1. Phase diagram of results of simulation with three alleles after 50 generations. The occasional equilibria with a frequency of one allele above 0.999 are treated as fixation of that allele, as would commonly occur in nature. Broken lines mean that more than 50 generations are needed to decide whether polymorphism or fixation will result, or (for $w_1 = 1$ and $g \leqslant 0.01$) that fixation of B will occur but the population will be effectively polymorphic for hundreds or thousands of generations.

runs used about an hour on the IBM 360-65, so extension was unfeasible. The three alleles started at equal frequencies.

Figure 1 gives an outline of the results for equal v_m. The equilibrium frequencies cannot be shown. For $w_1 = 1$ the equilibrium for allele B is $(1-g)/w_2$; for other values of w_1 an algebraic result is not apparent. The equilibria are obviously stable, if permanent, but they are not necessarily permanent; a few may be temporary plateaus. Oscillation in gene frequency before reaching equilibrium is rather common, and gene frequencies may rise as high as 0.999996 (for A, $w_1 = 1$, $w_2 = 0.01$, $g = 10^{-6}$, generations 4 and 5) before going to elimination of the abundant allele. In the latter case fixation of the abundant allele would sometimes be expected in a real species, and it might even be regarded

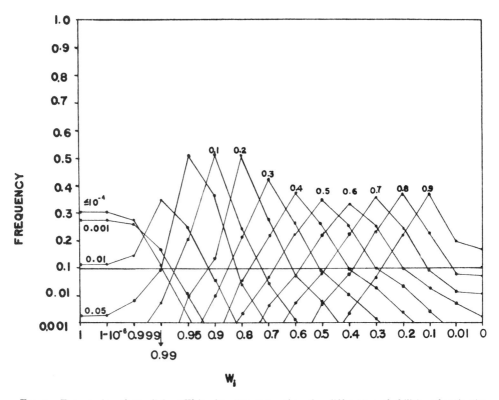

Fig. 2. Frequencies of 16 alleles (W_i) after 100 generations for different probabilities of extinction (g). Each line has its value of g marked on it. The horizontal line separates an arithmetic scale (above) from a logarithmic one (below).

as a case of fixation already with allowance for reverse mutation.

We see that it is possible to have not merely the occurrence of genotypes with intermediate dispersal, but in fact to have polymorphism, and even polymorphism for three alleles. The higher the probability of local extinction, the more scope there is for polymorphism. Even an allele which disperses all the offspring can be maintained with a probability of local extinction of only 0.05 per generation. There is often a major difference between an allele which has no dispersal and one which disperses at a rate of 10^{-6}. Some combinations of alleles cannot be maintained together by this model, but it is unclear how robust this conclusion is to changes in the assumptions.

Relaxation of Assumptions

Some of the assumptions of the model are artificial and others restrict its scope, so it is useful to see to what extent they are necessary for the conclusions.

Given not merely three alleles, but an indefinite number of alleles with different dispersal probabilities, what will be the outcome for a given probability of local extinction? I made a supplementary run with 16 alleles simultaneously for 100 generations. Figure 2 shows the frequencies at that time. The change in frequencies in the later generations gives no indication in any case of polymorphism, except that the most extreme pairs or triplets of alleles tend to behave similarly. The optimal probability of dispersal is about the same as the probability of local extinction.

If dispersal ability is controlled polygenically, the result would clearly be similar (except in rate of change) to that with many alleles at one locus, because the single optimal parental phenotype could then also be reached, although in a different way. Mutation acts similarly.

If the fertility or offspring viability of different genotypes differ, this will obviously affect the resulting frequencies, normally in the direction against dispersal. However, this modifies only the within-population selection and so does not affect the existence of a balance between the two kinds of selection. Variation in time or space on the probability of local extinction or of dispersal has an analogous effect. The effect of immigration (nondiscrete populations) was considered in the section on theory. Sexuality affects only the rate of change. Restriction to small or few populations introduces stochastic effects which I cannot predict without further simulation and which may be important in species for which such restriction is appropriate. The wide fluctuation in gene frequencies during selection indicates that equal initial frequencies of the alleles are unimportant. Discreteness of generations has no importance at the level I am considering.

The treatment of all populations as of equal size and with equal probability of extinction is not something that can obviously be ignored. Newer populations are likely both to be smaller and to be in greater danger of extinction than are older populations. Dispersal tends to be density-dependent, so older populations would tend to contribute more to the founding of new populations. It was to evaluate these effects roughly that I made the run with unequal v_m, weighting the contribution of each population directly by its age. The results are sufficiently similar to those with equal v_m that it is unnecessary to give them separately. There is some effect in the expected direction, and for a much greater assumed difference among populations it is of course possible that an appreciable effect would occur. The range of cases covered, however, seems realistic although not exhaustive.

Discussion

Unless extinction of local populations, species, and higher taxa occurs randomly, group selection occurs. It is obvious that extinction is nonrandom,[1] although also not completely selective, and the question we face is the importance of group selection relative to individual, kin, and progeny selection. Maynard Smith (1964) made some calculations which are appropriate and others have considered the subject also (e.g., Williams, 1966). It is clear that group selection is relatively weak, but this is not to say that it is always negligible. The special nature of dispersal provides the first reasonably general case where it seems to be important.

The reason for the occurrence of polymorphism for dispersal is clear. New populations tend to be founded by dispersers, but in each population the frequency of dispersers declines. With some frequencies of local extinction it is not surprising that more than one allele can sometimes be maintained, although this result is not obvious *a priori*. When an optimal allele is available it becomes fixed, but when one is not available a polymorphism can result. When dispersal ability depends on such a property as the possession of functional wings, as in many beetles, a single intermediate allele may not be possible, and intermediate phenotypes may be deleterious. (An allele controlling an environmentally triggered switch mechanism, however, could sometimes become fixed, as

---

[1] Dinosaurs became extinct worldwide, and large mammals have had a much larger probability of extinction than small ones in the Pliocene and Pleistocene (see Van Valen, 1970, for probabilities in the latter case at the species-lineage level.) Local populations with genotypes fitted for ephemeral habitats, such as active volcanoes, are clearly more susceptible to extinction than other populations of the same species.

in aphids.) An allele here is equivalent to an alternative genotype over many loci.

The theory requires dispersal to be inherited and this is known to be true (Wolda, 1963; Udvardy, 1969; Ogden, 1970).

A clone of an obligately asexual organism as used in the model, is obviously equivalent genetically, although perhaps not ecologically, to a species, so the theory applies to the maintenance of colonizing species in communities. In this form the theory is well known in outline (Andrewartha and Birch, 1954; Baker and Stebbins, 1965), and a homomorphism is easy to establish. It is interesting that there is not merely a homomorphism but a causal relationship between two variants. Group selection for dispersal creates colonizing species, and they are maintained in communities by the same dispersal ability.

Application of the theory to individual species requires knowledge of two rarely estimated parameters, probability of local extinction and variation in the probability of dispersal. I have not come across any species where both are known with much accuracy. However, to some degree we can use Figure 2 to predict the mean dispersal frequency from the probability of extinction, and vice versa. Figure 1 applies to polymorphism in a similar way. From Figure 2 a species with a dispersal of 10 or 20%, like the house mouse (*Mus musculus*; very rough estimate from data of DeLong, 1967; Levin et al., 1969; and Reimer and Petras, 1968) would be expected to have a probability of local extinction of about 0.1 or 0.2, which is not unreasonable (Lewontin, 1962). From Figure 1, a species with morphs that disperse with probabilities of 0.4 and 0.01 ($W_1 = 0.99$, $W_2 = 0.6$) would have an expected probability of local extinction near 0.05 or 0.1. These values may be possible for dimorphic beetles, but data seem unavailable. The numerical results depend on the assumptions in ways discussed in the preceding section.

Various authors have given aspects of the present theory (e.g., Darlington, 1943, 1957; Lidicker, 1962; Carlquist, 1965; MacArthur and Wilson, 1967; Łomnicki, 1969; Udvardy, 1969), but as far as I know not as a unified explanation. The only previous quantitative theory I have seen is that of Levins (1965), who shows that gene flow damps short-term geographic variation in selection and so is desirable as a biotic adaptation. He does not, however, show that natural selection can give such a result. In fact individual selection should be even less favorable to dispersal than in a constant environment, because the dispersers will by hypothesis tend to be poorly adapted to the environment where they arrive. This is, indeed, just how the gene flow is optimal, by reducing a too close tracking of small environmental fluctuations.

The fact that populations often undergo evolution for low dispersal, even frequently to the point of loss of wings, shows that for an important class of cases the Andrewartha and Birch (1954) model of population control is inapplicable. This model regards density-dependent regulation as unimportant and therefore requires, as an implicit corollary, that frequent local extinction occur. These populations do not seem to have had such an experience. Unfortunately the converse is untrue and we cannot say that populations which do disperse considerably have no density-dependent regulation.

SUMMARY

Selection within populations is normally against dispersal. New populations, however, are founded by dispersers, and if there is extinction of local populations there will be a balance between these modes of selection (group and individual). There are single-allele equilibria for given probabilities of extinction, the optimal values of dispersal equalling these probabilities, but polymorphism can occur if these equilibria are unattainable. The theory is homomorphic to that of the maintenance of colonizing species in a community. Group selection need not involve biotic adaptations and seems to be both common and, for dispersal ability, important.

ACKNOWLEDGMENTS

I am grateful to members of the population biology lunch group at the University of Chicago (Spring, 1970) for comments, and to the U. S. Public Health Service for a Research Career Development Award. Grants from the National Science Foundation and the Ford Foundation partly supported this work.

ADDENDUM

I understand that Levins has just published a rather similar theory in a book edited by M. Gerstenhaber and sponsored by the American Mathematical Association. I have not yet (November, 1971) been able either to obtain a full reference or to see the paper.

LITERATURE CITED

ANDREWARTHA, H. G., AND L. C. BIRCH. 1954. The distribution and abundance of animals. Chicago: Univ. Chicago Press. 782 pp.

BAKER, H. G., AND G. L. STEBBINS (eds.), 1965. The genetics of colonizing species. New York: Academic Press. 588 pp.

CARLQUIST, S. 1965. Island Life. Garden City: Natural History Press. 451 pp.

DARLINGTON, P. J. 1943. Carabids of mountains and islands. Ecol. Monog. 13:37–61.

———. 1957. Zoogeography. New York: Wiley. 675 pp.

DELONG, K. T. 1967. Population ecology of feral house mice. Ecology 48:611–634.

LEVINS, R. 1965. The theory of fitness in a heterogeneous environment. IV. The adaptive significance of gene flow. Evolution 18:635–638.

LEVIN, B. R., M. L. PETRAS, AND D. I. RASMUSSEN. 1969. The effect of migration on the maintenance of a lethal polymorphism in the house mouse. Amer. Natur. 103:547–661.

LEWONTIN, R. C. 1962. Interdeme selection controlling a polymorphism in the house mouse. Amer. Natur. 96:65–78.

LIDICKER, W. Z. 1962. Emigration as a possible mechanism permitting the regulation of population density below carrying capacity. Amer. Natur. 96:29–34.

ŁOMNICKI, A. 1969. Individual differences among adult members of a snail population. Nature 223:1073–1074.

MACARTHUR, R. H., AND E. O. WILSON. 1968. The theory of island biogeography. Princeton: Princeton Univ. Press. 203 pp.

MAYNARD SMITH, J. 1964. Group selection and kin selection. Nature 201:1145–1147.

OGDEN, J. C. 1970. Artificial selection for dispersal in flour beetles (Tenebrionidae: *Tribolium*). Ecology 51:130–133.

REIMER, J. D., AND M. L. PETRAS. 1968. Some aspects of commensal populations of *Mus musculus* in southwestern Ontario. Canadian Field-Nat. 82:32–42.

UDVARDY, M. D. F. 1969. Dynamic zoogeography. New York: Van Nostrand Reinhold. 445 pp.

VAN VALEN, L. 1970. Late Pleistocene extinctions. Proc. N. Amer. Paleont. Conv.:469–485.

WILLIAMS, G. C. 1966. Adaptation and natural selection. Princeton: Princeton Univ. Press. 307 pp.

WOLDA, H. C. 1963. Natural populations of the polymorphic landsnail *Cepaea nemoralis* (L.) Arch. Neerlandaises Zool. 15:381–471.

LARVAL DISPERSAL AND SPECIES LONGEVITY IN LOWER TERTIARY GASTROPODS

Thor A. Hansen

Department of Geology and Geophysics, Yale University, New Haven, Connecticut 06520

Abstract. *Species longevity in Lower Tertiary volutids (Gastropoda) is primarily controlled by a combination of developmental type and environmental tolerance. Larval dispersal may be an important factor in molluskan evolutionary rates.*

Ecologic factors affecting evolutionary rates in invertebrates have been the subject of a great deal of discussion (*1*). Various rate controls have been proposed, including feeding type (*2*), environmental tolerance (*1, 3, 4*), and population size (*5*). We discuss here how evolutionary rates may be influenced by larval dispersal, in particular for Lower Tertiary Volutidae (Gastropoda).

Larval dispersal has a significant effect on the geographic distribution of mollusks (*6, 7*). Living species with long-lived planktonic larvae may regularly cross the Atlantic Ocean while those with short or no planktonic stages are unable to cross any oceanic basins. Even local geographic irregularities such as brackish water coves and inlets may be a barrier to dispersal (*8*). The pronounced effect of larval type on species biogeography has generated studies relating the dispersal of fossil invertebrates to modes of speciation (*9*) and evolutionary rates (*4, 7, 10*).

The Paleocene-Eocene outcrops of the North American Gulf Coast provide a suitable framework for testing the effect of dispersal on species longevity. The stratigraphy has been extensively studied and molluskan fossils are generally well preserved.

In order to minimize the effect of factors other than dispersal (factors such as feeding type or morphologic complexity), a single family of gastropods, the Volutidae, was chosen for detailed analysis. Modern volutids are ecologically and morphologically a relatively homogeneous group (all burrowing carnivores-scavengers), and fossil species have an adequate proportion of both planktonics and nonplanktonics (living species have only nonplanktonic development).

Larval development for each species was determined by the criteria of Shuto (*9, 11*), and of the 42 taxa in the study

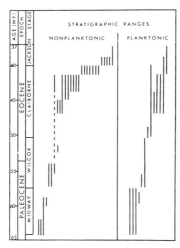

Fig. 1. Geologic ranges of species with non-planktonic or planktonic larval stages.

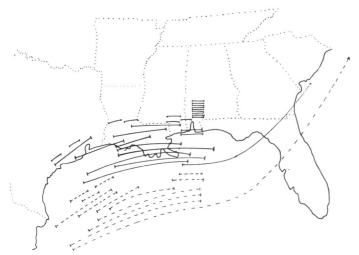

Fig. 2. Geographic distributions of all species, using maximum range in one interval for each. Solid line, nonplanktonics; dashed line, planktonics.

area (12), 29 were found to have nonplanktonic larvae, and 13 to have planktonic larvae. The stratigraphic range for each species was quantified by summing the durations of the geologic formations in which each was present (13, 14). In order to measure geographic ranges, a Paleocene-Eocene outcrop distribution map of the Gulf Coast was divided into geographic units approximately 75 km wide (± 5 km, in order to keep major collecting areas within one unit). Each species was mapped as continuously present between its two most distant localities within time-equivalent formations. This range was quantified by counting the intervening geographic units.

Under the above criteria, planktonics were found to have a mean species duration of 4.4 million years, compared to 2.2 million years for nonplanktonics. Geographic distribution for a species covered an average of 7.8 units for planktonics and 3.5 units for nonplanktonics. Values for longevity and distribution for planktonics are significantly higher as determined by the Mann-Whitney U test (two-tailed test for large samples, $z < .01$ and $.05$ respectively; see Fig. 1). Paleogeographic maps compiled for maximum marine transgressions and regressions show recurrent delta formation during regressive periods in southeast Texas and the Mississippi River Valley (15). During periods of regression, nonplanktonics tended to have their ranges restricted by these deltaic areas, while planktonics were generally able to disperse around them. In periods of delta submergence, both developmental types tended to be widespread, although nonplanktonics retained a higher proportion of restricted ranges. When all ranges are combined (Fig. 2), it can be seen that nonplanktonics have a restricted species distribution well within the limits of their overall range, while planktonics tend to have restricted distributions toward the edges of their normal range. This seems to corroborate Shuto's (9) observation that planktonic forms speciate near the edges of their distribution while nonplanktonics speciate within their total range.

The broad distributions of nonplanktonics during periods of submergence and the restricted ranges of some planktonics indicate differences in environmental tolerance among the volutids. Tolerance to environmental variations (eurytopy) has been suggested as having an effect on species longevity and distributions (3, 4). Species that are highly tolerant of changes in environment (eurytopic) are able to cross barriers because of their adaptability to extreme conditions. Such eurytopic forms tend to have wide ranges and can cope with environmental disturbances, which result in greater species longevity. If some of the volutids are more eurytopic than others, then the simple model of larval duration, biogeographic spread, and evolutionary rate is complicated by at least one other ecologic variable (eurytopy) that may affect species longevities.

To include this variable, environmental tolerance was estimated from relative abundance data that I collected, and from geographic distribution within a single 1- to 2-million-year time interval. The interval of maximum geographic distribution for each species was used as a measure of its tolerance. A species present in several different sedimentary regimes, or distributed over more than four geographic units in a single time horizon, was labeled eurytopic. A species present, at its maximum interval of distribution, in only one sedimentary type or in four geographic units or less was called stenotopic. A graph of larval type and environmental tolerance on longevity (Fig. 3) reveals that nearly all long-lived species are both planktonic and eurytopic (16). The majority of the short-lived forms are both nonplanktonic and stenotopic. Intermediate species are mainly nonplanktonic-eurytopic or planktonic-stenotopic. This arrangement corresponds to a model wherein environmental tolerance and dispersal combine to produce very long, very short, or intermediate species durations. In other words, two forces are at work: high dispersal and eurytopy both tend to extend species longevities, and low dispersal and stenotopy tend to curb them. When planktonic larvae and eurytopy occur together, species durations are greater than that expected from the independent influence of each factor. Similarly, when the two restraining forces, stenotopy and nonplanktonic larvae, are in operation together, species exhibit high turnover. Other combinations result in intermediate species durations.

If planktonic-eurytopic species of volutids are long-lived, why are there no planktonic volutids today? Figure 1 shows that while at the beginning of the Paleocene planktonics and nonplanktonics are about equally represented, the nonplanktonics steadily increase in number with time, particularly during rapid sea-level fluctuations in the Upper Middle Eocene. Environmental changes of this sort may favor species groups that can respond faster evolutionarily. It is

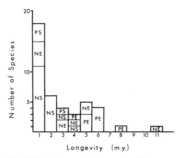

Fig. 3. Developmental type and environmental tolerance of each species with longevity. N, nonplanktonic; P, planktonic; S, stenotopic; E, eurytopic; m.y., million years.

possible that the planktonics were outcompeted by the nonplanktonics.

Evidence presented here supports the hypothesis that larval ecology has an effect on evolutionary rates. Along continental shelves, nonplanktonic, low dispersal species are easily isolated by local barriers during periods of regression. The subsequent increase in rates of extinction and speciation decreases average species longevity. Planktonic, high-dispersal species are less frequently isolated and tend towards long species duration. In any group of organisms, however, evolutionary rates will be influenced by a number of factors. Within the ecologically and morphologically uniform group of Lower Tertiary volutids, species longevities are primarily controlled by a combination of two factors, dispersal and environmental tolerance. Whether these factors control mollusks in general or even other families of gastropods is yet unknown, because many other ecologic controls must be taken into account.

References and Notes

1. E. G. Kauffman, in *Concepts and Methods of Biostratigraphy*, E. G. Kauffman and J. E. Hazel, Eds. (Dowden, Hutchinson & Ross, Stroudsburg, Pa., 1977), p. 109.
2. J. S. Levinton, *Palaeontology* 17, 579 (1974).
3. E. G. Kauffman, in *International Geological Congress, 24th Session*, J. E. Gill, Ed. (Harpell's Press Cooperative, Garden Vale, Quebec, 1972), section 7, p. 174.
4. J. B. C. Jackson, *Am. Nat.* 108, 541 (1974).
5. A. J. Boucot, *Evolution and Extinction Rate Controls, Developments in Paleontology and Stratigraphy* (Elsevier, Amsterdam, 1975), vol. 1.
6. G. Thorson, *Biol. Rev.* 25, 1 (1950); R. S. Scheltema, *Biol. Bull. Woods Hole Mass.* 140, 284 (1971).
7. R. S. Scheltema, in *Concepts and Methods of Biostratigraphy*, E. G. Kauffman and J. E. Hazel, Eds. (Dowden, Hutchinson & Ross, Stroudsburg, Pa., 1977), p. 73.
8. R. v. Cosel and M. Blöcher, *Arch. Molluskenkd.* 107, 195 (1977).
9. T. Shuto, *Lethaia* 7, 239 (1974).
10. Jackson (*4*) and Scheltema (*7*) presented a model for the effect of larval dispersal on biogeography and evolutionary rates of transoceanic species. Species with long-lived planktonic larvae easily maintain gene flow between populations, which suppresses geographic isolation. Moreover, local environmental disturbances have little effect on the entire species population because of its wide distribution, hence extinction rates are lower. The result is that long-lived planktonic species have high longevity but low speciation rates. On the other hand, species with short-lived planktonic larvae may occasionally traverse a barrier such as an ocean basin, but are generally unable to maintain genetic communication. Thus, populations diverge and geographic speciation may result. Local environmental disturbances are likely to affect the entire species, giving rise to high extinction rates. In this case, short-lived planktonic species have high extinction rates and high speciation rates (low longevity).
11. Shuto's (*9*) criteria are size of embryonic whorl and shape and ornamentation of protoconch whorls. Primarily, a small and pointed apex indicates a planktonic larval stage while a large and blunt apex is characteristic of nonplanktonic forms.
12. The volutid nomenclature of Palmer and Brann (*14*) was adopted. Only species that passed the following criteria were used: (i) a part of their range must include Alabama, Mississippi, Louisiana, or Texas, (ii) only fully named species were used (for example, not *Athleta* sp.), (iii) species based on a single unique specimen or species poorly described and in which the sole type has been lost were disqualified, and (iv) all subspecies were included under the specific name.
13. Published reports on nannofossils and planktonic foraminifera allow correlation of Gulf Coast Paleocene-Eocene stratigraphy with the new Paleocene time scale of J. Hardenbol and W. A. Berggren (*Bull. Am. Assoc. Pet. Geol.*, in press). For a similar scale, see W. A. Berggren, *Lethaia* 5, 195 (1972).
14. Species occurrences were taken from K. V. W. Palmer and D. C. Brann [*Bull. Am. Paleontol.* 48 (1965–66)] and L. Toulmin (*Ala. Geol. Surv. Monogr. 13*, in press).
15. Maps were drawn from W. L. Fisher [*Trans. Gulf Coast Assoc. Geol. Soc.* 19, 239 (1969)], C. J. Mann and W. A. Thomas [*ibid.* 18, 187 (1968)], and data compiled by the author from county geological reports.
16. The single long-lived nonplanktonic-eurytopic species is a problematical form present in one formation in the Upper Paleocene and one formation in the Upper Middle Eocene. Whether it is truly a single species is questionable, but it passed all the criteria of (*12*) and so is included.
17. I thank R. Dodge, E. Kauffman, D. Rhoads, N. Sohl, V. Tunnicliffe, and K. Waage for comments and criticism. Thanks go to J. B. C. Jackson and R. Scheltema for discussion.

20 October 1977

AUTHOR CITATION INDEX

Abbott, C. H., 62
Able, K. P., 5
Adams, P. A., 223
Agur, Z., 16, 82
Albrecht, F. O., 188, 224
Alden, B., 205
Alexander, R. D., 62
Allan, P. B. M., 62
Allen, A. A., 147
Allen, D. H., 247
Allen, H. W., 62
Allen, N., 69
Amanshauser, H., 62
Ambrose, H. W., III, 128, 170
Andersen, N. M., 180
Anderson, L. D., 180
Anderson, P. K., 128, 131, 286
Andersson, M., 257
Andrewartha, H. G., 14, 62, 64, 128, 188, 247, 294
Andrzejewski, R., 128, 132, 170, 171, 205
Archer, J., 238
Armitage, K. B., 138
Arora, G., 171
Ashby, K. R., 138
Atkins, M. S., 143
Aumann, G. D., 138
Austin, C. R., 129
Ayala, F. J., 171
Ayroza Galvão, A. L., 62

Bailey, E. D., 250
Bainbridge, R., 161
Baker, H. G., 5, 294
Baker, J. R., 147
Baker, R. R., 5, 266
Bakker, K., 16
Balch, R. E., 62
Balfour-Browne, F., 62
Balkau, B. J., 276
Ballard, E., 62
Banks, C. J., 70
Barnes, M. M., 62
Barness, L. A., 223
Bastock, M., 143
Bateman, A. J., 46, 238

Batzli, G. O., 128, 138
Beall, G., 62
Beckham, C. M., 62
Beddington, J. R., 244
Beebe, W., 62
Beer, J. R., 41
Beetsma, J., 143
Bellis, E. D., 276
Bengtsson, B. O., 257
Bennett, S., 276
Benson, R. B., 62, 63
Bent, A. C., 197
Berry, R. J., 128
Berthold, P., 32, 100
Betts, E., 224
Bhatnagar, P. L., 223
Bigelow, R. S., 62
Binsztein, N., 132
Birch, L. C., 14, 62, 128, 188, 224, 247, 257, 294
Bird, R. D., 63
Birdsell, J. B., 100, 128
Bishop, S. C., 276
Bishopp, F. C., 63
Blackith, R. E., 188, 224
Blackwall, J., 63
Blackwell, T. L., 14
Blair, W. F., 41, 46, 197
Blais, J. R., 63, 143, 188
Blakeman, J., 68
Blakley, N. R., 171
Blanchard, B. D., 197
Blem, C. R., 6
Blest, A. D., 143
Blöcher, M., 297
Blunck, H., 63, 188
Blyumental, T. I., 32
Bodenheimer, F. S., 63, 143
Boiry, L., 132
Bolduan, J., 82
Bond, F., 63
Bonnemaison, L., 63, 143, 245
Boorman, S. A., 245, 276
Boot, L. M., 133
Booth, C. O., 66, 143, 223
Botkin, D. B., 138
Boucot, A. J., 297

Boyce, A. J., 244
Boyko, H., 63
Bradt, G. W., 100, 128
Braendegaard, J., 63
Brann, D. C., 297
Brazendale, M. G., 15
Briese, L. A., 132, 138, 206
Brinkhust, R. O., 63, 180
Bristowe, W. S., 63
Brito da Cunha, A., 63
Bronson, F. H., 128, 130, 131, 133
Brown, B. A., 100
Brown, C. E., 63
Brown, C. K., 205, 266
Brown, E. S., 15, 63, 223, 257
Brown, L. E., 15
Brown, R. Z., 128, 138
Browne, L. B., 188
Bruce, H. M., 128
Bruehl, G. W., 223
Brussard, P. F., 257
Buckner, C. H., 129
Bueno, J. R. de la Torre, 180
Bujalska, G., 129, 132, 205
Burla, H., 63
Burmann, K., 63
Burt, W. H., 15, 41, 147, 197
Burton, J. F., 63
Busch, C., 132
Buxton, P. A., 63
Buzzati-Traverso, A. A., 286

Caldwell, R. L., 100, 171, 223, 224, 257
Calhoun, J. B., 129, 205, 250
Callahan, J. R., 180
Callan, E. McC., 63
Camerano, L., 63
Campbell, C. A., 171
Campbell, K. G., 63
Campbell, I. M., 224
Carl, E. A., 129
Carlisle, D. B., 224
Carlquist, S., 294
Carmon, J. L., 133
Carpenter, C. R., 46
Carter, A., 171

Author Citation Index

Casperson, K. B., 138
Caswell, G. H., 63, 143, 188, 224
Cavalcanti, A. G. L., 63
Cavalli-Sforza, L., 245
Cavicchi, S., 245
Cerkasov, J., 143
Chamberlain, J. C., 143
Chapman, J. A., 63, 143, 188
Chapman, R. K., 223
Chipman, R. K., 129
Chitty, D., 87, 138, 147, 197
Chitty, H., 129
Chiykowski, L. N., 223
Chnéour, A., 63
Christenson, L. R., 63
Christian, J. J., 129, 138, 197, 205, 250
Clark, A. M., 223
Clark, L. R., 63
Clarke, B. C., 180
Cloudsley-Thompson, J. L., 63, 64
Cockbain, A. J., 223
Cockrum, E. L., 15
Cohen, D., 266, 281
Cole, F. R., 171
Cole, L. C., 223, 245
Collins, J. P., 239
Colwell, R. K., 171
Comins, H. N., 244, 257
Common, I. F. B., 64, 70, 143
Conaway, C. H., 132
Conway, G. R., 244
Cook, E. F., 223
Cook, L. M., 277
Cook, W. C., 223
Corbet, P. S., 64, 143, 188
Corbett, L. K., 132, 258
Cornell, J. A., 245
Cornwell, P. B., 64
Cosel, R. v., 297
Cotterell, G. S., 64
Coulson, J. C., 15
Couturier, A., 64, 143
Cowan, E. T., 188
Cox, G. W., 224, 266
Crawford, D. E., 161
Crawford, G. I., 64
Crawley, L., 100, 245
Crawley, M. C., 129
Creaser, E. P., 161
Crichton, M. I., 64
Cross, E. A., 281
Crow, J. F., 277
Crowcroft, P., 129
Crowell, K. L., 129
Crumpacker, D. W., 245
Cummings, W. C., 161
Curtis, J., 64
Cushing, J. E., 197

Dana, S., 138
Danilevskii, A. S., 224

Dannreuther, T., 64
Darlington, P. J., 64, 294
Darlow, H. M., 64
Davenport, C. B., 41
Davey, J. T., 64
Davidson, J., 64
Davies, D. E., 193, 224
Davis, D. E., 129, 138, 205, 250
Davis, E. W., 64
Davis, M. A., 257
Dawson, C. E., 161
Dawson, P. S., 73, 82, 180
de Kort, C. A. D., 171
DeLong, K. T., 129, 205, 294
Dempster, J. P., 224
Denning, D., 63
Denno, R. F., 171
Derr, J. A., 171, 205
de Ruiter, L., 143
DeSmidt, W. J. J., 161
Dethier, V. G., 247
de Wilde, J., 143
DeYoung, C. A., 100
Dice, L. R., 41, 46, 197
Dickson, R. C., 64
DiHus, W. P. J., 171
Dingle, H., 100, 161, 171, 205, 223, 224, 245, 266
DiPace, M., 132
Dixon, A. F. G., 171
Dobrzhanski, F. G., 64
Dobzhansky, T., 63, 64, 245
Dolnik, V. R., 32
Dominski, A. S., 138
Donnelly, J., 67
Donnelly, U., 193
Dorsett, D. A., 223
Dorst, H. E., 64
Dorst, J., 223
Downes, J. A., 143
Dubinin, N. P., 64
Duffey, E., 64
Dunning, J. W., 64
Dymond, J. R., 129

Eastwood, E., 245
Ehrlich, P. R., 100, 257, 277
Eickwort, K. R., 180
Eimer, T., 64
Eleftheriou, B. E., 130
El-Fandy, M. G., 193
Eliot, N., 64
El Khidir, E., 143
Ellis, P. E., 224
Elton, C. S., 6, 64, 147, 188, 205
Emlen, J. T., Jr., 42, 138
Emlen, S. T., 15
Epling, C., 67
Erickson, M. M., 46
Errington, P. L., 130, 171, 197, 238
Essenberg, C., 180
Evans, D. M., 138

Evans, D. R., 188
Evans, F. C., 130, 286
Evans, M. G., 62
Ewing, K. P., 65

Fairbairn, D. J., 100
Falconer, D. S., 224
Farner, D. S., 224
Farr, J. A., 171
Farris, S. H., 143
Faure, J. C., 143
Fay, R. W., 68
Fedotov, D. M., 64, 143
Feldman, M. W., 276
Felenstein, J., 277
Felt, E. P., 193
Fernando, C. H., 64, 65, 180
Fisher, C. K., 66
Fisher, J., 46
Fisher, R. A., 15, 147, 223, 244
Fisher, W. L., 297
Fisler, G. F., 100
Fitch, H. S., 42
Fitting, L., 73
Fivizzani, A. J., 6
Foote, R. H., 63
Ford, E. B., 147, 245
Ford, R. G., 100
Forsdyke, A. G., 193
Foster, W. A., 15, 100, 171
Fox, K. A., 129
Fraenkel, G., 65, 188, 223
Frampton, V. L., 46
Fraser, F. C., 65
Free, C. A., 244
Freeman, J. A., 65
French, N. R., 15, 130, 138
French, R. A., 65, 70, 245
Fritzche, R., 65
Frost, W. E., 180
Fullagar, P. J., 130
Fye, R. E., 65

Gadgil, M., 130, 138, 245, 257, 266, 281
Gage, S. H., 277
Gaines, J. C., 65
Gains, M. S., 87, 131, 257, 266
Galbraith, D., 180
Gallopin, G., 132
Gambale, S., 101, 206
Gambles, R. M., 65
Gardner, A. D., 147
Gardner, A. E., 65, 67
Garrett-Jones, C., 143
Gates, D. L., 171
Gaunitz, D., 180
Gause, G. F., 65
Gentry, J. B., 130, 133, 138
Gerhold, H., 224
Getz, L. L., 16, 138, 171
Ghent, A. W., 73

Author Citation Index

Ghesquière, J., 65
Ghiselin, M. T., 15
Gibb, J., 46
Gilbert, L. E., 15, 100, 205, 277
Gill, D. E., 277
Gillespie, J. H., 205, 215, 257
Gillie, O. J., 87
Gilmour, D., 65
Gilpin, M. E., 238
Glick, P. A., 65
Gliwicz, J., 129, 132
Goldstein, A., 73
Golley, F. B., 138, 250
Goszczyński, J., 138
Gough, L. H., 193
Gradwell, G. R., 244
Graham, C. L., 223
Graham, K., 143
Grant, B., 173
Grant, P. R., 130, 138
Grant, W. E., 138
Grasse, P., 67
Gray, L. L., 172
Greathead, D. J., 65
Green, G. W., 100, 143
Green, J., 65
Green, R. G., 147
Greenbank, D. O., 65, 143, 188
Greenberg, G., 205
Greenberg, R., 171
Greensted, L. W., 65
Greenstein, J. A., 133
Greenwood, P. J., 15, 172, 245, 257
Grinnell, J., 15, 46, 100, 245
Grissell, E. E., 171
Grodziński, W., 138
Gunn, D. L., 193
Guthrie, D. M., 180
Gwinner, E., 15, 32, 100

Haeger, J. S., 143
Hagen, K. S., 143
Hägerstrand, T., 244
Haine, E., 65, 66
Hairston, N., 238
Haldane, J. B. S., 245
Hamilton, W. D., 245, 257, 266, 281
Hamilton, W. J., III, 245
Hammel, E. A., 205
Hans, H., 65
Hansing, E. D., 46
Hanson, W. D., 277
Hansson, L., 138, 257
Harden-Jones, F. R., 224
Hardy, A. C., 65
Harling, J., 138
Harper, J. L., 245, 257, 281
Harris, H., 205
Harrison, G. A., 244
Harrison, R. G., 257
Harvey, P. H., 172, 245, 257
Haskell, J. B., 223

Haskell, P. T., 143
Haskins, C. P., 286
Haskins, E. F., 286
Hassell, M. P., 244, 245
Hatt, R. T., 130
Hawkes, O. A. M., 65
Hayden, P., 15, 130
Hayward, K. J., 65
Healey, M. C., 130, 206
Healy, W. R., 277
Heape, W., 65, 223
Heath, J. E., 223
Heed, W. B., 65, 258
Hegmann, J. P., 205, 224, 266
Henson, W. R., 65, 143, 188
Herrnkind, W. F., 100, 161
Hilborn, R., 87
Hill, W. G., 277
Hill, W. H., 15
Hiraizumi, Y., 16, 29, 69, 158, 286
Ho, F. K., 180
Hocking, B., 65, 188, 223
Hodek, I., 65, 143
Hodson, A. C., 223
Hoffman, R. A., 67
Holdenried, R., 130
Holling, C. S., 244
Holman, J. A., 15
Holt, J. A., 129
Hood, R. E., 100
Hopkins, A. R., 65
Houlihan, R. T., 130
Howard, L. O., 193
Howard, W. E., 41, 42, 46, 130, 197, 286
Howell, A. B., 42
Hubby, J. L., 206
Huff, F. H., 223
Huffaker, C. B., 238, 245
Hungerford, H. B., 180
Hunter-Jones, P., 224
Huntington, C. E., 42
Hutchinson, G. E., 138, 257
Huxley, J., 29

Ibbotson, A., 188
Idyll, C. P., 161
Imler, R. H., 42
Immelmann, K., 32
Imms, A. D., 143
Incerti, G., 15, 130
Inglis, J. M., 100
Inouye, N., 73
Isaacson, A. J., 238
Istchenko, V. G., 132
Isted, G., 245
Istock, C. A., 266
Ivanova, L., 188
Iverson, S., 138
Iwao, S., 143
Iyama, S., 16, 29, 69, 158, 286

Jackson, D. J., 65, 66, 223
Jackson, J. B. C., 257, 297
Jacquard, A., 32
Jamieson, G. S., 180
Janion, S. M., 130, 172
Jannone, G., 66
Järvinen, O., 180, 181
Jensen, J. A., 69
Jepson, W. F., 69
Jewell, P. A., 130
Johannigsmeier, A. G., 138
Johansson, A. S., 143
Johnson, B., 66, 143, 188, 224
Johnson, C. G., 15, 66, 69, 82, 143, 180, 188, 223, 245
Johnson, M. S., 147
Johnson, N. K., 15, 172
Johnson, S. A., 188
Johnston, J. S., 258
Johnston, R. F., 42, 46, 130, 286
Jones, F. R. H., 245
Jones, I. B., 224
Jones, S. C., 66
Jones, T., 70
Jordan, P. A., 138
Joyce, R. J. V., 66

Kacher, H., 138
Kajak, A., 128, 205
Kalabukhov, N., 147
Kalela, O., 130, 131, 138
Kalmus, H., 69
Karlin, S., 82, 245
Karr, J. R., 266
Kauffman, E. G., 297
Kaufman, D. W., 138
Keeton, W. T., 6
Keilholz, 66
Keller, B. L., 87, 131, 138, 266
Kelly, E. O. G. L., 143
Kendrew, W. G., 193
Kennedy, J. S., 15, 66, 100, 143, 180, 188, 193, 223, 224, 245
Kershaw, W. J. S., 66, 223
Key, K. H. L., 66, 143
Keys, J. E., Jr., 138
Khaz'minskii, R. Z., 215
Kikkawa, J., 131
Kilpatrick, J. W., 68
Kimura, M., 29, 245, 277
King, C. E., 82
King, J. A., 131
Kinghorn, J. M., 143, 188
Kipeläinen, T., 131
Kirata, J., 143
Klein, H., 32
Kligler, I. J., 188
Kluyver, H. N., 46, 101
Knight, F. B., 206
Koch, M., 224
Kock, L. L., 138
Köhler, K., 66

301

Author Citation Index

Kolodziej, A., 131
Koponen, T., 131, 138
Krajewski, S., 180
Krebs, C. J., 15, 16, 87, 131, 133, 138, 206, 245, 266
Krogh, A., 223
Krull, J. N., 131
Krumme, D. W., 100
Küchemann, C. F., 244
Kurn, N., 281

Laake, E. W., 63
Lack, D., 42, 66, 224, 266
Lack, E., 66
Laird, E. F., 64
Lamb, K. P., 66
Lamond, D. R., 131
Lane, C., 66
Langley, C., 215
Larson, A. O., 66
Larson, C. L., 147
Lashley, K. S., 42
Lavabre, E. M., 66
Lavie, B., 101, 258
Lawson, F. R., 143
Lawton, J. H., 244
Lees, A. D., 66, 143, 180, 224, 245
Legett, W. C., 15
Lemmon, W. B., 173
Lempke, B. J., 66, 67
Lerner, I. M., 73, 286
Lestage, J. A., 67
Leston, D., 67
Leth, K. O., 180
Levene, H., 215
Levin, B. R., 294
Levin, S. A., 238
Levins, R., 180, 206, 224, 238, 245, 258, 277, 281, 294
Levinton, J. S., 297
Levitt, P. R., 245, 276
Lewis, T., 67, 143, 245
Lewontin, R. C., 180, 206, 223, 244, 258, 277, 294
Li, C. E., 277
Lichtenstein, J. L., 67
Lidicker, W. Z., Jr., 6, 73, 82, 101, 131, 138, 157, 206, 245, 286, 294
Lindquist, A. W., 67
Lindroth, C. H., 67, 180
Lindzey, G., 16
Linn, M. B., 46
Lipa, J. J., 180
Locket, G. H., 67
Lockley, R. M., 130
Lofts, B., 224
Łomnicki, A., 15, 101, 238, 258, 294
Long, D. B., 143
Longfield, C., 67, 188
Louch, C. D., 131
Lovell, P. H., 245, 281

Lowendorff, H. S., 138
Lund, M., 131
Lunson, E. A., 193

Macan, T. T., 67, 180
MacArthur, R. H., 180, 223, 224, 238, 245, 277, 294
Macaulay, E. D. M., 245
McCabe, T. T., 197
McClenaghan, L. R., Jr., 257
McCollum, F. C., 131
McDaniel, C. K., 205
McDonald, D. J., 73
McDonogh, R. S., 67
Macevicz, S., 281
McGregor, D., 138
McIntyre, G. A., 65
McKechnie, S. W., 277
Mackillop, A. T., 193
McLachlan, R., 67
MacLeod, J., 67
McMillan, W. W., 65
McPherson, A. B., 131
Macy, R. W., 67
Maher, W. J., 138
Mail, G. A., 69
Mainardi, D., 172
Makings, P., 143
Mann, C. J., 297
Marriner, T. F., 67
Marsan, M., 172
Marsden, H. M., 128, 131
Marshall, W., 67
Mathis, W., 68
Matthews, G. V. T., 224, 245
Matthey, W., 180
May, R. M., 238, 244, 257, 266, 281
Maynard, D. M., 161
Maynard Smith, J., 266, 281, 294
Mayr, E., 32
Mazurkiewicz, M., 131
Medler, J. T., 188, 223
Meier, A. H., 6
Mer, G., 188
Mercer, E., 16
Meredith, D. H., 15
Merrell, D. J., 277
Mertz, D. B., 180
Metzgar, L. H., 132
Meyer, J. R., 16
Meyer, R. K., 41
Miller, C. A., 188
Miller, E. R., 171
Miller, G. R., 138
Miller, R. S., 64
Millidge, A. F., 67
Milne, A., 67, 188
Milne, P. S., 65
Mirenda, J., 101
Missonier, J., 143
Mitchell, D. F., 67
Mitchell, R., 172

Mitis, H. von, 180
Mittler, T. E., 188
Moericke, V., 67
Mohr, E., 42
Moncreaff, H., 67
Moore, K. G., 245, 281
Moore, N. W., 188
Moorhouse, J. E., 143
Moran, P. A. P., 215
Moreau, R. E., 223
Moreton, B. D., 67
Morley, C., 67
Morris, R. D., 138, 188, 244
Morton, N. E., 277
Moser, J. C., 281
Moss, R., 101, 138, 245
Mott, D. G., 188
Muller, H. J., 143
Mundt, A. H., 67
Murphy, H. M., 128
Murray, B. G., Jr., 258
Murton, R. K., 224, 238
Muspratt, V., 70
Myers, J. H., 15, 87, 132, 138, 245, 266
Myers, K., 138
Mykytowycz, R., 101, 206

Narise, T., 16, 29, 69, 101, 158, 172, 286
Naumov, A. E., 132
Naumov, N. P., 147
Nayar, K. K., 143
Naylor, A. F., 16, 73
Needham, J., 193
Neville, A. C., 223
Newman, L. H., 67
Newsome, A. E., 132, 258
Nice, M. M., 46
Nichiporick, W., 223
Nicholson, A. J., 67, 198, 238
Nielsen, A. T., 67, 68, 188
Nielsen, E., 67
Nielsen, E. T., 67, 68, 143, 188, 223
Nijveldt, W., 68
Nordman, A. F., 68
Norris, M. J., 143, 188, 224
Northcott, T. H., 16
Nygren, J., 266

O'Donald, P., 172
Ogden, J. C., 73, 74, 82, 294
Olenjev, V. G., 132
Oliver, F. W., 68
Olszewski, J., 132
Orr, R. T., 16
Osborne, D. J., 224
Osten-Sacken, C. R., 68
Oster, G. F., 281
Ovtschinnikova, N. A., 132
Owen, D. F., 68, 133

Author Citation Index

Palmer, K. V. W., 297
Parker, G. A., 266
Parks, T. H., 143
Parrott, D. M. V., 128
Parshley, H. M., 68
Parsons, P. A., 132
Pasquali, A., 15, 130, 172
Paton, C. I., 68
Patterson, J. T., 158
Paul, L. C., 68
Pavan, C., 63
Payne, N. F., 16
Payne, R. S., 198
Pearson, A. K., 198
Pearson, K., 68
Pearson, E. O., 68
Pearson, O. P., 101, 132, 198
Penchaszadeh, P., 132
Pennycuick, C. J., 266
Perrins, C. M., 172
Perry, F. C., 193
Peto, R., 87
Petras, M., 294
Petrusewicz, K., 128, 132, 170, 206, 238
Peuman, H. L., 188
Piantanida, M., 132
Pickering, J., 16
Pieczyńska, E., 128, 205
Pielou, E. C., 239
Pielowski, Z., 132, 172
Pienkowski, R. L., 223
Pimentel, D., 68, 138
Pinowski, J., 16, 206
Piper, C. V., 68
Pitelka, F. A., 46, 128, 132, 138, 198
Pjastolova, O. A., 132
Poisson, R., 180
Pokrovski, A. V., 132
Pomianowska, I., 131
Ponting, P. J., 143
Popham, E. J., 68
Popov, G. B., 143
Powell, J. R., 245
Preston, F. W., 245
Price, G. R., 281
Provost, M. W., 68, 143, 188
Prus, T., 16, 74, 82
Pucek, Z., 132
Puppi, G., 245
Putnam, L. G., 68

Quarterman, K. D., 68

Radzievskaya, S. B., 68
Raevskii, V., 147
Rainey, R. C., 68, 70, 143, 188, 193, 223
Rajska, E., 131
Ramchandra Rao, Y., 193
Ramsey, P. R., 14, 132, 206

Rankin, M. A., 6, 171, 223
Rao, Y. R., 68
Rappaport, A., 266
Rappole, J. H., 172
Rasmuson, B., 266
Rasmuson, M., 266
Rasmussen, D. I., 294
Ray, T. S., 173
Redman, R. E., 138
Reimer, J. D., 294
Richard, G., 68
Richards, O. W., 68, 245
Richmond, M., 132
Rider, G., 245
Riggall, E. C., 68
Riggs, L. A., 16
Riley, C. F. C., 180
Ritte, U., 16, 82, 101, 258
Robert, P., 64, 143
Roberts, R. C., 206
Robertson, F. W., 133
Robertson, J. R., 180
Rockstein, M., 223
Roer, H., 68, 143, 188
Roff, D. A., 82, 172, 181, 258, 266, 281
Rogers, D. J., 245
Romer, J. I., 63
Root, F. M., 69
Ross, K. F. A., 69
Rostowzew, S., 245
Roth, P., 69
Rothschild, M., 69
Rowe, F. P., 129
Rowlands, I. W., 130
Rowley, W. A., 223
Rygg, T., 223
Ryszkowski, L., 132, 138

Sadleir, R. M. F. S., 138
Saitoh, T., 206
Sakai, K., 16, 29, 69, 158, 258, 286
Salisbury, E., 172
Sammeta, K. P. V., 224
Sanders, C. J., 206
Santibañez, S. F., 172
Saunt, J. W., 69
Savory, T. H., 69
Sawyer, J. S., 193
Sawyer, W. H., 223
Scheltema, R. S., 297
Scherbinovskii, N. S., 69
Schmidt-Koenig, K., 6, 16
Schneider, F., 69, 143, 188
Schneirla, T. C., 188
Schoof, H. F., 69
Schurr, K., 82
Scossiroli, R. E., 245
Scott, H., 69
Selye, H., 250
Sélys Longchamps, E. de., 193
Semeonoff, R., 133

Severin, H. H. P., 223
Shannon, H. J., 69
Shapiro, A. M., 101
Shapiro, L., 281
Sheppard, P. M., 69
Sheppe, W., 133, 206
Shipman, H. J., 188
Shvarts, S. S., 132
Shuto, T., 297
Siegel, S., 198
Sinclair, A. R. E., 138
Singer, M. C., 15, 100, 205, 277
Siverly, R. E., 69
Skellam, J. G., 245
Slater, J. A., 223
Slobodkin, L. B., 101, 224, 238, 266
Slonaker, J. R., 42
Smirnov, V. S., 138
Smith, C. E., 69
Smith, F. E., 238, 247
Smith, K. G., 69
Smith, M. H., 133, 138
Smith, W. W., 69
Smythe, M., 133
Snow, D. W., 69
Snyder, D. P., 198
Soans, A. B., 138
Sokoloff, A., 74
Solbreck, C., 266
Solot, S. B., 193
Southwick, C. H., 133
Southwood, T. R. E., 66, 69, 74, 143, 188, 223, 244, 245, 258, 266, 281
Spector, W. S., 138
Stańczykowska, A., 101
Stark, J., 63
Stebbins, G. L., 5, 281, 294
Stenseth, N. C., 258
Stern, C., 32
Stevenson, D. E., 172
Stoddart, D. M., 133, 138
Stokoe, W. J., 69
Stone, W. S., 158
Storm, R. M., 161
Stovin, G. H. T., 69
Stower, W. J., 224
Strangeways-Dixon, T., 143
Strecker, R. L., 133, 286
Street, P., 6
Strickland, A. H., 69
Stroyan, H. L. G., 66, 188, 224
Stuart, R. A., 266
Summerlin, C. T., 101
Sumner, F. B., 147
Svärdson, G., 16
Svihla, A., 198
Sweeney, R. C. H., 69
Swift, D. M., 138

Tagami, T. Y., 15, 130
Takada, H., 69

Author Citation Index

Tamarin, R. H., 87, 131, 133, 138, 266
Tardif, R. R., 172
Tast, J., 131, 133
Taylor, D., 133
Taylor, L. R., 66, 69, 143, 188, 239, 245, 266
Taylor, R. A. J., 239, 266
Temple, V., 69
Terman, C. R., 206
Theobold, F. T., 69
Thoday, J. M., 245
Thomas, E. S., 62
Thomas, W. A., 297
Thompson, A. L., 223
Thompson, A. T., 69
Thompson, D. Q., 138
Thompson, W. L., 46
Thorson, G., 297
Tilden, J. W., 143
Timofeeff-Ressovsky, N. H., 69
Timofeeff-Ressovsky, R. A., 69
Tinbergen, L., 101
Tinbergen, N., 42
Tiniakov, C. G., 64
Tinkle, D. W., 138
Tischler, W., 70
Tokmakova, S. G., 138
Topoff, H., 101
Toulmin, L., 297
Tower, W. L., 70, 143
Travis, D., 161
Trivers, R. L., 281
Truszkowski, J., 138
Tschinkel, W. R., 101
Tucker, V. A., 266
Tulloch, J. B. G., 70
Turner, B. N., 138
Tutt, J. W., 70
Tuttle, M. D., 172

Udvardy, M. D. F., 294
Urquhart, F. A., 16, 70, 101, 143, 188, 223, 245
Utida, S., 70, 143, 188, 224
Uvarov, B. P., 70, 172, 224

Vachon, M., 70

Valle, K. J., 70
van Belle, G., 101
van der Lee, S., 133
van der Pijl, L., 6
Van Emden, H. F., 69
van Soest, P. V., 138
Van Valen, L., 82, 133, 239, 277, 281, 294
Van Vleck, D. B., 133
Varley, G. C., 244, 245
Vepsäläinen, K., 173, 181, 258, 266
Verdier, M., 188
Verrier, M. L., 70
Vesey-Fitzgerald, D. F., 70
Vodjdani, S., 70
Volkonsky, M., 143
Von Bonde, C., 161
von Haartmann, L., 42

Wachter, K. W., 205
Waddington, C. H., 172
Wagner, H. O., 46
Wainwright, C. J., 70
Wakeland, C., 46, 188
Walker, R. L., 65
Walkowa, W., 128, 170
Wallace, B., 173
Wallace, L., 66
Waloff, N., 16, 68
Waloff, Z., 68, 70, 143, 173, 188, 193
Warner, D. W., 172
Warnock, J. E., 138
Warriner, C. C., 173
Wasilewski, A., 101
Waterhouse, D. F., 65
Watson, A., 101, 138, 245
Watt, K. E. F., 245
Watzl, O., 70
Way, M. J., 70
Webb, W. E., 70
Webb, W. L., 133
Weintraub, J., 143
Weise, C. M., 16
Weis-Fogh, T., 143, 188, 193, 223
Weiss, G. H., 245
Wellington, W. G., 16, 74, 143, 188
Westwood, N. J., 238

White, H. C., 173
White, J. A., 281
White, R. R., 277
Whitt, G. S., 16
Whitten, W. K., 133
Wiegert, R. G., 133
Wiener, J. G., 138
Wiens, J. A., 239
Wigglesworth, V. B., 143, 223
Wilbur, H. M., 239
Williams, A. B., 161
Williams, C. B., 6, 70, 143, 188, 223, 245
Williams, C. G., 245
Williams, C. M., 223
Williams, G. C., 296
Williams, M. C., 70
Williams, J. S., 245
Williams, M. C., 70
Wilson, E. O., 223, 245, 258, 294
Wiltshire, E. P., 70
Wodzicki, K., 258
Wolda, H. C., 294
Wolfe, J. L., 101
Wolfenbarger, D. A., 245
Wolfenbarger, D. O., 173, 245
Woolf, C. M., 87
Woyciechowski, M., 101
Wright, S., 29, 64, 245, 277
Wrocławek, M., 128, 171
Wunder, B. A., 138
Wynne-Edwards, V. C., 101, 133, 239, 245

Yates, W. W., 67
Yeaton, R. I., 16
York, G. T., 143
Young, E. C., 173
Yurkiewicz, W. J., 223

Zaher, M. A., 143
Zarrow, M. X., 130
Zeuthen, E., 223
Ziegler, J. R., 82, 266
Zimmerman, E. G., 138
Zohary, M., 281
Zverev, M. D., 147
Żyromska-Rudzka, H., 16, 82, 239

SUBJECT INDEX

Aphid. *See* Homoptera, Aphididae
Apodemus, 126, 144, 146
   flavicollis (Yellow-necked mouse), 108
   sylvaticus (Wood mouse), 113, 146
Arvicola terrestris (Water vole), 109, 118
Ascia monuste (Florida saltmarsh butterfly). *See* Lepidoptera, Pieridae

Bacteria
   Coulobacter crescentus, 280
   halophilic bacteria, 287
Bats (Chiroptera), 262, 264
   Myotis grisecens (Gray bat), 168
   Tadarida brasiliensis (Free-tailed bat), 12
Beaver *(Castor canadensis),* 95, 124, 126
Blackbird, European *(Turdus merula),* 240, 242, 243
Boundary layer, 173, 184, 186
Bruce effect, 115
Burt, W. H., 13, 33
Butterfly. *See* Lepidoptera

Carrying capacity
   definition of, 119, 134
   equilibrium density, relation to, 95, 120, 121, 122, 123, 126, 134-138, 177-178, 246-247, 282-286
Chitty, Dennis, 13, 83
Clethrionomys (Red-backed voles)
   glareolus (Bank vole), 94, 107, 108, 109, 113, 114, 118, 144-147, 204, 283
   rufocanus, 204
Coleoptera (Beetles), 13, 50, 51, 52-56, 58, 139, 218, 219, 292, 293
   Apionidae, *Apion,* 55
   Bruchidae, 56, 59
      Acanthoscelides obtectus, 56
   Carabidae, 53, 166
      Bambidion, 53
   Chrysomelidae, 139, 141
      Coelaenomenodera elaeidis (Hispid leaf-miner), 55
      Diabrotica undecimpunctata (Spotted cucumber beetle), 55
      Halticini, 55, 58
      Labidomera clivicollis, 176
      Leptinotarsa (Colorado potato beetle), 55, 141, 165-166
   Cincindelidae (Tiger beetles), *Cincindela,* 53
   Coccinellidae (Lady beetles), 47, 54, 55, 58, 139
      Adalia, 54
      Coccinella, 54
   Curculionidae (Weevils)
      Anthonomus grandis (Boll weevil), 58
      Bruchus pisorum, 44
      Callobruchus maculatus (Cow-pea weevil), 56, 142, 186, 221
      C. quadrimaculatus, 142
      Sitona, 55
   Dyticidae
      Agabus articus, 53
      A. biguttatus, 53
      A. didymus, 53
      A. sturmi, 53
      Hydroporus lepidus, 53
   Heteroceridae, 54
   Nitidulidae, *Meligethes,* 55
   Ptiliidae, 54
      Ptinella errabunda (Sub-cortical beetle), 280
   Scarabaeidae
      Melontha melontha, 55
      Phyllopertha horticula (Garden chaffer), 56, 185, 187
   Scolytidae, 54, 139, 141
      Dendroctonus, 140
      Trypodendron, 141
      T. lineatum, 54
   Staphylinidae, 54
   Tenebrionidae
      Tribolium (Flour beetle), 12, 13, 176, 234, 263
      T. castaneum, 10, 13, 71-74, 77-82, 92, 255
      T. confusum, 71-74
      Zophobus, 94
Colonization, 115
   definition, 5
Competition
   interspecific, 91, 117, 167, 169, 253, 254-255, 262
   intraspecific, 19-27, 148-158, 169, 278-281
Cormorant *(Phalacocorax carbo),* 12
Cricetid rodents. *See Arvicola terrestris; Clethrionomys; Lemmus lemmus; Microtus; Peromyscus; Reithrodontomys magalotis; Sigmodon hispidus*
Crustacea, 262
   copepod, 260
   Panulirus argus (Spiny lobster), 99, 159, 161

Subject Index

Ctenomys talarum (Tuco tuco), 108
Cynomys ludovicianus (Black-tailed prairie dog), 109, 124

Deer (Cervidae), 262. *See also Rangifer tarandus*
   Odocoileus virginianus (White-tailed deer), 96
Deme, definition of, 4
Diapause, 57-58, 174, 176, 179, 185, 186, 219, 264
Dinosaurs, 292
Dipodomys heermanni (Kangaroo rat), 36
Diptera (flies), 50, 52, 139, 218
   Calliphoridae (blowflies), 185
      Calliphora, 142, 185
      Lucilia, 60
   Chironomidae (midges), 141, 184
   Chloropidae, *Oscinella frit* (fritfly), 141, 217
   Cordyluridae (dung flies), 261
   Culicidae (mosquitoes), 139, 140, 142, 184, 218
      Aedes taeniorhynchus, 142
      Anopheles sacharovi, 185
   Drosophilidae (fruit flies)
      Drosophila, 57, 59, 96, 169, 170, 227, 240, 243, 244, 283
      D. ananassae, 11, 148-158
      D. funebris, 217
      D. pseudoobscura, 242
   Muscidae (houseflies), 140, 142
   Oestridae (bot flies), *Hypoderma,* 140
   Simulidae (black-flies), *Simulium,* 141
   Syrphidae, 54, 139, 141
Dispersal
   age-biased, 10, 11, 43, 92, 93, 95, 106, 108-109, 139, 202, 217-218, 254, 295-297
   behavior, heritability of, 13, 75-76, 220-221
   behavioral modifications, 48-49, 134, 137, 159-161, 167, 182-188, 205, 216-224. *See also* Social behavior
   costs and benefits, 90-92, 164-170, 252, 263, 282-284
   crowding effects on, 59, 60, 94, 134-138, 142-143
   definition of, 3, 103
   frustrated, 113, 116-119, 120-121, 126, 240
   gross mortality, effects on, 107-108, 168, 178-179
   growth rates, effects on, 109-114, 202, 221-222
   motivations for, 90-97, 116, 139-143, 219-220
   plant, 2, 165, 254, 278, 280, 287
   pre-saturation, 92-96, 97, 104-107, 109, 113, 116, 120, 125, 126, 127, 137-138, 139-143, 201, 202, 216, 246-247, 284
   reproductive interactions, 115-116, 118-119, 139-143, 165, 175, 185, 202-204, 221-222, 248-250, 262-264, 269-272
   saturation, 92, 104-105, 120, 125, 201
   selection for, 26-28, 71-74, 77-82, 90, 91, 99, 105-106, 107, 165, 177, 179, 202, 214-215, 225-226, 231-234, 244, 252-256, 260-265, 267-276, 278-281, 282-286, 287-294, 295-297
   sex-biased, 10, 92, 96, 106, 108-109, 142-143, 178, 202, 254

   sink, 95, 116, 126
   stress, effects on, 115-116, 248-250
   weather, effects of, 167-168, 175, 178, 179, 190-193, 217, 219
Dispersal-migration axis, 3, 4, 5, 9, 14, 97
Dispersion, definition of, 3, 43
Disseminules, 2, 10, 165

Elton, Charles, 1, 182, 187, 202
Emigration, definition of, 5, 103
Endocrine effects (on dispersal and migration), 115-116, 140-142, 161. *See also* Dispersal, reproductive interactions
Epideictic displays, 94, 244
Eruptions (population), 14, 178
Eutamias minimus (Least chipmunk), 11
Evolution
   dispersal. *See* Dispersal, selection for
   flightlessness, 10, 53, 201, 244, 293
   migration, 30-32, 47-48, 220-221, 252-256, 260-265
Exploratory excursions, 4, 12, 95, 96-97, 103, 107

Fish. *See* Guppy; Migration, fish
Flight-muscle histolysis, 165-166, 183, 220
Flour beetle *(Tribolium). See* Coleoptera, Tenebrionidae
Flycatcher
   Pied *(Muscicapa hypoleuca),* 35
   Western *(Empidonax difficilis),* 168
Food, relation to dispersal and migration, 93, 134-138, 141-142, 167, 175, 176-177, 225-229
Fruit fly. *See* Diptera, Drosophilidae
Fugitive species, definition of, 254

Gene flow, 1, 102, 103, 200-202, 267-276, 293
Genetics. *See also* Evolution
   behavioral, 9-14, 30-32, 77-82, 220-221
   drift, 28, 178, 267-276
   gene flow. *See* Gene flow
   inbreeding, 20-26, 91, 201, 255, 270, 274, 280
   intrapopulation variation, 12-14, 33-42, 77-82, 83-87, 179, 201, 207-215, 225-239. *See also* Polymorphism
   intraspecific variation, 11-12, 17-18, 19-27, 30-32, 43-46, 93, 106, 148-158, 165
   population, 91, 102, 200-202, 207-215, 237, 252-256, 267-276, 278-281, 287-294
Ground squirrel. *See Spermophilus*
Grouse, Red *(Lagopus lagopus),* 241
Guppy *(Poecilia reticulata, Lebistes),* 170, 283

Habitat heterogeneity, 59-61, 103, 144-147, 168, 175-177, 203, 207-215, 225-239, 240-245, 253-254, 262, 267-276, 285
Heteroptera (Bugs), 50, 139, 218
   Corixidae (Water-boatman), 12
      Sigara scotti, 166
   Gerridae (Water-striders), 166, 167, 262
      Gerris, 174-181
         G. argentatus, 174, 175, 178-179
         G. lacustris, 174, 175, 176, 178, 179

Subject Index

G. *lateralis*, 174, 175, 178, 179
G. *najas*, 174, 175, 178, 179, 180
G. *odontogaster*, 174, 175, 178
G. *paludum*, 175, 178
G. *rufoscutellatus*, 174, 175
G. *sphagnetorum*, 175, 178, 179, 180
G. *thoracicus*, 174, 175
Largidae *Iphita*, 141
Lygaeidae
   Lygaeus kalmii (Lesser milkweed bug), 13, 75-76, 95, 219, 220, 221, 223
   Oncopeltus fasciatus (Greater milkweed bug), 93, 142, 165, 167, 217, 218, 219, 220, 221, 222, 223
Miridae, *Lygus rugulipennis*, 58
Pentatomidae, *Murgantia histrionica* (Harlequin bug), 217
Pyrrhochoridae
   Dysdercus (cotton-stainers), 221, 222
   D. fasciatus, 220, 221
   D. nigrofasciatus, 220, 221
   D. sidae, 57, 59, 166
   D. superstitiosus, 220, 221
Scutelleridae (Shield-backed bugs), *Eurygaster integriceps*, 57, 58
Homing, 44, 96, 268
Homoptera
   Aleyrodidae, 139, 140
      Aleyrodes brassicae, 141
   Aphididae (aphids), 49, 54, 57, 61, 139, 140, 141, 142, 166, 182, 183, 185, 186, 187, 203, 218, 219, 222, 240, 241, 280, 293
      Aphis craccivora, 220
      A. fabae (Bean aphid), 54, 139, 183, 184, 216, 218, 221, 222
      Drepanosiphum platanoides (Sycamore aphid), 94
      Megoura viciae, 220
      Pemphigus trehernei, 167
      Tuberolachnus salignus, 186
   Cicadellidae (Leafhoppers), 139, 166
      Circulifer tenellus, 217
      Empoasca fabus, 217
      Macrosteles divisus, 44
      M. fascifrons, 217
      Prokelisia marginata, 166
House finch *(Carpodacus mexicanus)*, 44
Humans *(Homo sapiens)*, 36, 92-93, 107, 204, 242, 243, 244
Hydra, 92, 234
Hymenoptera
   ants, 139
   army ants, 93, 184
   Bombyx, 141
   honey bees, 140

Immigration, definition of, 5, 103

Jaeger, Parasitic *(Stercorarias parasiticus)*, 170

Krebs, C. R., 83, 87

Leafhopper. See Homoptera, Cicadellidae
Lemming. See *Lemmus lemmus*
Lemmus lemmus (Norwegian lemming), 113, 138
Lepidoptera (butterflies and moths), 47, 50, 51, 55, 96, 139, 140, 182, 183, 184, 187, 216, 218, 219, 241, 262
   Arctiidae, *Panaxia* (=*Callimorpha*) *dominula*, 48
   Eucosmidae, *Rhyacionia buoliana* (European pine-shoot moth), 142
   Geometridae
      Epirrhoe alternata, 242
      Xanthorhoe ferrugata, 242
      X. fluctuata, 241
   Lasiocampidae
      Malacosoma disstria, 220
      M. neustrie, 242
      M. pluviale, 220
   Noctuidae, 140
      Agrochola lychnidis, 241
      Agrotis infusa, 58
      Barathra brassicae, 142
      Laphygma exigua, 142
      Leucania unipunctata, 142
      Lycophotia varia, 242
      Plusia gamma, 142
      Xestia xanthographa, 242
   Nymphalidae
      Danaus plexippus (Monarch butterfly), 10, 57, 58, 98, 140, 141, 142, 217, 240, 242, 243
      Euphydryas editha (Checker-spot butterfly), 12, 95, 203, 204, 267
      Ford's marsh fritillary, 241
      Melitaea harrisii (Checker-spot butterfly), 246-247
   Pieridae
      Ascia monuste (Florida saltmarsh butterfly), 49, 60, 139, 140, 141, 142, 183, 184, 217
      Pieris brassicae, 142
   Plutellidae, *Plutella maculipennis* (Diamond-back moth), 220
   Pyralidae, *Achroia* (wax moth), 140
   Saturniidae
      Automeris, 140
      Cecropia, 140
      Cerodirphia speciosa, 141
   Sphyngidae, Eyed-hawkmoth, 141
   Tortricidae, *Choristoneura fumiferana* (Spruce bud worm), 59, 141, 142, 185, 220
Life-history strategies, 10-11, 205, 216-224, 253, 254, 260, 265, 276
Locust. See Orthoptera, Acrididae

Marine mammals, 98, 262, 263
Mate choice, 169-170
   rare-male advantage, 170
Microtus (Voles), 124, 126, 134, 135
   agrestis (Short-tailed vole), 115, 125
   californicus (California vole), 13, 36, 83-87, 96, 107, 110, 111, 112, 117, 123, 126, 284
   ochrogaster (Prairie vole), 14, 83-87, 106, 107, 109, 112, 115, 117, 118, 122, 123, 125
   oeconomus (Root vole), 109, 113

Subject Index

pennsylvanicus (Meadow vole), 14, 83-87, 106, 107, 109, 112, 117, 118, 122, 123, 125, 134-138, 168
pinetorum (Pine vole), 112, 123
townsendii (Townsend's vole), 83-87
Migration
 amphibians, 98
 birds, 30-32, 43-46, 97, 98, 106, 159, 161, 168, 204-205, 216, 223, 240, 243, 260, 262, 263, 264, 265
 caribou, 98, 99
 definition of, 3, 97
 fish, 98, 161, 216, 223, 240, 262, 264
  Pacific salmon, 260
 mammals, 262, 264
  desert, 262
  marine, 98, 262, 263
 motivations for, 97-99, 139-143
 pressure, 200
 turtles, marine, 98, 161
Migratory restlessness (zugenruhe), 30, 31, 99, 161, 223, 265
Milkweed bugs. See Heteroptera, Lygaeidae
Mite
 Hummingbird flower, 165
 Pygmephorus, 280
Mollusca, 295, 297
 Helix pomatia (Roman snail), 96
 marine bivalves, 256
 Volutidae (Gastropoda), 256, 295-297
Monarch butterfly. See Lepidoptera, Nymphalidae
Monkey, Toque *(Macaca sinica)*, 169
Mosquito. See Diptera, Culicidae
Mouse. See *Apodemus; Mus musculus; Peromyscus; Reithrodontomys magalotis*
Murid rodents. See *Apodemus; Mus musculus; Rattus*
Mus musculus (House mouse), 12, 95, 107, 113, 114, 115, 117-118, 120, 126, 169, 170, 204, 236, 248-250, 284, 293
Mustela vison (Mink), 11
Myocaster coypus (Nutria), 107

Narise, Takashi, 11, 96, 169
Natrix sipedon (Water snake), 14
Nomadism, definition of, 4

Odonata (dragonflies), 57, 139, 141, 185
Ondatra zibethica (Muskrat), 35-36
Oogenesis-flight syndrome, 139-143, 222
Orthoptera
 Acrididae (grasshoppers, locusts), 57, 218
  locust, 139, 140-141, 142, 143, 167, 182, 183, 184, 185, 186, 187, 222, 262
  Locustra migratoria (Migratory locust), 14, 140, 141, 185, 186, 220, 221
  L. migratoriodes, 142
  Melanoplus bilituratus, 59
  Nomadacris septemfasciata (Red locust), 220
  N. septempunctata, 142
  Schistocerca gregaria (Desert locust), 49, 59, 94, 139, 141, 142, 166 182, 183, 190-193, 217, 220

Tettigoniidae, *Anabrus simplex* (Morman cricket), 185
Oryctolagus cuniculus (European rabbit), 35, 38, 95, 204
Oryzomys palustris (Rice rat), 118

Perognathus (Pocket mice)
 formosus, 14, 109
 inornatus, 36
Peromyscus (Cricetidae), 126
 boylii (Brush mouse), 36
 leucopus (White-footed mouse), 115, 117, 167, 194-198, 204
 maniculatus (Deer mouse), 34, 36, 37, 38, 44, 95, 96, 115, 117, 118
 polionotus (Old-field mouse), 13, 106
 truei (Pinyon mouse), 36, 118
Phasmids (Walking-sticks), 141
Phoresy, 164, 165
Photoperiod, effects on dispersal and migration, 99, 141, 161, 174, 219-220, 221, 263
Pigeon, domestic *(Columba livia)*, 170
Polymorphism, 9, 12, 13, 33-42, 43-46, 58-59, 61, 106, 125, 142, 207-215, 220, 253, 255, 280, 284, 291, 292, 293. See also Genetics, intrapopulation variation
 wing-length, 9, 53, 58, 142, 166-167, 174-181, 264
Population regulation, 103, 111-113, 117, 119-124, 126-127, 144-147, 175, 176, 177, 202-205, 221-222, 225-239, 240-245, 246-247, 265, 282-286
 anti-regulating effect, 204
 Chitty hypothesis, 13
 contributing factor, 203
 key factor, 203
 migration, influences of, 168, 204-205
 regulating factor, 203
Predation, 91, 93-94, 169, 194-198, 254-255

Quail, Valley *(Lophortyx californicus)*, 34

Rabbit. See *Oryctolagus cuniculus*
Rana pipiens (Leopard frog), 273
Rangifer tarandus (Caribou), 98, 99
Rattus (Muridae)
 norvegicus, 115
 villosissimus, 107
Reithrodontomys magalotis (Western harvest mouse), 36
Reproductive value, 10, 216, 222

Salamanders
 Ambystoma maculatum (Spotted salamander), 227
 Notophthalmus viridescens (Red-spotted newt), 267-277
Sceloporus olivaceus (Fence lizard), 44
Screech owl *(Otus asio)*, 194-198
Shrew (Soricidae), 197
 Sorex cinereus (Masked shrew), 107
Sigmodon hispidus (Cotton rat), 95, 107, 115
Social behavior, 94, 95, 96, 99, 107, 114-116, 159-161, 169, 170, 202, 230-231, 248-250

Subject Index

Sparrow
    European Tree *(Passer montanus)*, 204
    House *(Passer domesticus)*, 35, 44
    Song *(Melospiza melodia)*, 34, 44
Speciation, 253, 256, 295-297
Spermophilus (Ground squirrels)
    beecheyi (California ground squirrel), 35, 36, 107
    richardsoni (Richardson's ground squirrel), 11
    undulatus (Arctic ground squirrel), 119
Starling *(Sturnus vulgaris)*, 35, 243

Termites (Isoptera), 139
Territoriality, 43, 45, 95, 167, 169, 204
Thomomys bottae (Pocket gopher), 36
Thrips (Thysanoptera), *Thrips imagines*, 47
Thysanura (Bristle-tails), 139
Titmice *(Parus)*
    P. major (Great tit), 93, 170
    P. rufescens (Chestnut-backed Chickadee), 93

Vagility, definition of, 165
Vole, 260. *See also Arvicola terrestris; Clethrionomys; Lemmus lemmus; Microtus*
Voltinism, 179

Warbler
    Blackcap *(Sylvia atricapilla)*, 12, 30-32
    Orange-crowned *(Vermivora celata)*, 12
    Phylloscopus, 30
Water-strider. *See* Heteroptera, Gerridae
Water-thrush, Northern *(Seiurus novemboracensis)*, 167
Weevil. *See* Coleoptera, Curculionidae
Whitten effect, 115

Zugenruhe. *See* Migratory restlessness

About the Editors

WILLIAM Z. LIDICKER, JR. is professor of zoology and curator of mammals at the University of California, Berkeley. He received the Ph.D. degree in 1957 from the University of Illinois. In addition to his position at the University, he has had one year associations with Sydney University (Australia), University College London, and the Royal Free Hospital of Medicine (London). His research centers on the ecology and evolution of mammals. Major themes have been population regulation, dispersal, population microstructuring, and social behavior.

ROY L. CALDWELL is associate professor of zoology and director of the Field Station for Behavioral Research at the University of California, Berkeley. He received the Ph.D. degree in 1969 from the University of Iowa and completed a year of postdoctoral research at Imperial College of Science and Technology (Silwood Park Field Station, United Kingdom) before taking a position at Berkeley. His research on insect dispersal focuses on the physiological and genetic bases of dispersal behavior.

THE LIBRARY
ST. MARY'S COLLEGE OF MARYLAND
ST. MARY'S CITY, MARYLAND 20686